Biomass Conversion Processes for Energy and Fuels

Biomass Conversion Processes for Energy and Fuels

Edited by

SAMIR S. SOFER
University of Oklahoma
Norman, Oklahoma

and

OSKAR R. ZABORSKY
National Science Foundation
Washington, DC

PLENUM PRESS • NEW YORK AND LONDON

Library of Congress Cataloging in Publication Data

Main entry under title:

Biomass conversion processes for energy and fuels.

Bibliography: p.
Includes index.
1. Biomass energy. I. Sofer, Samir S. II. Zaborsky, Oskar.
TP360.B587   333.95′3   81-15721
ISBN 0-306-40663-2   AACR2

© 1981 Plenum Press, New York
A Division of Plenum Publishing Corporation
233 Spring Street, New York, N.Y. 10013

Printed in the United States of America

Preface

Countless pages have been written on alternative energy sources since the fall of 1973 when our dependence on fossil petroleum resources became a grim reality. One such alternative is the use of biomass for producing energy and liquid and gaseous fuels. The term "biomass" generally refers to renewable organic matter generated by plants through photosynthesis. Thus trees, agricultural crops, and aquatic plants are prime sources of biomass. Furthermore, as these sources of biomass are harvested and processed into commercial products, residues and wastes are generated. These, together with municipal solid wastes, not only add to the total organic raw material base that can be utilized for energy purposes but they also need to be removed for environmental reasons.

Biomass has been used since antiquity for energy and material needs. In fact, firewood is still one of the most sought-after energy sources in most of the world. Furthermore, wood was still a dominant energy source in the U.S. only a hundred years ago (equal with coal). Currently, biomass contributes about 2 quadrillion Btu (1 quad = 10^{15} Btu) of energy to our total energy consumption of about 78 quad. Two quad may not seem large when compared to the contribution made by petroleum (38 quad) or natural gas (20 quad), but biomass is nearly comparable to nuclear energy (2.7 quad). Moreover, from silvicultural energy farms alone, the contribution from biomass could be as high as 5–10 quad by the end of this century. Without getting involved in controversy or perhaps a numbers game as to how much energy can be derived from biomass, one fact is clear. Biomass contributes to our present energy sources and could contribute even more so if appropriate actions are taken now and if a better understanding of its potential, as well as its limitations, is achieved.

This book brings together in one volume a description of the rather diverse elements of biomass sources and the currently more promising conversion processes for energy and fuels purposes. From the outset, we recognized that many topics would have to be excluded and that no single volume can adequately describe all the current processes being used or proposed. Moreover, the exciting developments occurring in research make it extremely difficult to project into the future with any certainty. Clearly, the biomass-to-energy area is subject to change, and many exciting advances in research and development will be made in the next few years. Concurrently, institutional, economic, and political issues need clarification, and sound policies need to be established for developing biomass energy.

This book should be considered as an introductory text to the topics of converting biomass to energy and fuels: namely, biomass sources, conversion processes, and technical and economic considerations. We have emphasized conversion processes, and the book provides chapters on both fundamental principles of the conversion processes and actual conversion systems. A unique feature of the volume is a description of all major conversion processes—direct combustion, thermochemical and biological—in one book. Also, an effort was made to describe the various conversion processes in a consistent format and to provide similar information to the reader so that intelligent comparisons can be made. In particular, we desired to obtain complete mass and energy balances for each conversion system. It is our hope that these chapters provide a beginning to more rigorous energy and mass balance analyses and of their more frequent reporting in the literature.

We express our gratitude to all the authors for their valuable contributions to this complex and, at times, controversial subject. We also thank Ms. Margaret Williford and Dr. Nancy Roundy for their able assistance. In particular, Ms. Williford's untiring assistance with the authors was most appreciated.

Samir S. Sofer
Norman, Oklahoma
Oskar R. Zaborsky
McLean, Virginia

Contents

PART II. CONVERSION PROCESSES
Section A. Direct Combustion Processes

Section B. Thermochemical Processes

Section C. Biochemical Conversion Processes

PART III. TECHNICAL AND ECONOMIC CONSIDERATIONS

PART I
BIOMASS SOURCES

1

Residues and Wastes

LUIS F. DIAZ AND CLARENCE G. GOLUEKE

1. Introduction

Judging from the literature and the many proposed schemes, great expectations exist regarding the recovery of useful items from residues—items which range from the lowly refillable container to the now highly prized energy sources, namely, the carbonaceous wastes. Whether or not the expectations will be realized with respect to energy from residues remains to be seen. Most likely, the realization will be somewhere between the extremes, i.e., between the trivial foreseen by the pessimist and the panacea predicted by the superoptimist. Regardless of position in the spectrum between extreme pessimism and superoptimism, the reality is that if all organic residues (euphemism for "wastes") were converted into energy, the resulting output would fulfill about 5–10% of the nation's total energy requirements. While these percentages may seem small, it should be remembered that any contribution, however small, has a significance.

An important factor, which stems from the geographical pattern of residue production, is the fact that the energy generated from those residues is concentrated in highly localized areas scattered throughout the nation. The significance of this fact is that the energy thus generated is an amount sufficient to meet a significantly large fraction of the energy needs of those areas. Thus, what is a quite small fraction when related to the needs of the nation as a whole, becomes a significantly large one when related to an individual community.

Upon making projections regarding the energy potential of residues, cer-

LUIS F. DIAZ and **CLARENCE G. GOLUEKE** • Cal Recovery Systems, Inc., Richmond, California 94804.

tain fundamental facts must be kept in mind. First, of course, is the obvious one that only the organic fraction of wastes can serve as an energy source. Here, "organic" is taken in the broad chemical sense and is not restricted to materials of biological origin; perhaps a more appropriate term might be "combustible." This limitation is not a major one in that at least 70% of the total mass of agricultural and municipal residues generated in the United States is combustible.

A second fundamental fact is that only a fraction of the combustible residues are available for conversion to energy. The constraints imposed by this fact originate from the scattered distribution of the residues. The implication of these constraints is that care must be taken not to exceed the limits of economic and energetic feasibility in collecting and transporting the residues to processing centers. This latter limitation largely restricts the conversion of residues to energy to large population centers and makes conversion from certain agricultural residues only marginally feasible.

A third fact is an absence of firm numbers on rates and amounts of waste generation. This lack applies equally to numbers reported for both agricultural and municipal waste generation. The problem is especially characteristic of municipal waste generation and in that instance, owes its origin to a singular dearth of studies involving firsthand observation. The result is a literature replete with "educated guesses" that, in turn, lead to a gross exaggeration of the actual rates of generation. It is only within the past few years that attempts have been initiated to determine the true rate of generation in a statistically sound manner. The ramifications of the lack of realistic estimates are many, not the least of which is the need for great caution in developing plans for energy recovery from a particular residue at a given site.

A fourth fact is that use of certain combustible materials for energy generation will have to compete with other uses that under certain circumstances may be deemed to be of higher priority. Two examples come to mind: (1) the use of reclaimed fiber to make paper; and (2) the return of manures and crop residues to the soil to increase the productivity of the soil. Increasing the productivity of the soil will be the most serious competing use because it is an essential element in food production. In the final analysis, food production would take precedence over energy production.

2. Municipal Solid Wastes

The combination of domestic, light industrial, and demolition solid wastes generated within a community usually is referred to as "municipal soild waste" (MSW). The proper management and disposal of MSW is considered to be among the more pressing of the environmental problems confronting the United States today, particularly the waste generated in the large metropolitan areas.

2.1. Quantity

Although MSW is of all wastes the most concentrated, the most visible, and perhaps the subject of most study, the precise quantities of MSW generated each day remain to be determined. The reason for the absence of firm figures is that attempts heretofore to determine quantities have consisted in merely making sporadic vehicle counts and then multiplying the number of vehicles by the assumed weight of the vehicle contents as judged from the nominal capacity of the vehicles. In the absence of even this minimal effort, estimates were based upon reported nationwide averages, the validity of which was even more tenuous. That this state of affairs is unfortunate and unsatisfactory need not be dwelt upon. Suffice it to state that it is imperative that a fairly accurate knowledge of the quantities generated in each area be had, not only for purposes of resource recovery, but also to provide and plan optimum collection and disposal services. Unfortunately, it has been only recently that when faced with seeking alternatives to land disposal, those involved with refuse collection and disposal are making an attempt to accurately determine the quantities of wastes handled each day.

A nationwide average frequently quoted is one based on surveys conducted by the United States Environmental Protection Agency (EPA).[1] Through the application of input/output relationships, an attempt was made in the EPA study to estimate and to project waste generation in the United States. A summary of the results is given in Table 1. According to the table, the total amount of post-consumer solid wastes generated in the United States was expected to reach 159 million tonnes/yr by 1980. That amount is equivalent to a daily per capita generation of about 1.94 kg. These estimates generally are considered to exceed the true amount. Furthermore, given the present economic situation in the United States, it is questionable whether or not the daily per capita generation of residential wastes will continue to increase as predicted. Records kept by the City of Los Angeles for almost a decade indicate that the per capita generation in that city is only about 1.02 kg/day. Consequently, the generation of MSW probably is from 1.4–1.8 kg per capita per day. The Los Angeles numbers have the advantage of validity in that they

TABLE 1. Baseline Estimates and Projections of Post-Consumer Solid Waste Generation from 1971 to 1990[a]

	Estimated				Projected		
Total waste generated	1971	1973	1974	1975	1980	1985	1990
Tonnes/yr (in millions)	121	131	131	123	159	182	204
Kilograms per capita per day	1.60	1.70	1.68	1.54	1.94	2.12	2.27

[a]Based on Reference 1.

were obtained through actual determinations rather than solely through vehicle counts. On the basis of the Los Angeles findings, it would appear that the EPA projections should be revised such that they show that the total amount of MSW generated in the United States in 1980 probably will vary between 110 and 148 million tonnes/yr.

2.2.   Characteristics

The heterogeneity of MSW is indicated by the data in Table 2, in which are listed the principal components of MSW as well as their percentages. According to the table, about 80% of the MSW stream is combustible, and about 82% of the combustibles are of biological origin, namely, paper, garbage, and yard wastes. When considering the data in Table 2, it should be kept in mind that they represent averages and that the amounts of the individual components not only vary seasonally, but also from region to region. This variation adds up to the constitution of a major problem in the reliance upon MSW as a source of energy.

2.3.   Fuel Value

The results of the ultimate analysis of "typical" MSW are given in Table 3. The "higher heat value" of the sample was 10,236 kJ/kg. Ultimate analysis together with the heat values and proximate analyses are used in determining the energy characteristics of a solid waste. Proximate analysis is the determination of moisture, volatile matter, fixed carbon, and ash as expressed as per-

TABLE 2.   Typical Composition of Municipal Solid Waste[a]

Component	Percent by weight
Paper	43.2
Metals	8.0
Glass, ceramics, rocks	10.8
Plastic, rubber, rags	4.5
Garbage, yard wastes	23.5
Miscellaneous	10.0
Total	100.0

[a]L. F. Diaz, G. M. Savage, R. P. Gobel, C. G. Golueke, and G. J. Trezek, Market Potential of Materials and Energy Recovered from Bay Area Solid Wastes College of Engineering, University of California, Berkeley. A report submitted to the State of California Solid Waste Management Board (March, 1976), p. 352.

TABLE 3.   Typical Ultimate Analysis and Heating Value for Municipal Solid Waste[a]

Component	Percent by weight (as received)
Moisture	26.04
Carbon	27.23
Hydrogen	3.85
Oxygen	21.49
Nitrogen	0.28
Chlorine	0.20
Sulfur	0.26
Ash	20.63

[a]Horner and Shifrin, Inc., Refuse as Supplementary Fuel for Power Plants, City of St. Louis, MO (March, 1970), p. 142.

centages of total weight of the sample. Ultimate analysis is the determination of moisture content, noncombustibles, and the carbon, hydrogen, oxygen, nitrogen, and sulfur content. "Higher heat" value means the "gross heat" value of the sample and is the total heat release per unit of material burned. Although as indicated by the data in Table 3, the combined carbon and hydrogen content of MSW may be substantially large (here, 31%), its heating value, even with moisture and ash removed, does not reflect the high percentage because the oxygen content also is large (here, 21.5%). Generally, the higher heat value will range from 9300 to 11,600 kJ/kg. The low sulfur content of MSW, only 0.26% of the sample described in Table 3, is an attractive feature of MSW in terms of the latter's use as a fuel. Such a sulfur content is much lower than that of high-grade fuel oil and low-sulfur coal. The nitrogen concentration is not enough to constitute a nitrogen oxide problem in conventional incineration.

Assuming a heating value midway in the range of higher heat values, namely, 10,500 kJ/kg and the revised projection for rate of MSW generation in 1980, it can be stated that MSW could be a source of 1.16 to 1.55×10^{15} kJ/yr. Of course, this is only an exercise in theory inasmuch as not all of the collected wastes would be available for energy production.

Findings made during the course of studies conducted by the authors indicate that to maintain economic viability, a waste-to-energy facility should accommodate at least 270 tonnes MSW/day. Taking into consideration this constraint on minimum capacity together with those factors that place a practical upper limit on capacity, namely, population density and land availability, one is drawn to the conclusion that the best location for a waste-to-energy facility would be in a metropolitan area that has a population ranging from about 150,000 to 200,000 people. This population size restriction and other constraints could reduce the amount of energy that can be feasibly recovered from MSW to about 0.74 to 1.06×10^{15} kJ/yr.

2.4. Upgrading the Fuel Value

The heterogeneous nature of MSW together with a variety of environmental, social, technical, and other constraints make it necessary in most cases to subject the waste to some type of segregation before it can be used as a fuel. The segregation can be done manually or mechanically.

In modern mechanical segregation, the waste stream usually is passed through a succession of processes that include size reduction, air classification, magnetic separation, and screening. The order of succession varies as does the number of times a particular type of process occurs in the succession. The result of the passage is that metals, glass, and a mostly organic fraction that is commonly termed "refuse-derived fuel" (RDF) are individually separated from the incoming mass of wastes. Each piece of equipment (unit process) is designed to perform in a prescribed manner under a given set of conditions,

TABLE 4. Comparison of RDF with Coal[a]

	Moisture (%)	Ash[b] (g/MJ)	Heating value, (J/g)	
			Dry solids	As received
Unscreened RDF	22.5	17.3	15,627	12,111
Screened RDF	16.3	7.8	18,568	15,541
Orient coal	12.5	2.8		26,875

[a]G. M. Savage, L. F. Diaz, and G. J. Trezek, *Waste Age* 9(3): 100–106, (1978).
[b]Based on as-received heating value.

which chiefly include flowrate, size distribution, and moisture content. The sequence and the degree of processing are altered to meet the specific needs of each plant. For example, the addition of a rotary screen (trommel) greatly improves the quality of refuse-derived fuel, as is demonstrated by the data in Table 4. In the table, the fuel properties of a sample of screened RDF and one of unscreened RDF are compared with those of coal. As the data show, screening the RDF increases its heating value and reduces its moisture content and ash content. In this particular case the ash content of the RDF was reduced from 17.3 to 7.8 g/MJ.

The wide variation in the physical characteristics of raw refuse may lead to a less than desired or at least unexpected performance by a unit process. Such an unpredicted performance aggravated by a lack of understanding of the basic principles that take place in each unit process constitutes a major reason for the poor functioning of certain refuse-processing plants now in operation.

Despite the errors made in designing refuse-processing plants, the overall knowledge regarding refuse processing and resource recovery continues to increase. The progress is due largely to the support provided by the federal government, especially the EPA. Progress in the technology is important because as the cost of conventional fossil fuel continues to escalate and the supply diminishes, the alternative of producing energy or fuels from MSW is approaching the point at which it will be economically competitive with other solid-waste-management options.

3. Municipal Sewage Sludges

A phenomenon of recent times is the trend to group wastewater sludges in a class with domestic wastes, light industrial wastes, and demolition debris when it comes to assigning responsibility for disposal. Thus, sewage sludge is more and more frequently being grouped with the three latter wastes, which conventionally bear the collective term "municipal solid wastes." Regardless of category or position assigned to them, the nature and quantity of sludges being generated at present and projected for the future are such that the disposition of the wastes is becoming a problem of the utmost gravity.

3.1. Types

"Sludges" is used in the plural sense because there are a number of types that differ among themselves not only in origin, but also in nature. Space allows only a brief cataloguing and description of the various sludges. Conventionally, the sludges produced in municipal wastewater treatment are grouped into four broad categories reflecting the step in which they are produced. They are raw or primary sludge, secondary sludge, tertiary sludge, and digested (anaerobically) sludge. Secondary sludge may be either activated sludge or trickling filter humus. Raw primary sludge consists of the solids settled out of the incoming raw sewage. It is very unstable and aesthetically objectionable. Secondary sludge is, as one would suspect, the collection of solids settled out of suspension in secondary treatment. The solids, while being relatively unstable, nevertheless do not have the objectionable aesthetic qualities common to raw-sewage sludge. Tertiary sludge is produced in the tertiary-treatment step. It has a very substantial chemical content (e.g., lime and aluminum) since it is the result of the chemical precipitation of the solids remaining in the treatment stream after the secondary step. However, the organic fraction of tertiary sludge is relatively unstable.

3.2. Quantities

In the primary sedimentation and activated-sludge steps of conventional wastewater treatment, the combined removal of incoming solids can average as much as 3.6 kg/m^3 wastewater treated.[2] Assuming an average production of wastewater of 0.379 m^3/person-day, the sewage produced by one million people each day will contain approximately 98 tonnes of dry solids. Therefore, the sludge contained in the entire daily discharge of sewage in the United States adds up to 8.64 × 10^6 tonnes dry wt/yr.

Sludge production as expressed in terms of dry weight is not nearly as impressive as when expressed in wet weight, the form in which it must be handled. This is true because of the high moisture content of the sludges. Thus, immediately upon removal from the settling basin, sludge contains about 95% water. After being subjected to some degree of stabilization, typically accomplished through digestion, the sludge is dewatered to approximately 30% solids. Consequently, since it is the wet sludge (70% moisture content) that constitutes the residue, a more realistic quantity as far as handling is concerned would be 28.2 million tonnes wet sludge/yr. The amounts are expected to increase drastically, largely due to the enactment of the Federal Water Pollution Control Act (P.L. 92-500). The act mandates that the level of wastewater treatment be upgraded to secondary treatment throughout the country. The expected result is an increase of two to five times the current generation of wastewater solids.

3.3. Characteristics

Municipal sludges contain organic matter, nutrients, and trace elements that can be utilized in the production of energy. The organic matter may range from about 50% in digested sludge to approximately 70% in raw sludge. The concentration of nutrients, of course, varies according to the type of inputs to the treatment plant. Typically, the concentration of nitrogen (N), phosphorus (as P_2O_5), and potassium (as K_2O) is on the order of 2% N, 4% P_2O_5, and 0.5% K_2O. Trace elements include Cd, Cu, Ni, Zn, Cr, Hg, and Pb.

Since the energy content of raw municipal sludge is about 16,284 kJ/kg of dry solids, at present nationwide generation rates, the potential energy production from municipal sludges could be about 1.41×10^{14} kJ/yr. However, the relatively large moisture content of sludge places a sharp limitation on its use as a fuel for combustion processes at this time, since the energy released in the combustion would about equal that required to dry the sludge. Another constraint on combustion, almost equally as serious as that of the unfavorable energy return, is the high cost of meeting air-pollution-control regulations currently in force or proposed for the future.

3.4. Utilization of Sludges

One of the best methods of recovering energy from municipal sludges is the production of methane through anaerobic digestion. Although a large number of treatment plants currently include anaerobic digesters, many are not functional, while the others are operated with the major objective being to accomplish a certain amount of treatment and volume reduction. Indeed, much of the methane is being wasted, and in only a few plants is some use made of the gas, usually as a fuel used in heating the digesters.

The utilization of sludges produced chemically in tertiary treatment is questionable at this time due to the presence of the noncombustible metallic compounds which become a part of the sludge mass.

In summary, it may be stated that the prospects for the utilization of sludges as an energy source seem dim at present. The reason is not confined to the unfavorable economics and energetics in so doing, but rather to the far more important factor of the utility of sludge in agriculture. Taking advantage of that utility involves a much less economic and energetic effort than would be required to transform the sludge into energy or into an energy source. Certainly, the environmental impact would be much less unfavorable. However, the positions would be reversed if it were found that the heavy-metal and hazardous-substance content of the sludges was sufficiently great as to rule out their being spread on the land.

4. Animal Wastes

In general, animal rearing takes place either under unconfined or under confined conditions. Wastes from unconfined animals constitute neither an environmental problem nor a practical source of organic matter for energy production. The number of animals per unit of area is so small, and the consequent dispersion of the animal wastes so great, that it becomes economically and energetically impractical to gather and transport the wastes to an energy-production site.

The situation is quite the opposite when the animals are reared under confined conditions, such as in a feedlot. Under these conditions, the amount of wastes generated per unit of area is sufficiently large as to lower collection and transport costs to a level at which the conversion of the wastes into an energy source becomes feasible. An added incentive for converting the wastes into an energy source is that by so doing, a substantial contribution is made to the solution of a disposal problem that rivals that of MSW in its gravity.

4.1. Quantity and Quality

Animals used for meat and dairy products make up the largest fraction of the animal population in the United States, and collectively they produce approximately 180 million tonnes (dry wt) of manure each year. According to a report issued in 1973 by the EPA[3] about 40 million tonnes (dry wt) of manure are generated by major farm animals under confined conditions. A more recent estimate indicates that the production of manure from confined livestock may now have reached 10.5 million tonnes (dry wt). All things considered, it may be assumed that about 50–80% of the 40.5 million tonnes, i.e., 20 to 32.4 million dry tonnes, can be collected within the constraints of economic feasibility.

Data pertaining to amount of manure produced by various animals and the composition of the manure are given in Table 5.[4–8] Ranges of values are given for each item so as to account for differences that arise from factors such as diet and degree of confinement. The data show that the heating value of the manures averages 17,450 kJ/kg dry matter. Of special importance as far as energy recovery is concerned is the high moisture content of the manures, namely from 60–85%. The relatively high moisture content of the manures limits the range of technology that can be used for energy recovery.

TABLE 5. Manure Production and Composition for Various Animals[a,b]

Animal	Animal wt (kg)	Production Volume (m³/day)	Wet wt (kg/day)	Composition (% wet wt) MC[c]	VS[d]	N	P	K
Beef cattle	500	0.028–0.037	27.7–36.6	85	9.33	0.47–0.70	0.09–0.25	0.14–0.28
Dairy cattle	500	0.031–0.036	30.2–35.0	85	7.98	0.38–0.53	0.06–0.1	0.13–0.3
Horse	500	0.025	28.0	60	14.3	0.86	0.13	
Swine	100	0.0056–0.0078	5.4–7.6	80	7.02	0.59–0.83	0.2–0.6	0.24
Sheep	50	0.002–0.003	1.9–3.0	70	21.5	1.0–1.9	0.3	0.78
Poultry	2.5	0.00014–0.00017	0.14–0.17	82	16.8	0.86	0.13	0.43

[a]Sources: References 4–8.
[b]Ranges are used in order to account for differences in quantity and composition due to various factors such as diet and degree of confinement.
[c]MC = moisture content.
[d]VS = volatile solids.

4.2. Utilization

Among the methods proposed for converting at least a fraction of the energy tied up in animal wastes into a useable form, one that seems to hold the most promise for success is methane production through anaerobic fermentation or "biogasification" as it often is termed. Biogasification is an attractive process because, among other reasons, the residue from the process (digested sludge) retains most of the plant and animal nutrients contained in the input wastes. The presence of the nutrients renders the residue suitable for use as a cattle feedstuff or as an organic fertilizer, or both. However, a disadvantage of biogasification, in terms of energy recovery from ruminant manures, stems from the nature of the digestive process in ruminants. Digestion in the ruminant alimentary tract results in a portion of the carbonaceous matter in the manures becoming unavailable for conversion into methane. The upshot is that instead of the 0.26–0.30 m³ methane/kg volatile solids introduced (expected on the basis of the digestion of raw sewage sludge), the actual yield is only 0.13–0.15 m³ methane/kg volatile solids.

Biogasification is being tried at a few large feedlots not only as a means of recovering energy, but also of alleviating the particularly complex problems that plague feedlot operations. The biogasification systems employed in the lots incorporate a variety of features designed to facilitate and maintain top quality sanitation and to prepare the wastes for efficient and effective processing. With biogasification as the process of choice, the second feature implies the production of a waste that is fresh, i.e., has not had time to age, and is free of inert materials such as dirt. Freshness is required because the extent of methane production is a function of the amount of undecomposed carbonaceous matter in the wastes. Obviously, the older the manure, the greater the degree of previous decomposition, and the less the amount of undecomposed matter available for conversion into methane. The significance of inert materials is mostly operational in nature, although, again, the larger the percentage of inert matter, the less will be the yield of methane per unit of wastes. The pertinent features are concrete floors and specially designed collection systems for the wastes. These systems include the installation of slatted flooring in a section of the lot such that the manure is deposited into a trough installed under the slats. The trough is equipped with a scraper blade activated at intervals to scrape the wastes from the trough into a collection sump. In the sump, the manure is slurried and then is introduced into the fermenter unit, i.e., the digester.

If a feedlot does not include the design features described in the preceding paragraphs and instead has a dirt floor and infrequent collection of manure, then biogasification definitely is not indicated. This conclusion is supported by a detailed analysis made by L. A. Schmid.[9] By his analyses, Schmid demonstrated that any attempt to biogasify wastes from a dirt floor feedlot managed according to conventional practice would be attended by a net loss rather than gain in energy.

5. Crop Residues

At first sight, crop residues seem to be an attractive energy source, and with good reason. They are primarily cellulosic (carbonaceous) in composition, and in general are relatively amenable to most energy-conversion processes. They are abundant, and very importantly, are beginning to pose a grave disposal problem now that burning in the field is being actively discouraged by regulatory agencies. Unfortunately, certain imposingly large obstacles must be overcome before the many advantages inherent in the production of energy from crop residues can be fully attained.

The feasibility of recovering and utilizing crop residues as an energy source depends upon a number of factors, among which are technical and economic constraints, social and environmental considerations, and number and stringency of regulations. The energy recovery potential of the utilization of a residue can be evaluated on the basis of certain parameters, the more important of which are quantity and quality of the residue, seasonality of generation, and location of the residue. Other parameters, nontechnical in nature but nonetheless important, include factors of an ecological, social, or political nature and competition from other uses for the residue. Competition from the use of residues to provide the organic matter essential to soil productivity is an exceedingly important factor because, for practical purposes, residues constitute the principal source of organic matter in the soil. The importance to soil productivity rests in the improvement in soil texture, increase in water-holding capacity, and increase in efficiency of chemical fertilizer utilization that result from the decomposition of organic matter in soil.

5.1. Types

In this chapter, crop residues are classified more or less into two major categories based on whether the residues are left on or in the soil after harvesting, or whether they are collected and removed from the field as a part of or along with the harvested crop. Residues in the second group constitute the greater part of the waste stream from the processing plants. Examples of the first category are wheat and corn stover; examples of the second include chaff, rice hulls, and fruit peelings. The first group is further subdivided into residues left on the soil surface and those in the soil. An obvious example of the latter are nonedible roots.

5.2. Quantity

Estimates as to the quantity of crop residues generated annually range from about 306 million dry tonnes[10] to 378 million dry tonnes.[11] In terms of economic and energetic feasibility, from 90 million to 243 million dry tonnes/yr are available for energy production.[10,12] In general, feasibility would demand that a large land area be cultivated with a crop from which the yield of residue per unit of area (tonnes/ha) is substantial. Typically, the yield is larger with field crops than with vegetable crops.

A broad approximation of the quantity of residue generated in the production of a given crop can be made by multiplying the numerical weight of the crop by the residue coefficient for that crop. The residue coefficient is the ratio of the dry weight of the above-ground residue to the weight of the harvested crop at field moisture. Coefficients for the following six major crops as reported in the literature are as follows: soybeans, 0.55–2.60; corn, 0.55–1.20; cotton, 1.20–3.00; wheat, 0.47–1.75; sugar beet, 0.07–0.20; and sugarcane, 0.13–0.25.[6] The wide range of numerical values of coefficients for individual crops is a function of differences in crop variety, growth conditions, and harvesting practices and in the procedures followed in estimating the coefficients. Generally, the higher the crop yield, the greater is the proportion of residue.

5.3. Quality

Usually, crop residues are fairly homogeneous in composition. Among their more important characteristics are particle size, moisture and ash content, and bulk density. Typically, cereal crop residues, with the possible exception of rice, tend to be relatively dry in that normally their moisture content is on the order of 15%. The energy or heating values of crop residues for the most part fall within the range of 11,500–18,600 kJ/kg. In making estimates, the value is assumed in this chapter to average 16,300 kJ/kg.

5.4. Seasonality of Generation

Seasonality of generation, and hence of availability, is related to the time at which a residue is generated and is determined mostly by type of crop, geographic location, and climate. An accurate determination of the time of availability is essential to the proper planning of the mechanical aspects of the activities that precede actual utilization. Such activities include collection, transport, and storage.

Seasonality is a disadvantageous factor because it necessitates a storage of the residues to ensure a constant supply of material throughout the year. At a minimum, storage involves tying up land and additional handling. Moreover, the storage must be such that access is had to stored waste throughout the year and the material does not deteriorate.

5.5. Location of Residues

The utilization of crop residues as sources of energy apparently is site-dependent in two respects: (1) Utilization is confined to agricultural areas, or reasonably closely thereto; (2) The energy production and consumption must take place relatively near the site of residue production. The location of a crop site, in terms of both distance from the point of use of the residue and the nature of the terrain at the site, can be the deciding factor with respect to the economic feasibility of use of a given crop residue. Distance dictates transportation requirements and their magnitude, while terrain exerts an influence on the mode and ease of collection.

5.6. Other Factors

Mentioned only in passing, but not to be overlooked, are the social, ecological, legal, and political aspects, because ultimately they determine the viability of a project. Among these aspects is the competition with other uses discussed earlier. Potentially competitive uses include utilization as a substitute ingredient in animal feeds, as a feedstock for the manufacture of paper, and as a soil conditioner. In terms of global approach, these latter uses may be more efficient than fuel production.

Based on the estimated quantities of residues generated in the United

States and their average heating value (16,300 kJ/kg), the total energy content of crop residues that could be feasibly used would be as much as 1.5–4.0 × 10^{15} kJ/yr. Of course, as emphasized earlier, a number of factors interfere with the attainment of that energy and hence determine the percentage of the energy that can be recovered. Not the least of the factors are moisture content, transportation, and efficiency of conversion process.

In summary, it can be stated that although the energy from the conversion of the entire output of crop residues to energy would account for but a very small percentage of the energy consumed in the United States, crop residues do constitute a renewable source of energy. As the need for more food production and the cost of conventional fossil fuel increase, the use of crop residues as a source of energy will become more attractive.

6. Industrial Wastes

Inasmuch as only organic residues have the potential to serve as an energy source, the major industrial residues of interest in a chapter on residues are those generated in the forest-products and the food-processing industries. The forest-products residues received attention in another section of this chapter; hence, in this section attention is focused on food-processing wastes. The wide variation of types of food processing wastes reflects the great diversity of foods that are processed. Each food has its peculiar residue. Thus, in residues from fruit packing, sugars and pectins are present in an appreciable concentration; whereas in food-processing grains, starch and cellulose are important constituents. Another important difference is in fiber content; fruit wastes have a far less fiber content than do those from cereal processing. As one would suspect, wastes from meat-processing plants characteristically have protein and fat contents far exceeding those in fruit and vegetable residues.

Because wet handling is the predominant mode of waste handling in food processing, the greater part of the solids are in a dissolved or suspended form, or both. "Slurry" would be a suitable descriptive designation. In vegetable processing, the solids residuals (as distinguished from suspended solids) range from a low 100 kg/tonne in tomato canning to 670 kg/tonne in pumpkin and squash canning.[13] The range in fruit processing is from 150 kg/tonne in cherry canning to 450 kg/tonne in pineapple canning.[14] From the 11,110,750 raw tonnes of the seven types of fruit processed in 1974, 4,244,760 tonnes of solid residues were generated, and about 3,038,450 tonnes of solid residue were produced from the 10,675,390 tonnes of vegetables (14 varieties) processed that year.[13,14]

The generally high moisture content of food-processing residues places a sharp restriction on the use of thermal-energy-recovery processes. Consequently, methane production becomes the principal candidate. Despite the fact that the water content of food processing residues is declining each year because of a greater degree of water recycling within the plant, a combination of logistical and economical problems together with a strong competition for use as an animal feedstuff takes away from the attractiveness of energy recovery through methane production, or any other method.

Two types of residues constitute exceptions to the negative prospects stated in the preceding paragraph. They are: (1) waste seed of peaches, nectarines, plums, apricots, olives and cherries, and shells of almonds, walnuts, and other nuts; and (2) bagasse. In California, some 136,000 tonnes of waste seed and shells were generated. In fact, a $2 million plant was built in the state in 1967 to convert waste seed and shells into charcoal briquets.

A great deal of attention is being directed to the energy potential of bagasse.[11] Bagasse is the product of the sugar cane extraction process and represents about 30% of the weight of the raw sugar cane. The amount of bagasse generated in the continental United States is on the order of 1.1×10^6 dry tonnes/yr, of which the greater part is concentrated in Louisiana, Mississippi, and Florida.[11] Hawaii and Puerto Rico could make sizeable contributions to this inventory. The quality of bagasse as an energy source is indicated by the data presented in Table 6. The carbon, hydrogen, and oxygen content is similar to that characteristic of plant organic matter. If one assumes a heating value of 18,930 kJ/kg, then the heating value of the total mass of bagasse generated in the nation (continental) would amount to 4.54×10^{13} kJ/yr.

The utilization of bagasse in the production of energy has become a very practical occurrence in Hawaii. Burning of crop wastes (mostly bagasse) accounts for approximately 13% of all electricity generated within the state. The percentage reaches 50% on the island of Hawaii (i.e., the "big island"). On the island of Kauai, a 21,650-kW electric power plant is presently under construction. The plant, when completed, will increase the island's existing electric generating capacity by 20% and will save the state the equivalent of 100,000 bbl oil/yr.[15]

TABLE 6. Typical Analysis of Bagasse

Carbon	46.2%
Hydrogen	6.4%
Oxygen	45.9%
Ash	1.5%
Moisture content	50%
Higher heating value	18,930 kJ/kg
Lower heating value	17,490 kJ/kg

7. Forest Products

The utilization of wood and wood wastes as sources of energy is by no means a newly discovered technology. Wood was the major fuel used in the generation of heat energy in the United States until the latter part of the nineteenth century, at which time it was replaced by fossil fuels—first by coal and then by petroleum and natural gas. The availability of alternate fuels together with the depletion of forests and the development of an ever-increasing demand for forest products in the form of raw materials for lumber, paper, and other items led to a decline in the use of forest products as a fuel to the point at which its use as a fuel is now negligible. However, the decline may be reversed by recent unfavorable developments in the area of energy availability and supply. These developments, particularly with respect to petroleum, are forcing the nation to look for alternate sources of energy. Wood, being a versatile and renewable resource, may once again become a major source of heat energy.

The percentages of the total volume of the timber felled in the United States that become logging residues, wood products, paper products, and fuel are listed in Table 7. According to the table, 85% by volume of the total round-wood trimmed logs harvested nationally is devoted to the manufacture of wood and paper products, and only about 4% becomes fuel. The remaining 11% is in the form of logging residue.

The steps involved from the harvesting to the utilization of forest resources are multiple and complex. The important fact is that in each step, a certain amount of residue is generated. In some steps the residue must be disposed of; in other steps, it may become a source of energy or a feedstock, or both, for another step (process). An example of the latter possibility is the manufacture of particle board from wood chips and sawdust.

In general, the wastes generated in lumbering and related industries can be classified into two fairly broad categories: (1) forestry wastes, and (2) manufacturing wastes.

TABLE 7. Approximate Disposition of Forest
Products[a]

Product	Percent by volume of total harvest
Logging residues	11.0
Wood products	57.0
Paper products	28.0
Fuel wood	4.0
Total	100.0

[a]Source: Reference 21.

The term "forestry wastes" is reserved for those wastes that are generated on the field. Types include annual litterfall, dead trees, forest-fire remains, and wastes generated in the course of culling and logging. Although sound estimates as to the overall amounts of forestry wastes are unavailable and difficult to make, it is certain that of the various types, the amounts of those from culling (which includes dead and over-age trees) and logging are the largest in volume and the most significant in terms of energy source.

Wastes generated as a direct result of an industrial (manufacturing) activity are classified as "manufacturing wastes." Two major industries serve as sources of manufacturing wastes, namely, wood products and pulp and paper. Although these wastes usually are placed in the broad category of industrial wastes discussed elsewhere, continuity and clarity are better served by speaking of them at this time.

7.1. Logging Residues

Debris from breakage, tree tops, tree limbs, small stems, and roots accumulated during a standard logging operation and typically considered to be nonusable, come within the category "logging residues." The qualification "typically" is made because certain of these materials may be harvested and chipped in the forest, and then be put to some use. It also should be noted that there are other residues that are normally left on the field but which are not a direct result of the logging operation *per se*. They include shrubs, saplings, and unsuitable trees.

7.1.1. Quantity

The amount of logging residues vary according to species, logging practices, geographical location, terrain, and other factors. The existence of the many variables render it quite difficult to arrive at an accurate estimate of total quantities. Current estimates range from 9 to 89 tonnes (dry wt)/ha.[16] The nationwide average is estimated as being approximately 20 tonnes/ha.[17] At an average generation of 20 tonnes/ha, the total amount of logging residues generated on a nationwide basis is about 180 million tonnes/yr (dry wt). Of the total, approximately 100 million tonnes are in the form of stumps and root systems,[12] and consequently only 70 million tonnes may be considered to be readily available.

7.1.2. Characteristics

Inasmuch as logging residues are primarily cellulosic and lignaceous in nature, carbon, oxygen, and hydrogen constitute the major elements in their molecular structure. Generally, carbon makes up about 50% of the mass, oxygen 40%, and nitrogen 5%. The heating value of logging residues generally is about 18,610 kJ/kg (dry weight). Ash content of the residues in excess of that normally characteristic of organic matter is a function of the method of collection. In certain operations, relatively large amounts of inert materials, in particular rock, dirt, and sand, are picked up along with the residues. The obvious result is a very sizeable percentage of ash. Generally, the ash content of wood is less than 1.0%, while that of bark may range from 2% to 10%.[18]

The moisture content of logging residues is comparable to that of hogged fuel, namely, about 40–60%.[19]

Using as a basis the numbers given in the preceding paragraphs, one could conclude that the total amount of recoverable energy from logging residues is about 1.3×10^{15} kJ/yr. This quantity is in agreement with more recent estimates.[11]

7.2. Residues from Wood Product Manufacturing

Since the lumber and plywood industry used approximately 105 million tonnes (dry wt) of raw material (roundwood) in 1970 to produce about 37 million tonnes (dry wt) of primary product, the processing wastes must have amounted to 68 million tonnes (dry wt) as residue.[20,21] The approximate disposition of the residue is indicated in Table 8. According to the table, approximately 18 million tonnes (dry wt), i.e., about 27% of the residues, were used as an energy source, while 36% remained unused. The energy content of the 18 million tonnes used as fuel amounted to a total of about 3.4×10^{14} kJ.

Residues from wood product manufacture that are used as sources of energy consist primarily of bark and "hogged" fuel.

Ultimate analyses and heating values of wood, bark, and hogged fuel are presented in Table 9. According to the table, carbon, oxygen, and hydrogen are the major elements in these residues and make up about 50%, 40%, and

TABLE 8. Approximate Disposition of Residues Generated in the Forest Products Industry in 1970

Disposition	Wood products (tonnes $\times 10^6$)	Pulp and paper products (dry wt)
Feedstock to other processes	25	0
Fuel	18	40[a]
Unused	24	0
Total	67	40

[a]Includes about 5 million tonnes of bark and 35 million tonnes of solids from spent black liquor.

TABLE 9. Ultimate Analyses of Wood, Bark, and Hogged Fuel[a]

Component	Yellow pine	Oak bark	Pine bark	Hogged Douglas fir
Carbon[b]	52.6	49.7	53.4	52.3
Oxygen[b]	40.1	39.3	37.9	40.5
Hydrogen[b]	7.0	5.4	5.6	6.3
Nitrogen[b]	—[c]	0.2	0.1	—
Sulfur[b]	—	0.1	0.1	—
Ash[b]	1.31	5.3	2.9	0.8
Heating Value (kJ/kg)	22,300	19,420	20,950	21,000

[a]Sources: *Steam, Its Generation and Use,* Babcock & Wilcox, New York (1972); Fryling, G. R., ed., *Combustion Engineering,* Combustion Engineering Inc., New York (1966).
[b]Values given as percent dry weight.
[c]Dashes indicate no data available.

6% of their dry weight, respectively. Heating values range from approximately 19,420 to 22,300 kJ/kg.

The moisture content of wood and bark varies considerably, not only because of the water bound in the cellular structure, but also because of the effect of time of year and geographic location on water content. Methods of storage and transportation also exert an influence on moisture content. For example, transportation of logs by water can bring the moisture content of the bark up to 80%.[19,22] Typically, the moisture of hogged fuel is about 50%.

Industries involved in the manufacture of wood products presently utilize 37% (dry wt) of their residues as a feedstock for other industries, primarily in the pulp and paper industry, and about 27% as a source of fuel. The increasing demand for both raw material and energy undoubtedly will lead to a reduction in the quantities of residues that presently are not being used. The deciding factor will be economics.

7.3. Residues from Pulp and Paper Manufacture

In 1972 the pulp and paper industry in the United States produced 37 million tonnes (dry wt) of wood pulp and about 49.5 million tonnes (dry wt) of paper and paperboard products. The total quantity of feedstock for this production amounted to approximately 76 million tonnes (dry wt), of which about 62% (47 million tonnes) came from virgin material (roundwood) and approximately 38% (29 million tonnes) from residues.[23] As shown in Table 8, production of the 76 million tonnes of feedstock resulted in a generation of about 40 million tonnes of residues (dry wt). The composition of the residues was approximately 34.6 million tonnes of solids from spent black liquor and 5.4 million tonnes of bark.

Two residues produced in pulp and paper manufacture are major sources of energy to the industry. They are bark and spent cooking liquor. Bark is the outer layer of the pulpwood and is removed at the pulpmill. Spent cooking liquor (black liquor) is a liquid waste generated in the pulping process and is characterized by a heavy concentration of dissolved organic chemicals. In order to recover the inorganic chemicals, the black liquor typically is concentrated and burned. Of the 2.7×10^{15} kJ energy utilized by the pulp and paper industry in 1972, approximately 35%, or 0.945×10^{15} kJ, was obtained from bark and black liquor.

The black liquor generated in the soda and sulfate process typically contains about 98% of the total alkali introduced into the digester to degrade and dissolve lignin, and thus free the wood fibers. If the liquor were discarded, it would be necessary to replace the alkali, and the cost of new chemicals combined with that of treating the waste would make the process prohibitively expensive.

Although the main function of black liquor treatment has been chemical recovery, energy production from the combustion of black liquor solids has assumed an equal importance. Black liquor contains about 23% total solids and a relatively high concentration of sodium carbonate. To recover the alkali, the black liquor must be evaporated. The evaporation increases the solids concentration from the initial 23% to a final 45–70%.[24]

Characteristics of typical black liquor solids are indicated in Table 10. As the data show, black liquor solids have a carbon content of about 43%, and a heating value of about 15,400 kJ/kg. The table also reveals the presence of a certain amount of impurities. These impurities can lead to the development of problems with the recovery equipment.

Because of the nature of the forest products industry as a whole, and particularly because of the integration between the various products and the indus-

TABLE 10. Composition and Heating Value of Black Liquor Solids[a]

Element	Percent
Carbon	42.6
Oxygen	31.7
Hydrogen	3.6
Sodium	18.3
Sulfur	3.6
Inert mineral oxides	0.2
Total	100.0
Heating value (kJ/kg of dry solids): 15,400	

[a]Source: Reference 24.

try's experience in utilizing wastes and residues, it is expected that (1) the amount of residues will be reduced, and (2) those that are generated will be fully utilized within the industry.

References

1. U.S. Environmental Protection Agency, Office of Solid Waste Management Programs, Resource Recovery and Waste Reduction, Fourth Report to Congress, EPA Publication SW-600, US EPA, Washington (1974), p. 142.
2. G. M. Fair, J. C. Geyer, and D. A. Okun, *Water and Wastewater Engineering, Vol. 2: Water Purification and Wastewater Treatment and Disposal,* John Wiley and Sons, Inc., New York (1968).
3. J. D. Denit, Development Document for Proposed Effluent Limitations Guidelines and New Source Performance for the Feedlots Point Source Category, EPA 440/1-73/004, Effluent Guidelines Division, Office of Air and Water Programs, U.S. EPA, Washington (1973).
4. E. G. Bruns and J. W. Crowley, Solid Manure Handling for Livestock Housing, Feeding, and Yard Facilities in Wisconsin, Bulletin No. A2418, University Extension, University of Wisconsin (1973).
5. R. C. Loehr, Animal wastes: A national problem, *J. Sanit. Eng. Div. Am. Soc. Civ. Eng.,* **85**, 181–221 (1969).
6. National Academy of Sciences, *Methane Generation from Human, Animal, and Agricultural Wastes,* NAS, Washington, D.C. (1977), p. 131.
7. L. L. Anderson, Energy Potential From Organic Wastes: A Review of the Quantities and Sources, Information Circular 8549, U.S. Department of the Interior, Bureau of Mines, Washington, D.C. (1972), p. 16.
8. B. L. Meck, Guidelines for Manure Use and Disposal in the Western Region, Washington Research Center Bulletin 814 (1975).
9. L. A. Schmid, Feedlot wastes to useful energy—fact or fiction, *J. Environ. Eng. Div.,* ASCE, *101* (EES), Proc. Paper **1647**:787–793 (October, 1975).
10. A. D. Poole and R. H. Williams, Flower power: Prospects for photosynthetic energy, in: *Toward a Solar Civilization* (R. H. Williams, ed.), MIT Press, Cambridge (1978), pp. 145–168.
11. J. M. Radovich, P. G. Risser, T. G. Shannon, C. F. Pomeroy, S. S. Sofer, and C. M. Sliepcevich, Evaluation of the Potential for Producing Liquid Fuels from Biomaterials, EPRI Report AF-974, Special Study Project TPS77-716, Electric Power Research Institute, Palo Alto (January, 1979), p. 127.
12. J. R. Beneman, Biofuels: A Survey, EPRI Report ER-746-SR, Electric Power Research Institute, Palo Alto (June, 1978), p. 88.
13. B. S. Luh and J. G. Woodroof, *Commercial Vegetable Processing,* The Avi Publishing Co., Westport (1975).
14. J. G. Woodroof and B. S. Luh, *Commercial Fruit Processing,* The Avi Publishing Co., Westport (1975).
15. P. M. Kohn, Hawaii: Alternative Energy Lab, *Chem. Engineering,* **86**(12), 86–90 (1979).
16. J. B. Granthan, E. M. Estep, J. M. Pierovich, H. Tarkow, and T. C. Adams, Energy and Raw Material Potentials of Wood Residue in the Pacific Coast States, Forest Service, USDA, Pacific Northwest Forest and Range Experiment Station, Portland, Oregon (1973), p. 52.
17. K. Howlett and A. Gamache, Silviculture Biomass Farms, IV: Forest and Mill Residues as a Potential Source of Biomass, Technical Report 7347, Georgia Pacific Corp., Mitre Corp. (May, 1977).

18. Chang Ying-Pe and R. A. Mitchell Chemical composition of common North American pulp-wood barks, *Tappi* **38** (5), 315–320 (1955).
19. G. R. Fryling, ed. *Combustion Engineering,* Combustion Engineering Inc., New York (1966).
20. T. H. Ellis, The Role of Wood Residue in the National Energy picture, in: *Wood Residue as an Energy Source,* Forest Products Research Society Energy Workshop, Denver (September, 1976), p. 118.
21. J. B. Grantham, Anticipated competition for available wood fuels in the United States, in: *Fuels and Energy from Renewable Resources,* (D. A. Tillman, K. V. Sarkanen, and L. L. Anderson, eds.) Academic Press, San Francisco (1977), pp. 55–91.
22. M. J. Leman, Air pollution abatement applied to a boiler plant firing salt water soaked hogged fuel, in: *Wood Residue as an Energy Source,* Forest Products Research Society Energy Workshop, Denver (September, 1976), pp. 60–65.
23. R. J. Auchter, Raw material supply, in: Future Technical Needs and Trends in the Paper Industry-II, Committee Assignment Report No. 64, Technical Association of the Pulp and Paper Industry, Inc., (1976).
24. R. G. MacDonald, ed. *Pulp and Paper Manufacture, I: The Pulping of Wood,* McGraw-Hill, New York, (1969).

2

Agricultural and Forestry Residues

PAUL G. RISSER

1. Introduction

Since the early 1950s, an interest has been demonstrated in largescale production of fuels from biomass.[1] Some industries are already utilizing biomass for energy on a limited scale. A Diamond-Sunsweet walnut factory near Stockton, California, has successfully tested a gasifier that converts walnut shells to low Btu gas. The estimated cost for the gas is about $1 per million Btu, less than half the cost of the natural gas it now burns.[2] The Andersons, a grain exporting firm, uses the cobs of the corn grains as a heating fuel.[3] The company assures a year-round supply of this heating–feedstock by offering higher prices to farmers during off-seasons.

The concept of energy farms, plants grown purposely for their fuel value, is often discussed,[4] and is covered in this book by Henry. It has been estimated that short rotation, close-spaced tree farms using high-yield species such as poplars or eucalyptus could produce as much as 4.5×10^{15} Btu annually on 10% of the currently idle forest and pasture land.[5] Some technical problems must be solved before large-scale production is possible, and environmental trade-offs will need to be considered.

But energy demands are increasing, and unless major changes are made in energy consumption patterns, it will be necessary to examine all the possibilities of energy sources currently available, including residues and wastes. Agricultural and forestry residues, municipal wastes, and animal manures can be used as feedstocks for conversion to liquid and gaseous fuels that could be

PAUL G. RISSER • Illinois Natural History Survey, 607 E. Peabody, Champaign, Illinois 61820.

available as peak load supplements to current energy sources. Municipal wastes and manures have been given much attention and are covered by Diaz and Golueke. Although agricultural and forestry residues have undesirable characteristics, such as high moisture content, bulky and inconvenient particle shapes, and large quantities being required for economically feasible conversion, they have many advantages. Residues are relatively abundant, available over a wide geographic area, and many types have little or no current economic value. Since the potential feedstock materials are available, renewable, and utilizable with current or soon-to-be-current technology, they deserve a close examination.

This chapter identifies and evaluates potential sources of biomass for energy feedstocks produced from residues by existing forestry and agricultural techniques. The abundance, seasonal and geographic availability, moisture content, mineral content, and energy equivalency values of the residues are tabulated and discussed. This information is then used for estimates of the amount of energy currently available from these sources on a state, regional, and national basis. The value of each type of residue as an energy feedstock is then determined on the basis of technology and transportation costs.

Several feedstock-oriented studies have been completed. These range from a detailed, county-by-county inventory of residues in the U.S.[6] to a comprehensive study of the potential for silvicultural biomass farms.[5] Such studies assist in providing necessary information for use in the future, when conversion technology is ready for wide-scale application. However, there is a need for a more detailed, comprehensive, inventory to indicate regions of the country that have a ready source of biomass. Also, there is a lack of specific literature on feedstock quality. This chapter intends to fill these needs by providing a current agricultural and forestry residue inventory in a readily interpretable form. Also, chemical, moisture, and energy content information on selected feedstocks is given.

2. Methods of Data Analysis

2.1. Species Selected for Study and Analytical Rationale

A "residue" is any material remaining after the desired portions of the plant have been removed. Forestry residues include wood and bark residues which accumulate at primary wood manufacturing operations during the production of lumber and other wood products, bark from pulp mills processing roundwood, and slash left on the forest floor following logging operations. Agricultural residues are crop plant parts left in the field after harvest, and residues accumulated at packing sheds during the sorting or cleaning of produce or initial nearsite processing of crop yields. Agricultural field residues are

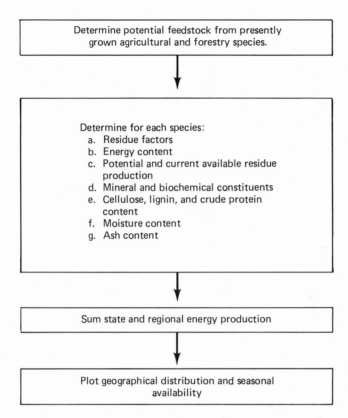

FIGURE 1. Analytical scheme used for data employed in the projection of potential biomass feedstocks.

usually plowed into the soil at zero disposal cost (not including the cost of plowing), while packing-shed residues are often sold as cattle feed, or disposed of at some economic loss.

The acquisition of information and subsequent analysis steps followed for this study are depicted in Figure 1. The species selected for evaluation (Table 1) were chosen because of previous bioconversion technology studies, or because they were judged to be crops which potentially could be grown specifically as feedstocks. Hay or forage crops were not considered because the whole plant is normally harvested and used.

Agricultural waste residues were estimated using the following calculations (Table 2):

$$\text{Crop yield} \times \text{Residue factor} = \text{Total residue}$$
$$\text{Total residue} \times \text{Availability percentage} = \text{Available residue}$$

TABLE 1. Species Selected for Evaluation as Feedstocks

Agricultural crop species		Forestry crop species	
Wheat	Sorghum	Douglas fir	Scotch pine
Grain corn	Rice	Loblolly pine	Western hemlock
Soybeans	Sugarcane	Longleaf pine	White fir
Oats	Cotton	Slash pine	Western red cedar
Potatoes	Peanut hay	Shortleaf pine	Redwood
Barley	Rye		
Sugarbeets	Seed grasses		

TABLE 2. Residue Factors, Availability Percentages, and Waste
Percentages Used in Calculating Agricultural Residues for Energy
Feedstock

Crop	Residue Factor	Availability Percentage	Waste Percentage
Wheat	2.53	85	99
Grain corn	1.10	90	45
Soybeans	2.14	80	100
Oats	3.01	85	75
Potatoes	1.40	90	100
Barley	2.50	85	99
Sugarbeet field	0.52	90	100
Sugarbeet pulp	0.07	100	0
Sorghum grain	1.57	90	40
Rice straw	1.43	90	100
Rice hulls	0.18	100	20
Sugarcane field	0.43	50	100
Sugarcane bagasse	0.38	100	0
Cotton	2.45	60	100
Peanut hay	1.48	95	90
Rye	2.50	85	99
Seed grasses	4.44	60	97

where

Crop yield = total yield of crop plant under consideration

Residue factor = a crop-specific estimator which when multi-
plied by the crop yield identifies actual amount
of residue produced by a unit of harvested crop
(Table 2)

Total residue = all residue generated by a crop, including
above-ground and below-ground portions

Availability percentage = the percent of the total residue either collected during normal harvesting procedures or considered to be collectable (Table 2)

Waste percentage = the percent of the available residue disposed of at zero or negative cost and not used for any other purpose, and/or left in the field and returned to the land (Table 2)

TABLE 3. Soybean Residue Availability by Season (Quarter) for Thirty-One Selected States

State producing crop	Percent total residue available per given quarter[a]			
	1st	2nd	3rd	4th
New York	0	0	0	100
New Jersey	0	0	0	100
Pennsylvania	0	0	0	100
Delaware	0	0	0	100
Maryland	0	0	0	100
Michigan	0	0	0	100
Wisconsin	0	0	0	100
Minnesota	0	0	50	50
Ohio	0	0	50	50
Indiana	0	0	50	50
Illinois	0	0	50	50
Iowa	0	0	0	100
Missouri	0	0	50	50
North Dakota	0	0	0	100
South Dakota	0	0	0	100
Nebraska	0	0	0	100
Kansas	0	0	50	50
Virginia	0	0	0	100
West Virginia	0	0	0	100
North Carolina	0	0	0	100
Kentucky	0	0	0	100
Tennessee	0	0	0	100
South Carolina	0	0	0	100
Georgia	0	0	0	100
Florida	0	0	0	100
Alabama	0	0	0	100
Mississippi	0	0	0	100
Arkansas	0	0	0	100
Louisiana	0	0	0	100
Oklahoma	0	0	50	50
Texas	0	0	0	100

[a]Quarter: 1st (winter), January–March; 2nd (spring), April–June; 3rd (summer, July–September; 4th (fall), October–December.

TABLE 4. Chemical Analysis of Selected Forestry Species in % Dry Weight[a]

Feedstock	Nitrogen		Sulfur		Hydrogen		Carbon		Oxygen		Ash		Calcium	Potassium
	wood	bark	wood	bark	wood	bark	wood	bark	wood	bark	wood	bark		
Douglas Fir *Pseudotsuga menziesii*	0.01	0.01	—	—	6.3	5.8	52.3	51.2	40.5	39.2	0.8	3.7	—	—
Loblolly pine *Pinus taeda*	0.01(sd)	—	—	—	6.3(sd)	5.6	51.8(sd)	56.3	41.3(sd)	37.7	0.5(sd)	—	—	—
Longleaf pine *P. palustris*	0.01(sd)	—	—	—	6.3(sd)	5.5	51.8(sd)	56.4	41.3(sd)	37.4	0.5(sd)	—	—	—
Slash pine *P. elliotti*	0.01(sd)	—	0.4	—	6.3(sd)	5.4	51.8(sd)	56.2	41.3(sd)	37.3	0.5(sd)	—	—	—
Shortleaf pine *P. echinata*	0.01(sd)	—	0.4	—	6.3(sd)	5.6	51.8(sd)	57.2	41.3(sd)	36.1	0.5(sd)	—	—	—
Scotch pine *P. silvestris* (80–100 yr old specimen)	1.18–1.56[12]b(n)		0.03–0.04[12](n)		—		—		—		—		0.14–0.31[12](n)	0.15–0.53[12](n)
	0.65[12](yb)		0.02(yb)		—		—		—		—		0.23[12](yb)	0.26[12](yb)
	0.15[12](tb)		trace[12](tb)		—		—		—		—		0.09[12](tb)	0.02[12](tb)

Western hemlock	0.1	0.0	0.1	5.8	6.2	50.4	53.0	41.4	39.3	2.2	1.5		
Tsuga heterophylla	1.48[12](n)	—	—	—	—	59.5[12](n)	—	—	—	—	—	0.38[12](n)	0.47[12](n)
White Fir													
Abies concolor	—	—	—	—	—	—	—	—	—	—	—	—	—
Grand Fir													
A. grandis	1.46[12](n)	—	—	—	—	62.8[12](n)	—	—	—	—	—	needles 0.91[12]	needles 0.46[12]
Western Redcedar													
Thuja plicata	—	—	—	—	—	—	—	—	—	—	—	—	—
Redwood													
Sequoia sempervirens	.01	—	—	5.9	—	53.5	40.3	—	0.2	—	—	—	—

[a] The data are for wood or bark as indicated in the column headings unless one of the following abbreviations appear in parentheses after a datum: n = needles, sd = sawdust, yb = young branch, tb = timber and bark.

[b] Superscript numbers in parentheses designate references.

TABLE 5. Mineral Content of Selected Crop Species[a]

Feedstock	Nitrogen	Sulfur	Calcium	Phosphorus	Potassium	Magnesium
Wheat						
grain	2.11[13]	—	0.028[14]	0.96[14]	0.48[14]	0.118[14]
stems	0.35[16]	—	0.0518–0.0721[15]	0.0259–0.0389[15]	—	0.00016–0.00019[15]
roots	0.82[13]	—	—	—	—	—
	0.89[17]	—	—	—	—	—
straw	0.5[7]	0.03–0.16[8]	—	—	—	—
leaves	1.13[16]	—	—	—	—	—
head	1.64[16]	—	—	—	—	—
Grain(field) corn						
stem	2.04[13]	—	—	—	—	—
Oats						
seeds	2.36[13]	—	0.09[18]	0.42[18]	—	—
straw	1.1[7]	—	—	—	—	—
whole plant	2.25[13]	—	—	—	—	—
Potatoes						
tuber	—	—	—	0.17–0.184[17]	0.185–2.84[17]	—
Barley						
grain	—	—	0.87[18]	0.45[18]	0.16–0.27[19]	—
straw	—	—	—	0.05–0.67[16]	0.93–2.78[19]	—
Sorghum						
hay	—	—	0.56[20]	0.31[20]	1.54[20]	0.4[20]
Rice						
grain	3.45–4.49[21]	0.152–0.197[21]	—	0.308–0.325[21]	2.51–3.05[21]	—
leaf	3.33–3.96[21]	0.075–0.103[21]	—	0.169–0.176[21]	0.675–0.898[21]	—
Peanut						
hay	1.696[22]	—	—	—	—	—
nut	—	—	0.0048[12]	0.0137[12]	0.0687[12]	0.0157[12]

Rye	—					
grain	—	—	0.07[14,]	0.38[14]	0.52[14]	0.13[14,]
Seedgrasses	—	—	0.41[20]	0.23[20]	2.2[20]	0.25[20]
Timothy hay	—	—	0.31[20]	0.19[20]	hay 1.66[20]	hay 0.17[20]
prebloom	1.648[20]		—	—	—	—
early bloom	1.088[20]		—	—	—	—
full bloom	0.992[20]		—	—	—	—
1st yr average	0.925[20]		—	—	—	—
2nd yr average	1.28[20]		—	—	—	—
average regrowth after harvest	2.496[20]		—	—	—	—
Orchard grass						
hay	3.728[20]		0.45[20]	0.32[20]	2.1[20]	0.32[20]
Kentucky bluegrass						
hay	1.312	2.5[20]	0.39[20]	0.27[20]	1.72[20]	0.21[20]
1st yr average	1.952[20]		—	—	—	—
2nd yr average	1.968[20]		—	—	—	—
average regrowth after harvest	2.88[20]		—	—	—	—
Chewings fescue(hay)	—		0.23[20]	0.19[20]	1.6[20]	0.21[20]
Red fescue(hay)	—		0.40[20]	0.24[20]	1.8[20]	0.16[20]
Tall fescue(hay)	—		0.30[20]	0.24[20]	2.0[20]	0.32[20]
1st yr average	1.84[20]		—	—	—	—
2nd yr average	1.664[20]		—	—	—	—
average regrowth after harvest	2.832[20]		—	—	—	—
Rye grass						
hay	—		0.46–0.49[20]	0.25–0.32[20]	1.96–2.0[20]	0.21–0.34[20]

[a] Expressed in percent dry weight.
[b] Computed from protein content.
[c] Superscript numbers in parentheses designate references.

TABLE 6. Composition and Energy Content of Selected Forestry Species[a]

Feedstock	Moisture content (% fresh wt)[a]		Volatile matter (% dry weight)[a]		Fixed carbon % (dry weight)[b]		Ash content % (dry weight)[a]		Cellulose/ hemicellulose content (% dry weight)[a]	Lignin content (% dry wt)[b]	Energy equivalency (Btu/lb dry wt)[b]		Magnesium content (% dry wt)[c]	Phosphorous content (% dry wt)[c]
	wood	bark	wood	bark	wood	bark	wood	bark			wood	bark		
Douglas fir	43.0	—	86.2	70.6(o)	13.7	27.2(o)	0.1	2.2(o)	—	—	8,890.0	9,790.0	—	—
Pseudotsuga menziesii		50.0		73.0(y)		25.8(y)		1.2(y)	—	—			—	—
Loblolly pine	51.0													
Pinus taeda				65.7		33.9		0.4	—	—	8,600.0	9,360.0	—	—
Longleaf pine														
P. palustris				67.0		32.3		0.7	—	—	8,600.0	9,360.0	—	—
Slash pine	61.0	53.0												
P. elliotti				65.9		33.4		0.7	d	c	8,600.0	9,360.0	—	—
Shortleaf pine														
P. echinata				65.5		33.8		0.7	—	—	8,600.0	9,360.0	—	—
Scotch pine														
P. silvestris									d	—	—	—	0.08–0.09(n)	0.03–0.12(n)
80–100 yr old											—	—	0.08(yb)	0.06(yb)
specimen											—	—	0.01(tb)	trace(tb)

Western Hemlock

Tsuga heterophylla	56.0	—	84.8	74.3	15.2	24.0	0.2	1.7	—	—	8,410.0	9,400.0	0.13(n)	0.12(n)
White fir														
Abies concolor	—	—	84.4	74.9	15.1	22.6	0.5	2.5	—	—	8,210.0	—	—	—
Grand Fir														
A. grandis	—	—	84.4	73.4	15.1	24.0	0.5	2.6	—	—	8,657.0(n)	—	0.16(n)	0.11(n)
Western Redcedar														
Thuja plicata	—	—	77.0	86.7	21.0	13.1	2.0	0.2	—	—	—	8,790.0	—	—
Redwood														
Sequoia sempervirens	—	—	83.5	71.3	16.1	27.9	0.4	0.8	—	—	—	—	—	—

[a] The data are for wood or bark as indicated in the column headings unless one of the following abbreviations appear in parentheses after a datum: n = needles, yb = young branch, tb = timber and bark, o = old, y = young.

[b] Data from Reference 2.

[c] Data from Reference 23.

[d] Cellulose content for *Pinus elliotti*: needles 42.5, top 41.5, branches 36.9, stem 51.1, bark 23.7, roots 44.6. Hemicellulose content for *Pinus sylvestris*: Needles 22.3, top 31.2, branches 33.7, stem 26.8, bark 24.9, roots 25.6.

[e] Lignin content for *Pinus elliotti*: needles 37.3, top 32.5, branches 35.1, stem 27.8, bark 49.9, roots 31.3.

TABLE 7. Composition and Energy Content of Selected Crop Species

Feedstock	Moisture content (% fresh wt)	Ash content (% dry wt)	Cellulose content (% dry wt)	Lignin content (% dry wt)	Crude protein content (% dry wt)	Energy equivalency 10^6 Btu/ton (dry wt)
Wheat	28.0[7],a	14.0[7]	—	—	—	15.0
bran	—	4.35[14]	—	—	—	—
grain	—	2.02[14]	—	—	16.1–19[23]	—
flour	—	0.59[14]	—	—	9.84–10.73[15]	—
straw	—	—	10.5[20]	—	3.9[22]	—
hay	—	—	—	—	6.1[22]	—
Grain corn (field)	47.0[7]	10.0[7]	—	4.2–8.8[20]	—	15.0
leaf blade	71.3[20]	—	20.6–23.2[20]	—	—	—
leaf sheat	67.0[20]	—	—	—	—	—
stem	74.8[20]	—	22.0[20]	—	12.75[13];b	—
Oats	10.0[7]	15.0[7]	—	—	15.21–19.93[25]	—
	—	3.22[18]	—	—	whole plant 14.06[18];b	15.0
Potatoes	89.0[7]	10.0[7]	—	—	—	15.0
tuber	—	—	4.2–4.35[7]	—	—	
stem	—	—	—	12.5[20]	—	
Barley	9.0[7]	15.0[7]	—	—	9.8–14.3[19]	—
grain	—	2.1–2.9[19]	—	—	—	—
grain fiber	—	2.98[18]	4.4–6.0[13]	—	—	—
straw	—	—	—	—	2.3–7.4[19]	—
whole plant	—	—	—	—	14.0[18];b	15.0
Sugarbeet						
field	80.0[7]	20.0[7]	—	—	—	15.0
pulp	7.0[7]	7.0[7]	—	—	—	15.0
Sorghum	60.0[7]	16.0[7]	—	—	—	15.0
Rice						
straw	20.0[7]	18.0[7]	—	—	—	13.1
hulls	4.0[7]	22.0[7]	—	—	—	15.6
bran	—	8.2–9.4[24]	—	—	13.1–15.4[24]	—
germ	—	9.1[24]	—	—	21.3[24]	—
whole plant	—	—	—	—	21.56–28.06[21];b	—
Sugarcane						
field	60.0[7]	16.0[7]	—	—	—	15.0
bagasse	50.0[7]	25.0[7]	—	—	—	16.6
Cotton	50.0[7]	19.0[7]	—	—	—	15.0
gin trash	8.0[7]	10.0[7]	—	—	—	—
Peanut(hay)	20.0[7]	9.0[7]	—	—	10.6[26]	15.0
Rye						
grain	28.0[7]	15.0[7]	—	—	—	15.0
flour	—	1.93[14]	—	—	—	—
bran	—	0.82[14]	—	—	—	—
middlings	—	2.63[14]	—	—	—	—
straw	—	0.83[14]	—	straw 14.0[20]	—	—
Seedgrasses (combined)						
Timothy(hay)	20.0[7]	4.0[7]	—	—	—	16.0
pre bloom	13.6[20]	6.0[20]	—	3.3–13.8[26]	10.3[20]	—
early bloom	21.2[20]	4.9[20]	—	—	6.8[20]	—
full bloom	23.9[20]	4.4[20]	—	—	6.2[20]	—

TABLE 7. (*continued*)

Feedstock	Moisture content (% fresh wt)	Ash content (% dry wt)	Cellulose content (% dry wt)	Lignin content (% dry wt)	Crude protein content (% dry wt)	Energy equivalency 10^6 Btu/ton (dry wt)
1st yr average	—	—	—	—	5.7[20]	—
2nd yr average	—	—	—	—	8.0[20]	—
average regrowth after harvest	—	—	—	—	15.6[20]	—
Orchard grass(hay)	11.4[20]	6.8[20]	—	—	7.7[20]	—
Kentucky blue grass(hay)	11.4[20]	6.8[20]	—	12.6–15.5[20]	8.2[20]	—
leaf	—	—	—	4.55[20]	—	—
stem	—	—	—	5.7[20]	—	—
1st yr average	—	—	—	—	12.2[20]	—
2nd yr average	—	—	—	—	12.3[20]	—
average regrowth after harvest	—	—	—	—	18.0[20]	—
Chewings fescue						
Red fescue	—	—	—	—	—	—
Tall fescue						
leaf blade	71.5[20]					
leaf sheat	75.4[20]	—	17.0–24.0[20]	—	—	—
stem	—	—	27.0[20]	—	—	—
flower	—	—	20.0[5]	4.5–7.5[20]	—	—
Rye grass						
leaf	—	18.9–26.0[20]	3.17.8[20]	—	—	—
blade	74.0[20]	—	—	—	—	—
sheath	77.0[20]	—	—	—	—	—
stem	—	—	22.0[20]	—	—	—
flower	—	—	23.0[20]	—	—	—

[a]Superscript numbers in parentheses designate references.
[b]Computed from various nitrogen content.

It is these wasted residues that are considered to be available for use as an energy feedstock. The factors and percentages used in these calculations were developed by Stanford Research Institute (SRI); the "waste percentage" is a combination of "percentage returned to the land" and "percentage wasted".[6]

The use of agricultural residues for energy feedstock depends in part on the seasonal availability. Seasonal availability of residues must be considered since long-term storage of these high-bulk, low-density products is expensive and results in deterioration of feedstock quality. Seasonal distributions for waste residues from the selected crops were estimated by multiplying waste residues by the seasonal availability percentage[6] in each quarter of the year (January–March; April–June; July–September; October–December). A sample of residue seasonality for soybean residue is given in Table 3.

2.2. Sources of Information

Forestry residue figures are those from the SRI residue inventory.[6] Since trees can be stored "on the stump" and for the most part, harvested as needed, seasonality was not considered to be an important factor influencing the use of forestry residues.

The value of a specific crop or forestry residue as an energy feedstock depends not only on its availability and abundance, but on its energy content. The estimated energy available from each crop[6] was then combined with chemical content data. Energy estimates from forestry residues were computed using the energy content of loblolly pine wood, because a detailed analysis of residues produced for all tree species was not available. Energy contents of other tree species are presented with the mineral content data for forestry species. All the estimates were calculated on the geographical scale of the state, and summed to give regional and national totals.

Data used in this study were obtained from existing literature. Values for forestry residue production were obtained from the SRI's Crop, Forestry, and Manure Residue Inventory for the continental United States.[6] The crop annual yields and per acre yields were from United States Department of Agriculture's Agricultural Statistics, 1976. Factors used in computing residue amounts and energy equivalents for the crops were published in 1976,[7] and energy equivalents for forestry species were published in 1977.[5] Information used in compiling the mineral content tables was obtained from various articles and publications indicated in the references.

3. Chemical Quality of Agricultural and Forestry Residues

The nitrogen and sulfur content of energy feedstock is important because of possible pollution which would be generated upon combustion. Relatively few data are available on nitrogen and sulfur contents of trees, but the nitrogen content is less than 1.5%, and the sulfur is less than 0.1% (Table 4). Nitrogen content in agricultural species ranges from about 0.4% in wheat to as high as 4.5% in rice. The sulfur content of rice is less than 0.2% and is 0.16% or less in wheat. These concentrations are lower than those for coal. In general, plants contain about the same amount of sulfur as phosphorus.[8] As is characteristic of these data, considerable differences in nutrient concentration arise in different laboratories, and the concentration varies with plant part, age, and growing conditions. The concentrations of calcium, phosphorus, potassium, and magnesium (Table 5) appear not to be present in amounts which would negatively affect the proposed conversion processes, and this ash might be useful as a fertilizer.[9] Conversion processes are more efficient when the feedstock used has a low moisture and ash content.[5] Trees have moisture contents of around

55% and ash contents of about 2.5% or less (Table 6). On the other hand, the moisture content of agricultural crops is frequently well above 55%, especially in succulent plants and plant parts such as corn leaves, potatoes, sugar beets, and sugar cane (Table 7). Coal itself has a moisture content of approximately 30%.[10]

The ash content of crop plants ranges from about 4% to as high as about 20% in rice and sugarcane. In summary, tree species, as a group, have lower concentrations of nitrogen, sulfur, moisture, and ash than agricultural crops, but wood, with its high proportion of lignin, has a much higher energy conten.[11]

4. Estimates of Agricultural and Forestry Residues

4.1. Regional Distribution

As much as 80% of the biomass is left on the forest floor following conventional logging practices in southeast pine forests.[27] Combined annual

TABLE 8. Regional and Total U.S. Agricultural Available Residues: Production [in 10^6 tonne (10^6 ton)][a] for Sixteen Selected Agricultural crops

Crop	Region[b]					U.S. Total
	I	II	III	IV	V	
Wheat	33.6 (37.0)	53.1 (58.5)	17.9 (19.7)	2.0 (2.2)	17.2 (19.0)	123.7 (136.7)
Grain corn field	1.46 (1.5)	33.9 (37.4)	1.8 (2.0)	4.8 (5.3)	39.7 (43.8)	81.6 (90.0)
Soybeans	0.0 (0.0)	23.7 (26.2)	8.0 (8.8)	11.5 (12.7)	34.9 (38.5)	78.2 (86.2)
Oats	0.9 (1.0)	10.3 (11.3)	0.7 (0.8)	0.45 (0.5)	5.8 (6.5)	18.2 (20.1)
Potatoes	10.9 (12.0)	1.8 (2.0)	0.18 (0.2)	0.6 (0.7)	4.7 (5.2)	18.2 (20.1)
Barley	10.3 (11.3)	5.9 (6.5)	0.27 (0.3)	0.18 (0.2)	1.08 (1.2)	17.4 (19.2)
Sugarbeet field	8.2 (9.0)	3.0 (3.3)	0.18 (0.2)	0.0 (0.0)	1.08 (1.2)	12.4 (13.7)
Grain sorghum	0.7 (0.8)	4.1 (4.5)	5.8 (6.5)	0.18 (0.2)	0.09 (0.1)	10.9 (12.1)
Rice straw	1.7 (1.9)	0.09 (0.1)	5.3 (5.8)	0.36 (0.4)	0.0 (0.0)	7.4 (8.2)
Sugarcane-field	0.0 (0.0)	0.0 (0.0)	1.5 (1.7)	1.9 (2.1)	0.0 (0.0)	3.4 (3.8)
Cotton	0.8 (0.9)	0.09 (0.1)	1.2 (1.3)	0.6 (0.7)	0.0 (0.0)	2.7 (3.0)
Peanuts	0.0 (0.0)	0.0 (0.0)	0.36 (0.4)	1.6 (1.8)	0.18 (0.2)	2.2 (2.4)
Sugarcane bagasse	0.0 (0.0)	0.0 (0.0)	0.45 (0.5)	0.6 (0.7)	0.0 (0.0)	1.09 (1.2)
Rye	0.9 (0.1)	0.5 (0.6)	0.09 (0.1)	0.18 (0.2)	0.36 (0.4)	1.3 (1.4)
Seed grasses	0.45 (0.5)	0.09 (0.1)	0.0 (0.0)	0.0 (0.0)	0.0 (0.0)	0.5 (0.6)
Rice hulls	0.09 (0.1)	0.0 (0.0)	0.18 (0.2)	0.0 (0.0)	0.0 (0.0)	0.27 (0.3)
Total	69.19 (75.8)	136.48 (150.6)	43.41 (48.5)	24.0 (27.7)	259.89 (116.1)	279.46 (418.7)

[a]Dry wt.
[b]Designated regions are shown in Fig. 2.

TABLE 9. Regional and Total U.S. Forestry
Available Residue Production (10^6 tons)[a]

Region	Production
I	19.2 (21.2)
II	1.7 (1.9)
III	8.6 (9.5)
IV	19.5 (21.5)
V	12.5 (13.8)
U.S. Total	61.5 (67.9)

[a]Dry wt.

wood-mill and logging residues in the United States represent over 7×10^{14}
Btu, the equivalent of 111×10^6 barrels of oil or about 1.5–2.0% of our annual
oil consumption.[28] Nevertheless, the total amount of residue from crops (Table
8)* is more than six times that from forestry (Table 9) in the five regions of

*Throughout this report, both metric and English units are used. When both sets of units are
given, the metric appears first, with its English equivalent in parentheses. Metric ton (tonne)
= 1000 kg; U.S. short ton (ton) = 2000 lb.

FIGURE 2. Map of the continental United States depicting regions as designated for use in the
study.

TABLE 10. Regional and Total U.S. Energy Equivalency (10^{12} Btu) of Agricultural Available Residues for Sixteen Selected Crops

Crop	Region					U.S. total
	I	II	III	IV	V	
Wheat	554.3	877.3	295.0	33.5	285.2	2045.3
Grain corn	22.1	561.4	29.3	78.8	657.4	1349.0
Soybeans	0.0	393.4	131.7	190.3	456.8	1172.2
Oats	14.9	169.2	12.3	7.0	97.1	300.5
Potatoes	178.3	30.4	3.0	10.1	78.5	300.3
Barley	165.5	97.1	3.9	3.3	18.7	288.5
Sugarbeet field	135.1	49.5	3.1	0.0	17.9	205.6
Grain sorghum	11.7	66.9	97.4	2.5	1.7	180.2
Rice straw	29.0	0.7	87.0	6.4	0.0	123.1
Sugarcane field	0.0	0.0	25.1	30.7	0.0	55.8
Cotton	14.0	1.0	19.0	9.9	0.0	43.9
Peanuts	0.2	0.0	6.6	27.1	2.7	36.6
Sugarcane bagasse	0.0	0.0	8.0	9.8	0.0	17.8
Rye	0.7	8.2	1.3	2.3	5.2	17.7
Seed grasses	6.7	1.9	0.1	0.2	0.3	9.3
Rice hulls	0.8	0.0	2.4	0.2	0.0	3.4
Sugarbeet pulp	0.0	0.0	0.0	0.0	0.0	0.0
Total	1133.4	2257.0	725.2	412.1	1621.5	6149.2

the country (Fig. 2). Of the sixteen crop plants chosen for evaluation, wheat, field corn, and soybeans produce 35% of the total available waste residue.

If the crops are considered on the basis of energy equivalents, wheat, corn, and soybeans contribute 74% of the total from the selected species (Table 10). According to the calculations, the energy available from crop residues is a little more than five times greater than that available from forestry residues (Table 11 and Figure 3).

TABLE 11. Regional and Total U.S. Energy Equivalency (10^{12} Btu) of Forestry Available Residues

Region	Production
I	364.8
II	33.1
III	162.7
IV	369.5
V	238.0
U.S. Total	1168.1

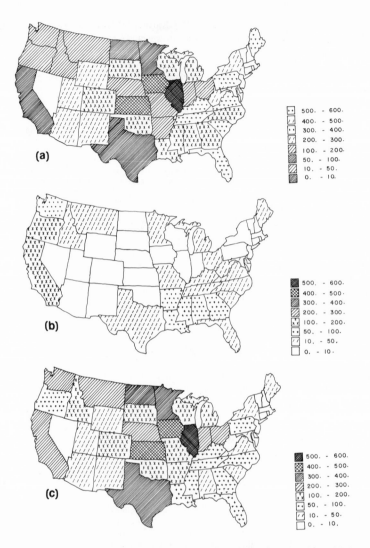

FIGURE 3. Agricultural crop residues (a), forestry residues (b), and total residues (c) by state in energy equivalents (10^{12} Btu).

Geographic considerations are important in determining the amount of area required for supplying a particular amount of biomass. For instance, the area required to supply 10^6 tons of corn silage is 956 square miles in Illinois, 1116 square miles in Iowa, 3003 square miles in Nebraska, and 20,833 square miles in the High Plains of Texas.[29]

Region II possesses the greatest amount of crop residues and the smallest amount of forestry residues. Conversely, Region IV has the greatest amount of energy in forestry residues, but the smallest amount in crop residues. Regions III and IV have about half the amount of total energy equivalents in the other three regions, and Region II has the greatest total energy equivalents, about 30% of the total in the continental United States (Table 12).

4.2. Seasonal Availability

Although forestry residues are largely available all year, there is a marked seasonality to the availability of agricultural crop residues (Fig. 4). In the win-

TABLE 12. Total Energy Equivalency (10^{12} Btu) of Agricultural and Forestry Available Residues, by State and Region

Region I		Region IV	
Washington	289.1	North Carolina	104.7
Idaho	205.0	South Carolina	70.8
Montana	260.0	Georgia	169.7
Colorado	107.3	Florida	74.0
Utah	19.0	Alabama	138.3
New Mexico	23.3	Mississippi	146.9
Arizona	45.0	Tennessee	76.7
Nevada	5.1	Total	781.1
California	351.1		
Oregon	201.4	Region V	
Total	1506.3	Maine	61.4
		New Hampshire	54.0
Region II		Vermont	4.0
Minnesota	364.4	Massachusettes	2.3
Iowa	479.0	Rhode Island	1.2
Missouri	205.5	Connecticut	16.0
Kansas	429.2	New York	37.6
Nebraska	291.2	New Jersey	7.3
South Dakota	148.8	Pennsylvania	54.1
North Dakota	372.4	Delaware	13.5
Total	2290.5	Maryland	41.9
		Virginia	60.2
Region III		West Virginia	48.7
Arkansas	204.4	Kentucky	76.5
Louisiana	149.3	Ohio	265.2
Texas	351.8	Indiana	298.7
Oklahoma	180.5	Michigan	143.7
Total	886.0	Illinois	599.0
		Wisconsin	104.7
		Total	1890.0
	U.S. Total	7353.9	

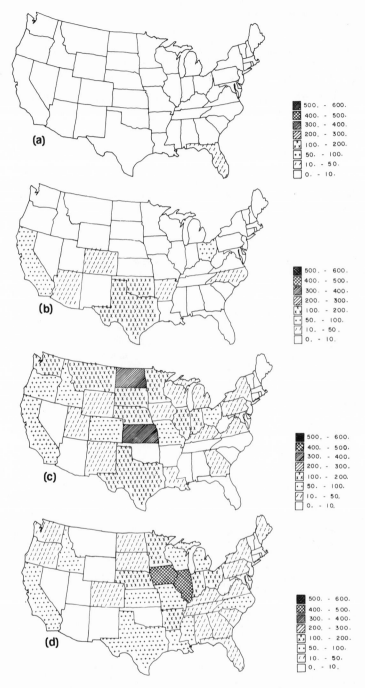

FIGURE 4. Agricultural crop residues in (a) winter, (b) spring, (c) summer, and (d) fall expressed as energy equivalents (10^{12} Btu).

TABLE 13. Agricultural Available Residue Production for Six Selected Species[a]

Crop	Region					U.S. average
	I	II	III	IV	V	
Wheat	5.53 (2.44)	4.03 (1.78)	3.45 (1.52)	4.08 (1.80)	5.72 (2.52)	4.56 (2.01)
Corn (grain)	3.42 (1.51)	2.77 (1.22)	3.45 (1.52)	2.06 (0.91)	3.52 (1.55)	3.04 (1.34)
Soybeans	0.00 (0.00)	3.33 (1.47)	2.79 (1.23)	.73 (0.32)	3.83 (1.69)	2.13 (0.94)
Oats	3.33 (1.47)	0.29 (0.13)	2.22 (0.98)	3.06 (1.35)	0.43 (0.19)	1.87 (0.82)
Potatoes	20.77 (9.16)	24.06 (10.61)	27.01 (11.91)	23.41 (10.32)	33.02 (14.56)	25.65 (11.31)
Barley	1.56 (0.69)	4.20 (1.85)	3.63 (1.60)	4.58 (2.07)	5.24 (2.31)	3.84 (1.70)

[a]Values in tonnes/hectare. The numbers in parentheses are the equivalents in tons/acre.

ter, some crop residues are available, but mostly in California and Florida. By spring, residues are available in most of the southern half of the United States. During the summer and fall, crop residues are available through virtually all of the country.

Individual crop species vary as to seasonality across the country. For example, 55% and 45% of the tomato crop in California is available in the first and third quarters of the year; in Florida, 80% in the second quarter, and 20% in the fourth; and in Washington 100% in the fourth quarter.[30]

The per hectare production for the top six residue producing crops is given in Table 13. Potatoes, the sixth ranked crop in available residue productions, produce the most residue per hectare; an average of 25.65 tonne/ha (11.31 tons/acre), with Region I having the highest yield of 20.77 tonne/ha (9.16 tons/acre) and Region V the second highest, 33.02 tonne/ha (14.56 tons/acre). Wheat, which is the most available waste residue produced among the crop species, has an average per hectare residue yield of 4.56 tonne/ha (2.01 tons/acre). Region V has the best per hectare yield of available waste residues, 5.72 tonne/ha (2.52 tons/acre), and Region I is next with 5.53 tonne/ha (2.44 tons/acre). The per hectare production values of available waste residues are useful in predicting the area of land required to furnish sufficient feedstock to support a conversion plant and in predicting potential residue production for future use.

References

1. C. Wilson, *Energy for Survival: The Alternative to Extinction,* Garden City, N.Y., Anchor Press/Doubleday (1975).
2. A. L. Hammond, Photosynthetic solar energy: Rediscovering biomass fuels, *Science* **197**, 745–746 (1977).

3. W. Hudson, Commercial use of corn cob residue, in: *Biomass—a cash crop for the future?*, Midwest Research Institute and Battelle Columbus Laboratories, Kansas City, Missouri (March 2–3, 1977), pp. 116–120.

4. Szego, G. Design, operation and economics of the energy plantation, Proceedings of a conference on capturing the sun through bioconversion, Washington, D.C. (March 10–12, 1976), published by Washington Center for Metropolitan Studies, Massachusetts Avenue, Washington D.C., pp. 217–240.

5. Silvicultural Biomass Farms, Mitre Corporation, Mitre Technical Report 6, 7347 (1977).

6. Crop, Forestry and Manure Residue Inventory—Continental United States, Vol. 1–8, Stanford Research Institute, Menlo Park, California (1976).

7. An Evaluation of Agricultural Residues as an Energy Feedstock, Vol. 1, 2, Stanford Research Institute, (1976).

8. B. B. Tucker and J. M. Baker, Jr., Sulfur content of wheat straw grown in Oklahoma, Agricultural Experimental Station, Oklahoma State University, Bulletin B-7 (1972).

9. J. R. Host and S. R. Pfenninger, Plant nutrients in flyash from barkfired boilers, U.S. Forest Service, Intermountain Forest and Range Experiment Station, Research Note INT-247 (1978).

10. D. T. McPhee, Report of panel discussion, in: *Biomass—a cash crop for the future?* Midwest Research Institute and Battelle Columbus Laboratories, Kansas City, Missouri (March 2–3, 1977), p. 337.

11. I. Falkehag, Waste in bioproduction, bioconversions, and bioconsumption, Proceedings of a conference on capturing the sun through bioconversion, Washington, D.C. (March 10–12, 1976), pp. 161–172.

12. A. D. Allen, Biomass potential from underexploited species, in: Proceedings of a conference on the production of biomass from grains, crop residues, forages, and grasses for conversion to fuels and chemicals, Midwest Research Institute, (March 2–3, 1977), pp. 27–33.

13. P. L. Altman and D. S. Dittmer, *Environmental Biology*, Federation of the American Society for Experimental Biology, (1966), Table 161, p. 489.

14. B. Singh and N. R. Reddy, Phytic acid and mineral composition of Triticales, *J. Food Sci.* 42(3), 1077–1083 (1977).

15. A. Nahapetian and A. Bassiri, Variations in concentrations and interrelationships of phytic phosphorus, magnesium, calcium, zinc, and iron in wheat varieties during two years, *J. Agriculture and Food Chem.* 24(5), 947–950 (1976).

16. L. A. Daigger, D. H. Sander, and G. A. Peterson, Nitrogen content of winter wheat during growth and maturation, *Agron. J.* 65(5), 815–818 (1976).

17. P. Muthusivomy, V. Rairkumar, V. Sundaramoorthy and K. K. Krishnamoorthy, Mineral content of potato tubers as influenced by application of industrial and mineral wastes, *J. Indian Soc. Soil Scien.* 24(4), 421–426 (1976).

18. I. R. Sibbald and K. Price, Relationships between metabolizable energy values for poultry and some physical and chemical data describing Canadian wheats, oats, and barleys. *Canad. J. Anim. Sci.* 56(2), 255–268 (1976).

19. G. Fadak and A. R. Mack, Influence of soil moisture levels and planting dates on field and chemical fractions in two barley cultivars, *Canadian J. Plant Sci.* 57(1), 261–268 (1977).

20. G. W. Butler and W. Bailey, eds., *Chemistry and Biochemistry of Herbage*, Vol. 1–3, New York, Academic Press, (1973).

21. P. B. Pillai and H. G. Singh, Effect of different sources of sulfur and iron of flag leaf composition and grain field of rice on calcareous soils, *Indian J. Agricultural Sci.* 45(8), 340–343 (1975).

22. H. D. Hughes, M. E. Heath, and D. S. Metcalf, *Forages: The Science of Grassland Agriculture*, 2nd ed., Iowa State University, Ames, Iowa (1973).

23. S. Dubetz, Effects of high rates of nitrogen on neepawa wheat grown under irrigation, part I: Yield and protein content, *Can. J. Plant Sci.* **57**(2), 331–336 (1977).

24. A. A. Betschart, R. Y. Fong and R. M. Saunders, Rice by-products: Comparative extraction and precipitation of nitrogen from U.S. and Spanish bran and germ, *J. Food Sci.* **42**(4), 1088–1093 (1977).

25. H. W. Ohm, Response of 21 oat cultivars to nitrogen fertilizer, *Agron. J.* **68**(5), 773–775 (1976).

26. T. H. Brown, Rate of loss of dry matter and change in chemical composition of nine pasture species over summer, *Austral. J. Exper. Agric. An. Husbandry* **17**(84), 75–79 (1977).

27. P. Koch and C. W. Boyd, Effect of harvesting method on biomass removal from southern pine sites, Mississippi State University, (March 21–22, 1978), pp. 54–59.

28. A. R. Ek and D. H. Dawson, Yields of intensively grown *Populus:* Actual and projected, pp. 5–9. Intensive Plantation Publication, U.S. Forest Service, North Central Experiment Station (1976), pp. 5–9.

29. T. A. McClure, Corn production practices, in: *Biomass—A Cash Crop for the Future?*, Midwest Research Institute and Battelle Columbus Laboratories, Kansas City, Missouri (March 2–3, 1977), pp. 145–177.

30. J. A. Alich, Crop, forestry, and manure residues: An energy resource. Proceedings of a conference on Capturing the Sun through Bioconversion, Washington, D.C. (March 10–12, 1976), Published by Washington Center for Metropolitan Studies, Massachusetts Avenue, Washington, D.C., pp. 127–138.

3
Aquatic Biomass

PAUL G. RISSER

1. Introduction

The possibility of using plant biomass as a fuel has been discussed for some time.[1-4] As discussed by Risser, it has been estimated that agricultural and forestry wastes could contribute as much as 2% of the country's current annual fossil energy use. In 1957, Tamiyo discussed the possibilities of using mass algal cultures as an energy supply and concluded that the future was very encouraging for these large-scale systems.[5] However, the following discussion will show that more recent authors have presented far less optimistic predictions.

The ocean has been proposed as a possible biomass energy source[6] particularly because of the apparently large production rates of giant kelp, *Macrocystis pyrifera*,[7,8] but also because of benthic plants[9] and algae.[10] Although the productivity of estuaries has been studied,[11] it is the salt marshes that have been considered to represent habitats of potential high biomass production.[12-15] Freshwater marshes are known to have relatively high rates of primary productivity.[16-18] In addition, selected species such as water hyacinths *(Eichhornia crassipes)* have been considered potential biomass sources because of their rapid growth rates.[16,19]

This chapter reviews the potential of aquatic biomass as a source for energy feedstocks. For biomass to represent a convenient feedstock, it must be available in adequate amounts and have sufficient energy content so that the energy conversion process can operate efficiently. In considering the potential

PAUL G. RISSER • Illinois Natural History Survey, 607 E. Peabody, Champaign, Illinois 61820.

use of aquatic biomass, nutrient content, especially nitrogen and sulfur, must be considered because the effluents from the conversion process may not meet existing air quality standards or the operations may require costly equipment needed to reduce the resulting nitrous and sulfur oxides.

Terrestrial and marine biomass are treated elsewhere in this volume, and this chapter is limited to potential feedstock production from freshwater lakes and streams and from wetlands. For convenience, the biomass sources are discussed by first considering algae and then aquatic macrophytes.

2. Algal Biomass

Algae represent the organisms that are the basis of most aquatic food chains and, as such, have received considerable study.[20,21] As a potential biomass source, these organisms have been regarded as particularly promising because the gross photosynthetic rate is low relative to the net photosynthetic rate. Unlike vascular plants, algae do not develop large amounts of nonphotosynthetic secondary tissue which then require energy for respiration.

2.1. Yield

Algae have been considered as a potential food or protein source, and various arrangements for their mass culture have been investigated. Tamiyo[5] reviewed the literature on algal mass cultures in 1957 and summarized the range of average yields from 5 to 20 g dry wt-m^{-2}-day^{-1}. Tamiyo also suggested that the potential was quite promising for algal production of biomass and as a source of protein. Since 1957, investigators have developed various mass culture systems in various parts of the world. The average yields are usually 15–25 g dry wt-m^{-2}-day^{-1} and the maximum rates are in the range of 30–40 g dry wt-m^{-2}-day^{-1}.[4] As Goldman discusses, these rates represent about 5% of the incoming solar radiation and are limited by the quantum yield of the photosynthetic process and the light saturation response characteristic of all plants which utilize the Calvin cycle of CO_2 fixation.[3] Further, additional energy is lost by reflective losses of sunlight and through respiration of the algal cells. These mass culture systems depend upon large quantities of sunlight since this is a diffuse form of energy even under conditions of full sunlight. As a consequence, the culture systems must be outdoors to take advantage of the natural energy, and large enough to produce significant biomass, and as such, require massive systems with reasonably sophisticated environmental controls.

2.2. Feasibility of Culture Systems

Algae use solar energy which is a "free" source of energy. The cells can be fermented to methane with conversion efficiencies of 50–70%.[4,22] Nevertheless, their potential as a source of biomass feedstock appears to be limited at best. Obviously, energy of some type is necessary to build and operate the mass culture systems. If algae were to be used as large sources of energy, extremely large amounts of water and nutrients, especially nitrogen and phosphorous, would be required. Furthermore, large land areas would be required if algal cultures were to be effective in supplying useful amounts of energy. It has been estimated that it would require over one-half million ha (2000 mi^2) to produce energy through algal mass cultures equivalent to that projected for the energy produced from one large coal gasification plant.[6] It has also been projected that an area the size ranging from parts of New England to the size of Texas would be required for algal mass cultures to produce 5% of the projected energy demands of 1990.[23] So in spite of promising inherent characteristics of algal cells and several decades of investigations, it appears unlikely that mass algal cultures will be a significant biomass feedstock source or even a net-energy producing system.[24] This negative assessment comes not so much from the inherent characteristics of the algae, but rather from the large amounts of space and nutrients that would be required for mass cultures producing significant amounts of feedstock material.

3. Aquatic Macrophyte Biomass

3.1. Yield of Selected Species

A number of aquatic weeds grow rapidly and are considered to be nuisance plants. Although these species may be controlled by water level changes,[25] they frequently produce large amounts of biomass under the nutrient-rich runoff conditions found in urban or agricultural areas.[26] As an example, water hyacinth *(Eichhornia crassipes)* grows rapidly and may choke waterways in southeastern United States.[19] However, because each kilogram dry weight of water hyacinth can produce almost 400 liters of biogas containing approximately 65% methane, these plants have been considered as an energy feedstock.[27] Available data[16] indicate that annual productivity of water hyacinth is only from 150 to 440 g dry wt-m^{-2}. Penfound[19] reports growth of 14.6 g dry wt-m^{-2}-day^{-1} from June 23 through August 6; higher rates in the order of 30 g dry wt-m^{-2}-day^{-1} might be possible over short time periods if only young, fast-growth plants are maintained in the system.[16] Under these conditions, significant amounts of nitrogen and phosphorus would be required to maintain this rapid growth system.[28]

Westlake[16] summarized the biomass standing crop and production rates of several aquatic plants (Table 1). Some of these measured values are questionable [e.g., giant bulrush *(Scirpus lacustris)*; although Auclair *et al.*[18]

TABLE 1. Maximum Standing Crop and Growing Season Productivities of Representative Aquatic Macrophytes[a]

Species	Maximum standing crop (g dry wt-m^{-2})	Growing season productivity (g dry wt-m^{-2}-day^{-1})
Marsh cordgrass *(Spartina alterniflora)*	4,200	10.0
Great bulrush *(Scirpus lacustris)* and giant reed *(Arundo donax)*	10,000	28.0
Coon tail *(Ceratophyllum demersum)*	710	2.5
Water parsnip *(Berula* species) and crowfoot *(Ranunculus* species)	500	4.2
Wapato *(Sagittaria latifolia)*	810	7.5
Water hyacinth *(Eichhornia crassipes)*	1,500	7.4–22.0

[a]Source: Reference 16.

showed approximately equivalent values for river bulrush *(Scirpus fluviatlis)*] and some measurements include other organisms in the ecosystem (e.g., measurements for marsh cordgrass, *Spartina alterniflora,* include the associated algae). Penfound[16] reported standing crop values of approximately 1500 g dry wt-m^{-2} for cattails, *Typha latifolia,* and 800 g dry wt-m^{-2} for lizards tail, *Saururus cernus.* Hopkinson et al.[14] measured the annual production of seven species in a Louisiana salt marsh (Table 2).

Inclusion of the weights of underground plant parts can have a significant influence on the amount of reported biomass. For example, Klopatek and Stearns[29] studied the emergent macrophytes in a Wisconsin freshwater marsh and found that annual net primary productivity ranged from 1181 g dry wt-m^{-2} for sedge, *Carex lacustris,* to 3200 g dry wt-m^{-2} for cattails, *Typha latifolia.* Penfound's[19] study in Oklahoma of the aboveground parts found the annual production of cattails to be only 800 g dry wt-m^{-2}.

TABLE 2. Annual Biomass Production of Seven Species in a Louisiana Saltmarsh[a]

Species	Annual biomass production (g dry wt-m^{-2}-yr^{-1})
Common salt grass *(Distichlis spicata)*	3237
Rush *(Juncus roemerianus)*	3416
Common reed *(Phragmites communis)*	2318
Arrowhead *(Sagittaria falcata)*	1501
Marsh cordgrass *(Spartina alterniflora)*	2658
Big cordgrass *(S. cynosuroides)*	1355
Saltmeadow cordgrass *(S. patens)*	6043

[a]Source: Reference 14.

3.2. Yield of Marsh Communities

In a large marsh complex in southern Quebec, Canada, during a 150-day growing season, the shoot net productivity averaged 6.1 g dry wt-m^{-2}-day^{-1} and the terminal standing crop was 845 g dry wt-m^{-2}. This community was dominated by river bulrush *(Scirpus fluviatilis)*, water horsetail *(Equisetum fluviatile)*, great bulrush *(Scirpus validus)*, common reed *(Phragmites communis)*, and spike-rush *(Eleocharis palustris)*. Although the average rate of productivity was around 6 g-m^{-2}-day^{-1}, the authors[18] noted that the seasonal productivity demonstrated a strong bimodal pattern with peaks in late July (12.9 g-m^{-2}-day^{-1}) and mid September (4.7 g-m^{-2}-day^{-1}).

Biomass production in freshwater wetland areas may be influenced by hydrological cycles,[25,30] biological controls[31] and nutrients.[18,32] However, rich swamps usually are characterized by annual dry weight production values of 2000–3700 g-m^{-2}. Salt marshes may produce 3500 g-m^{-2} or more each year, especially if the algae components are added. Phytoplankton biomass is usually highest where the water is enriched by nutrients and may range up to 700 g dry wt-m^{-2}.[16] Streams and rivers are quite variable, but may produce between 100 and 600 g dry wt-m^{-2}-yr^{-1}.[16,23,33] It is therefore clear that wetlands, but not aquatic habitats, are very productive and are capable, especially where nutrients are plentiful, of producing large quantities of biomass.

3.3. Energy Equivalents and Nutrient Contents

The energy equivalents, nutrient concentrations, and ash contents of the biomass are important considerations if the material is to be used as energy feedstock. Most plants contain 46–48% carbon in the organic matter constituents of their tissue, but some algae with high fat content (and, therefore, high energy equivalent) contain up to 54%. On the other hand, most terrestrial plants contain only about 5% ash content, but for calcareous aquatic plants, the ash content ranges from about 25% in some pondweeds (Potamogetonaceae) to 50% in some of the stoneworks (Characeae) and up to 90% in the corals (Corallinaceae).[16]

Some algae have energy equivalents significantly higher than terrestrial plants, but because of the relatively high ash content of most macrophytes (water hyacinth ranges from 10–30% ash), the energy equivalents relative to the dry weights are approximately equal to terrestrial plants.[34-36] In a study of 11 species of vascular aquatic plants, Boyd[34] found a range of 3906–4552 kcal-g^{-1} dry wt. Auclair[37] examined the tissue contents of five wetland species and found that the ash content ranged from 5.9% in common reed *(Phragmites communis)* to 15.6% in water horsetail *(Equisetum fluviatile)*, with an average of 8.5% dry weight. Nitrogen concentrations ranged from 1.2–2.1% dry weight; therefore, carbon and nitrogen ratios are usually between 20:1 and 30:1, and these ratios are appropriate for methane-producing bacteria.

4. Discussion

The advantages of mass algal cultures are that solar energy is used and that relatively small amounts of energy are lost as respiration since secondary tissue is lacking in these organisms. The rate of productivity averages 15–25 g dry wt-m^{-2}-day^{-1} with maximum rates as high as 40 g dry wt-m^{-2}-day^{-1}. Although this rate of productivity is adequate for consideration as feedstock production, large amounts of dedicated land and large quantities of both water and nutrients are required. For these reasons, mass algal cultures have never achieved the energy feedstock production levels which have been predicted from the algal growth rates in small cultures.

Current research is addressing questions about methods to harvest algae and to increase algal populations without the tremendously large demands for space, nutrients, and water. In some instances, nutrients are obtained from oxidation ponds used for wastewater treatment while space and water are conserved by the use of either shallow, aerated, high-flow ponds or recirculating systems. Under these specialized techniques, rates of approximately 60 g dry wt-m^{-2}-day^{-1} have been produced on a small scale. Whether these systems can be developed and are economically feasible has yet to be determined.

Wetland communities, with emergent aquatic macrophytes have very high rates of annual productivity (as much as 6000 g dry wt-m^2-yr^{-1}) and produce relatively large quantities of biomass. Therefore, their potential for producing feedstocks on a per unit area is favorable. However, there are relatively small areas of these isolated wetlands, and, like virtually all herbaceous sources of feedstock biomass, the rates of production vary seasonally. Exceptions may exist along the Gulf Coast, where growth occurs throughout the year, but the winter growing rates are very slow. Although nitrogen and sulfur contents are satisfactorily low, the energy content is not as high as for some terrestrial biomass sources. Also, wetlands are a fragile ecosystem, and any disruption would not only affect the subsequent regrowth capability of the plant community but would have an impact on the adjacent or associated wetland biological systems. Therefore, while aquatic macrophytes and wetland communities may be possible sources of biomass, the technical problems of converting these biomass types must be overcome. An evaluation must be made of the geographical and seasonal availability and the ecological consequences of biomass removal.

References

1. M. Calvin, Photosynthesis as a resource for energy and materials, *Am. Sci.* **64**, 270–278 (1976).
2. A. Mitsui, S. Miyachi, A. San Pietro, and S. Tamura (eds.), *Biological Solar Energy Conversion,* Academic Press, New York (1977).

3. J. A. Bassham, Photosynthesis: The path of carbon, in: *Plant Biochemistry* (J. Bonner and J. E. Varner, eds.), Academic Press, New York, NY (1965), pp. 875–902.

4. J. C. Goldman, Outdoor algal mass cultures, I: Applications, *Water Res.* **13**, 1–20 (1979).

5. H. Tamiyo, The mass culture of algae, *Ann. Rev. Plant Physiol.* **8**, 309–334 (1957).

6. M. S. Doty, Status of marine agronomy, with special reference to the tropics, in: *Proceedings of the Ninth International Seaweed Symposium,* (A. Jensen and J. Stein, eds.) Science Press, Princeton NJ (1978), pp. 35–38.

7. G. A. Jackson, Nutrients and production of giant kelp, *Macrocystis pyrifera,* off southern California, *Limnol. Oceanogr.* **22**, 979–995 (1977).

8. M. R. Hart, D. deFremery, C. K. Lyon and G. O. Kohler, Processing of *Macrocystis pyrifera* (Phaeophyceae), for fermentation to methane, in: *Proceedings of the Ninth International Seaweed Symposium,* (A. Jensen and J. Stein, eds.), Science Press, Princeton, NJ (1978), pp. 493–498.

9. C. W. Schneider, and R. B. Searles, Standing crop of benthic seaweeds on the Carolina continental shelf, in: *Proceedings of the Ninth International Seaweed Symposium,* (A. Jensen and J. Stein, eds.), Science Press, Princeton, NJ (1978), pp. 293–301.

10. R. W. Eppley, E. H. Renger and W. G. Harrison, Nitrate and phytoplankton production in southern California coastal waters, *Limnol. Oceanogr.* **24**, 483–494 (1979).

11. D. L. Correll, Estuarine productivity, *BioScience* **28**, 646–650 (1978).

12. A. A. de la Cruz, The role of tidal marshes in the productivity of coastal waters, *Assoc. Southeast Biol. Bull.* **20**, 147–156 (1973).

13. B. G. Hatcher and K. H. Mann, Aboveground production of marsh cordgrass *(Spartina alterniflora)* near the northern end of its range, *J. Fish. Res. Bd. Can.* **32**, 83–87 (1975).

14. C. S. Hopkinson, J. G. Gosselink, and R. T. Parrondo, Aboveground production of seven marsh plant species in coastal Louisiana, *Ecology* **59**, 760–769 (1978).

15. D. A. White, T. E. Weiss, J. M. Trapani, and L. B. Their, Productivity and decomposition of the dominant salt marsh plants in Louisiana, *Ecology* **59**, 751–759 (1978).

16. D. F. Westlake Comparisons of plant productivity, *Biol. Rev.* **38**, 385–425 (1963).

17. C. R. Goldman (ed.), *Primary Productivity in Aquatic Environments* (proceedings of an International Biological Programme Fresh Water Production symposium, Pallanza, Italy, April 26–May 1, 1965), University of California Press, Berkeley, CA (1966).

18. A. N. D. Auclair, A. Bouchard, and J. Pajaezkowski, Plant standing crop and productivity relations in a *Scirpus-Equisetum* wetland, *Ecology* **57**, 941–952 (1976).

19. W. T. Penfound, Primary production of vascular aquatic plants, *Limnol. Oceanogr.* **1**, 92–101 (1956).

20. M. Brylinsky and K. M. Mann, An analysis of factors governing productivity in lakes and reservoirs, *Limnol. Oceanogr.* **18**, 1–14 (1973).

21. J. R. Beneman, Biofuels: A survey, ER-746-SR Electric Power Research Institute, Palo Alto, CA (1978).

22. C. G. Golueke and W. J. Oswald, Power from solar energy via algae-produced methane, *Solar Energy* **7**, 86–92 (1963).

23. E. Ashare, D. C. Augenstein, A. C. Sharon, R. L. Wentworth, E. H. Wilson, and D. L. Wise, Cost analysis of algae biomass systems, Report No. 1738, Dynatech R/D Co., Cambridge, MA. (1978).

24. J. D. Goldman and J. H. Ryther, Mass production of algae: Bioengineering aspects, in: *Biological Solar Energy Conversion* (A. Mitsui, S. Miyachi, A. San Pietro and S. Tamura, eds.), Academic Press, New York (1977), pp. 367–378.

25. W. J. Bond and M. G. Roberts, The colonization of Cabora Bassa, Mozambique, a man-made lake, by floating aquatic macrophytes, *Hydrobiologia* **60**, 243–260 (1978).

26. E. Hesser and O. Gangstad, Nuisance aquatic macrophyte growth, *J. Aquat. Plant Manage.* **16**, 11–13 (1978).

27. National Academy of Sciences, Making aquatic weeds useful: Some perspectives for developing countries, NAS, Washington, D.C. (1976).

28. J. D. McCullough, A study of phytoplankton primary productivity and nutrient concentrations in Livingston Reservoir, *Texas J. Sci.* **30**, 377–388 (1978).

29. J. M. Klopatek and F. W. Stearns, Primary productivity of emergent macrophytes in a Wisconsin freshwater marsh ecosystem, *Am. Midl. Nat.* **100**, 320–332 (1978).

30. J. G. Gosselink and R. E. Turner, The role of hydrology in freshwater wetland ecosystems, in: *Freshwater Wetlands: Ecological Processes and Management Potential* (R. E. Good, D. F. Whigham, and R. L. Simpson, eds.), pp. 63–78, Academic Press, New York (1978).

31. A. A. de la Cruz, Primary production processes: Summary and Recommendations, in: *Freshwater Wetlands: Ecological Processes and Management Potential* (R. E. Good, D. F. Whigham, and R. L. Simpson, eds.), pp. 79–86, Academic Press, New York.

32. M. S. Adams, P. Guilizzoni, and S. Adams, Relationship of dissolved inorganic carbon to macrophyte photosynthesis in some Italian lakes, *Limnol. Oceanogr.* **23**, 912–919 (1978).

33. M. Owens and R. W. Edwards, The effects of plants on river conditions, III: Crop studies and estimates of net productivities of macrophytes in four streams in southern England, *J. Ecol.* **50**, 157–162 (1962).

34. C. E. Boyd, Amino acid, protein, and caloric content of vascular aquatic macrophytes, *Ecology* **51**, 902–906 (1970).

35. J. M. McClure, The secondary constituents of aquatic angiosperms, in: *Phytochemical Phylogeny* (J. B. Harborne, ed.), Academic Press, New York (1970), pp. 233–268.

36. W. J. Nelson and L. S. Palmer, Nutritive value and general chemical composition of *Elodea, Myriophyllum, Vallisineria* and other aquatic plants, *Minn. Agric. Expt. Sta. Tech. Bull.* **136**, 1–47 (1938).

37. A. N. D. Auclair, Factors affecting tissue nutrient concentrations in a *Scirpus–Equisetum* wetland, *Ecology* **60**, 337–348 (1979).

4

Marine Biomass

Ivan T. Show, Jr.

1. Introduction

In order to develop marine biomass as a source of raw materials, a large dependable and economical supply of suitable biomass must be developed; however, our ability to develop such a supply is largely unproven. Although terrestrial biomass has received considerable attention, the development of terrestrial biomass crops has been hampered by competition with food crops, other uses of land and water, and the cost of supplying nutrients.

To date, we lack the basic knowledge needed to develop a marine biomass source although investigations are proceeding.[1] The only marine organism to receive considerable attention has been the Pacific giant kelp, *Macrocystis pyrifera,* this kelp having been the subject of continuing efforts to investigate marine farming concepts. Early studies on *Macrocystic pyrifera* were performed by the Naval Weapons Center in their Ocean Food and Energy Farm Project during which several small ocean farms were installed and observed. Currently, the Institute of Gas Technology (IGT) has a test-scale *Macrocystis* farm in operation off Newport Beach, California.

As with research on production, almost all research on preprocessing and processing has concentrated on *M. pyrifera.* The most comprehensive study to date has been carried out by the Agricultural Research Service, USDA[2] to evaluate the production of methane from *M. pyrifera.*

The studies by Show *et al.*[3] and Radovich *et al.*[4] were intended to help fill one of the most obvious gaps in our knowledge: the lack of a comparative

IVAN T. SHOW, JR. • Science Applications, Inc., 464 Prospect Street, LaJolla, California 92038.

assessment of marine biomass sources and processing methods. There are other less obvious inadequacies in our knowledge, however. For instance, little is known about the growth dependence of potential marine biomass organisms on major variables such as temperature, salinity, nutrient levels, and population density. In addition, the roles of negative influence such as disease, predation, and weather are largely unknown.

2. Biological Characteristics of Marine Plants

Most marine plants are in a group known as the algae, most of which are primitive and relatively simple in structure. The majority are microscopic and unicellular; however, some grow to over 80 m in length. A few flowering plants have invaded the oceans from the land, but they are confined to the shallow margins of the sea.

Larger, multicellular marine algae lack the complex supporting structures of terrestrial plants; the high density of sea water (about 1000 times that of air) makes such support unnecessary. Also, the entire surface of marine algae, especially unicellular forms, is photosynthetically active while in terrestrial plants, photosynthesis is usually restricted to specialized structures, usually leaves.

Algae which live attached to the bottom have specialized structures called holdfasts. Holdfasts, however, serve none of the nutritive functions of roots; algae draw nutrients and water through their entire surface. This difference gives the larger marine algae an advantage over similar-sized terrestrial plants. Microscopic forms, of course, benefit from a very high surface to volume ratio.

An extremely important factor in the efficiency of marine plants involves a basic difference between the functioning of marine and terrestrial ecosystems. In marine ecosystems, especially when one considers microscopic algae, literally all of the plant production is eaten by animals. The abnormal event of a locust swarm attacking a field is analogous to the normal situation in the sea. It is apparent, therefore, that a marine farm must create an artificial ecosystem in which the critical plant-herbivore link is restricted or absent. It is also apparent that, if man is to remove a large proportion of this rapidly overturning plant material, then the effects on the natural ecosystem must be carefully considered.

Solar energy can be fixed by marine algae at relatively high efficiencies because of their simple structure and generally high surface-to-volume ratio mentioned above. Under proper conditions, they can be maintained in their growth phase year round making continuous harvest possible or perhaps even necessary. Productivity figures for some terrestrial and marine plants are given in Table 1. These values suggest high yields of marine plants relative to more complex terrestrial plants.

TABLE 1. Terrestrial and Marine Photosynthetic Productivity

Vegetation type	Production (kg/m²-yr)
Terrestrial	
Trees	0.9–2.8
Grasses	1.1–6.8
Marine	
Algae (waste treatment ponds)	4.5
Algae (laboratory culture)	6.8–13.5
Kelp (natural beds)	4.9

Ecologically, marine plants can be divided into two major categories: phytoplankton and benthos. Phytoplankton are mostly microscopic and unicellular and live floating unattached in the upper layers of the water column. The major taxa of phytoplankton are the diatoms, dinoflagellates, coccolithophorids, and certain blue-green algae. The major taxa of benthic plants are the blue-green algae (Cyanophyta), green algae (Chlorophyta), red algae (Rhodophyta), brown algae (Phaeophyta), and a very few flowering plants (Spermatophyta). Each taxa will be discussed with regard to its potential as a source of marine biomass.

Phytoplankton are appealing in many respects. Many diatoms would be good sources of fuels because of their high carbohydrate content. Both diatoms and dinoflagellates have been extensively cultured; natural population polycultures have shown remarkably high productivity. As mentioned earlier, phytoplankton populations can turn over very rapidly, a characteristic which could lead to high yields relative to net production. In addition, phytoplankton are found in almost all sunlit marine environments, and their life cycles are relatively simple, consisting of asexual fission and simple modes of sexual reproduction.

Two possible disadvantages of phytoplankton are their small size and their almost universal possession of a siliceous or carbonaceous exoskeleton. Small size adds to the complexity of containment and harvesting. Presence of an exoskeleton significantly increases the ash content of the plant.

Blue-green algae are among the simplest and most primitive of living creatures; they are generally microscopic and unicellular or form loose aggregates of cells. Most are tropical. Currently, they are receiving no serious consideration as a marine biomass source. Their small size would make them very difficult to harvest.

The red algae are a group with real immediate potential as a source of marine biomass. Of utmost importance is that a number of species have long been successfully cultured in Japan and China; therefore, a great deal is known about handling certain species. Also, yields of organic compounds tend to be relatively high and many species can be harvested by clipping without destroying the whole plant.

As a group, the red algae have several disadvantages. They are generally small, the largest being about one meter in length. This small size makes high biomass density difficult to achieve because most red algae live attached to a hard bottom; maximum density is therefore severely limited by the amount of hard "substrate" available. Also, red algae tend to be the deepest living of the marine algae. This means that most species live under conditions difficult to reproduce in culture. Lastly, the most complex life cycles among all living plants occur in this group; this increases the complexity of cultivation.

The brown algae also have some very attractive characteristics. Many species have exceedingly high growth rates and yields and have been cultivated or harvested for many years. We know more about the growth, ecology, and chemical makeup of a few genera of brown algae (*Macrocystis, Laminaria,* and *Fucus*) than any other algae. Also, many species tend to be very large, exceeding 80 m in length and can be harvested by clipping without destroying the entire plant.

The major disadvantage of the brown algae is that individual species tend to be intolerant to environmental variations. In particular, they are sensitive to ocean swell and often grow only in a very narrow temperature range.

Most species of green algae are small and delicate and grow in shallow coastal bays, sloughs, and estuaries. These quiet coastal areas might be operationally and economically attractive for the cultivation of marine biomass; Japanese commercial algae farming is successful in these kinds of areas.

Marine flowering plants have high productivity, comparable to the brown algae. They live almost exclusively in shallow, protected areas; the surf grasses *(Phyllospadix)* of the Pacific are an exception. Their most appealing characteristic is that most species have relatively simple reproductive patterns; many are monoecious (both sexes in the same individual). Typically, seeds are formed which simply drop to the bottom and germinate there while many reproduce by bud or rhizome (horizontal root) formation. Some can be propogated from cuttings.

Marine flowering plants have two major disadvantages. First, they are generally restricted to the very edge of the sea, thus restricting the areas in which they can be cultivated. Second, there is almost no information on our ability to culture or cultivate them.

General characteristics of the various broad categories of algae have been stated. However, each group within a given category is diverse in its own right. Because one can find numerous and often significant exceptions to nearly any generality, any choice of a marine biomass source will have to be made by selecting species suitable to particular geographic areas. No single species or even larger taxon is going to be universally applicable to all areas.

3. Geographical Distribution of Marine Plants

Marine plants are found in almost all sunlit regions of the ocean: phyto-plankton drifting in the upper portion of the water column and benthic algae and spermatophytes attached to the bottom near the shores. Some marine plants are found in highly unlikely places such as the diatoms that grow attached to the undersurface of polar sea ice or the brown alga, *Sargassum,* that floats unattached in the central portions of the subtropical Atlantic and Pacific Oceans. The latter is a significant phenomenon; in fact, so much *Sargassum* is found in the North Atlantic that the region is called the Sargasso Sea.

Although the distribution of marine plants is a very complex subject, some general patterns can be discerned. For instance, planktonic diatoms predomi-nate in subpolar and low temperature seas; dinoflagellates predominate in high temperate and tropical seas. The benthic algae and spermatophytes have the general patterns shown in Table 2. The pattern with increasing depth is par-ticularly evident, being controlled primarily by light intensity which decreases exponentially with depth. Latitudinal differences are controlled largely by water temperature. What is not shown in Table 2 is the taxonomic changes with latitude; for instance, spermatophytes change from marsh grasses in the temperate and subpolar regions to mangroves in the tropics.

4. Primary Production

Primary production refers to the amount of organic material produced or carbon fixed photosynthetically by a plant during a certain interval of time. It is extremely important that anyone involved in marine biomass production understand the concepts associated with primary production and the produc-tion patterns observed in the oceans.

TABLE 2. Distribution of Benthic Marine Algae

Algae	Surface	Shallow	Intermediate	Deep
Tropical	Spermatophytes Blue-green algae	Spermatophytes Green algae	Coralline red algae	—
Low temperate	Spermatophytes	Green algae Spermatophytes	Brown algae	Red algae
High temperate	Spermatophytes	Spermatophytes	Brown algae	Red algae
Subpolar	Spermatophytes	Spermatophytes	Brown algae	Red algae

4.1. Fundamental Considerations

Data on marine primary production is widely scattered in the literature. In addition, much of the data is not directly comparable for one reason or another. Comparability of data is adversely affected by two factors. First, the units of measurement used in the literature are not uniform; units one encounters are of the form mass area^{-1}-unit time^{-1}. Mass is most often given as dry weight or weight carbon; these are not comparable because carbon content varies considerably with environmental changes. Area measurements are not a problem. Units of time are, however, very much a problem; units most often encountered are days and years. However, one cannot simply multiply by 365 to convert days to years because of strong seasonal cycles in production seen in most oceanic areas and because most studies have been done during the highest productivity of the year.

The second and more serious problem is the variance in defining and measuring production, which is variously expressed as gross production, net production, or yield. Gross production is the total amount of energy bound or carbon fixed. Net production is total or gross production minus the metabolic consumption (primarily through respiration) and other needs of the plant. Finally, yield is the amount of plant material available to man for harvest. Production is measured by a variety of methods; principally O_2 uptake, biomass change over time, and radiocarbon uptake. These methods are not directly comparable *per se*. In practice, however, we are forced to accept the results as comparable.

4.2. Geographical Distribution of Primary Production

The most productive zones of the oceans are the mangrove swamps, coral reefs, kelp and seagrass beds, coastal areas, upwelling zones, and estuaries. In general, coastal areas are about as productive as some forests, moist grasslands, and lands under primitive cultivation (ca. 100 g C/m^2-yr); estuaries are about as productive as evergreen forests and land under intensive cultivation (ca. 2000 g C/m^2-yr). Most open ocean areas are comparable to terrestrial deserts (ca. 50 g C/m^2-yr).

Steeman-Nielsen[5] gives total oceanic production as 11–14 × 10^9 tonne C/yr; Koblentz-Mishke[6] gives it as 14–16 × 10^9 tonne C/yr. Based on a total oceanic production of 20 × 10^9 tonne C/yr, Ryther[7] estimated offshore production as 16.3 × 10^9 tonne C/yr, coastal production as 3.6 × 10^9 tonne C/yr, and upwelling zone production as 0.1 × 10^9 tonne C/yr.

Phytoplankton primary production in the open ocean generally ranges from 25 to 75 g C/m^2-hr with an average of 50 g C/m^2-yr and a maximum of 7300 g C/m^2-yr occurring in the Bering and Greenlands Seas. Open ocean regions comprise about 90% of the world ocean, but only a small proportion of this area has particularly high productivity. Highly productive offshore areas include, but are not limited to, polar seas, equatorial divergence, the Kuroshio Current, and the Gulf Stream.

Phytoplankton primary production in the shallow coastal areas generally ranges from 50 to 400 g C/m²-yr with an average of 100 g C/m²-yr. These areas comprise about 7.5% of the worlds oceans and are similar in productivity to the more productive offshore regions.

Upwelling zones are highly productive, phytoplankton production normally exceeding 300 g C/m²-yr. The most important upwelling zones are off Peru, Oregon, Southern California, northwest and southwest Africa, Southeast Asia, the Yucatan Peninsula, and Antarctica.

Phytoplankton production is usually controlled by the degree of nutrient enrichment; a classification given by Steeman-Nielsen,[5] identifying four production zone categories, is probably applicable to most oceanic areas. These categories, with a range of average production and examples of geographic location, are as follows:

1. Considerable enrichment
   average production = 0.5 − 3.0 g C/m²-day
   example: southern Benguela Current (west coast of Africa)
2. Steady but lesser enrichment
   average production = 0.2 − 0.5 g C/m²-day
   example: near equatorial convergences
3. Without pronounced enrichment and some turbulence
   average production = 0.1 − 0.2 g C/m²-day
   example: most tropical and subtropical offshore areas.
4. Old surface water with no effective enrichment or turbulence
   average production = 0.05 − 0.1 g C/m²-day
   example: central Sargasso Sea.

Not surprisingly, regions of high benthic algal production coincide with regions of high phytoplankton production with one major restriction: except for the Mediterranean Sea, the coasts of India, and the northwestern coast of Africa, high benthic algal production occurs where water temperature does not exceed 20°C.

Large brown kelps are the most productive benthic algae overall. In natural beds off California and India, they produce up to 2000 g C/m²-yr. Others produce as much as 1500 g C/m²-yr under cultivation in Japan. Some of the smaller brown algae produce as much as 6.9 kg dry wt/m²-yr in the Mediterranean and 3.13 kg dry wt/m²-yr off Hawaii.

The red algae generally rank next in production. High production values include 16.4 kg dry wt/m²-yr in the Canary Archipelago and 4.4 kg dry wt/m²-yr in culture in Florida. Although there is very little known about the production of green algae, they are probably comparable to red algae.

Marine spermatophytes are usually found in nutrient-rich coastal areas. It is not surprising, therefore, that their productivity is relatively high, production ranging from 400 to 1500 g C/m^2-yr. The highest production occurs in temperate salt marshes.

4.3. Some Highly Productive Plants

Show et al.[3] have developed a list of plants which have high potential as marine biomass sources. Table 3 shows this list and the maximum potential yield found in the literature for each taxa. Besides high organic yield, each plant in the list occurs over a wide range of geographical or ecological regions and is thought to be capable of being cultured or cultivated.

Pelvetia, Thalassia, and natural phytoplankton populations are seemingly placed out of order in Table 3. All of these are placed out of order because of lack of documentation of their having been cultured or cultivated. Enteromorpha and Monostroma are included because of well-known but undocumented high production values.

5. Chemical Properties of Marine Plants

The concentration of most chemicals in marine plants is related to the classification of the plants; the closer the relationship, the more similar the

TABLE 3. Maximum Productivity of Selected Plants

Taxa	Maximum production (kg dry wt/m^2-yr)
Laminaria (brown algae)	24.1
Macrocystis (brown algae)	17.2
Chondrus (red algae)	16.4
Gracilaria (red algae)	16.4
Fucus (brown algae)	15.7
Porphyra (red algae)	15.3
Eucheuma (red algae)	7.7
Gigartina (red algae)	7.6
Iridea (red algae)	7.6
Neoagardhiella (red algae)	5.6
Pelvetia (brown algae)	12.8
Hypnea (red algae)	4.4
Sargassum (brown algae)	3.1
Thalassia (spermatophyte)	7.9
Ruppia (spermatophyte)	2.3
Natural phytoplankton population	6.9

chemistry. Most of our discussion of chemistry is derived from data presented by Vinogradov,[8] Paine and Vadas,[9] Chapman,[10] Dawson,[11] Hoagland,[12] Mautner,[13] and Platt and Irwin.[14]

5.1. Inorganic Chemistry

Marine plants contain about 50–89% water in two forms: as a salt solution and bound onto colloidal materials. Variations in water content are fairly consistent from species to species. Some variation does occur within any taxomonic group, however, dependent on the season, geographical region, and the age and health of individual plants. The giant kelps (brown algae) contain the greatest concentration of water (as much as 89%); smaller brown algae and green algae typically contain about 80% water, while the red algae contain only about 70% water. Marine spermatophytes, particularly the turtle grass *Thalassia,* have very low water content (as low as 50–55%). In general, algae with simple broad fronds (leaf-like structures) contain greater than 80% water while bushy and feathery forms contain less than 80% water.

Ash is a collective term for all inorganic material in the plant with the exception of water. The majority of ash in marine plants contains alkali metals, chlorides, sulfates, phosphates, carbonates, and silicates. Ash typically makes up about five percent of the dry weight of the plant. Four of that five percent is in the form of water soluble compounds. Major elements which occur in the ash include bromine, calcium, iodine, iron, magnesium, phosphorus, potassium, silicon, and sulfur. Table 4 gives examples of the major chemical forms and their concentrations. Of particular importance is sulfur. Marine algae contain more sulfur (0.5-1.0% of dry weight) than any other living organisms except sulfur-fixing bacteria.

More than 90% of the nitrogen in marine plants occurs as protein; the remainder occurs as nitrates, nitrites, and ammonia. One-half to two-thirds of

TABLE 4. Concentration[a] of Major Trace Element Forms

Taxa	Na_2O	K_2O	MgO	CaO	SO_3^{2-}	P_2O_5	Cl^-	Fe_2O_3	SiO_2
Laminaria digitata	18.3	25.9	6.7	8.7	11.3	2.8	28.5	0.4	—
Laminaria saccharina	18.3	24.6	5.8	9.6	12.9	2.7	24.9	0.4	0.7
Macrocystis pyrifera	13.75	34.0	3.6	6.7	6.4	1.7	33.7	0.4	—
Condrus species	18.7	17.3	1.4	7.2	41.2	13.0	—	—	—
Fucus serratus	23.3	8.7	9.7	14.6	22.7	2.8	17.6	0.8	1.0
Fucus vesiculosus	20.4	12.0	8.8	14.0	25.2	3.0	15.0	2.7	4.5
Porphyra canaliculata	5.0	3.3	0.8	1.5	9.8	0.2	—	—	—
Iridea edulis	16.9	23.4	—	20.5	25.2	13.0	1.0	—	—
Sargassum vulgare	4.7	4.2	3.7	57.8	18.9	1.0	1.8	2.7	5.8
Enteromorpha species	20.9	7.1	3.3	16.6	27.9	2.2	14.2	0.8	10.3

[a]Given in % ash-free dry weight.

the protein and virtually all of the inorganic nitrogen are water soluble. Mean nitrogen concentration for all marine plants is about 3% of the dry weight. Red algae have the highest (3.0–7.0% of dry weight), green algae have the next highest (ca. 3.0%), and brown algae have the lowest (1.5–2.0%) concentration of nitrogen. Nitrogen concentrations generally decrease, especially among the brown algae, from high to low latitudes.

5.2. Organic Chemistry

The protein content of marine plants has been extensively studied by investigators interested in their use as human food or fodder. Lewis[15] reviewed this literature and reported wide variations from species to species in protein concentrations (7.4–41.1% of dry weight). Protein quality may vary from part to part of the same individual, but does not seem to vary with age. Although toxic compounds (e.g., 2-aminocaprylic acid) are present in some algae, others have all amino acids essential for human nutrition.

A number of other organic compounds have been studied in marine algae. These include linolenic and acrylic acids, sesquiterpenes, terpenoid lactones, phenols, and chlorophyll derivatives.[16] Several complex carbohydrates of commercial value have been particularly well studied. Alginic acid, a slightly water-soluble polysaccharide of brown algae, consists largely of the calcium and magnesium salts of mixed polymers of D-mannuronic and L-glucuronic acids. The sodium salts of these polymers are known as algin. Agar is a complex polysaccharide found primarily in red algae. It is similar to algin and contains a neutral gelling fraction, agarose, and a sulfated nongelling fraction, agaropectin.

Laminarins, found primarily in *Laminaria,* are polysaccharides consisting of β-D-glucose in a 1:3 linkage. Both water soluble and insoluble forms are known. Other of these so-called phycocolloids include fucoidin (containing L-fucose), carrageenan (a complex galactin), iridophycin, and funorin. Table 5 gives the chief sources of these polysaccharides.

Highly sophisticated liquid and gas chromatographic techniques were recently applied to a study of the hydrocarbons in *M. pyrifera.* Rossi *et al.*[17] found the hydrocarbon levels to be quite low: 2–9 μg/g dry wt saturated hydrocarbon and 2–5 μg/g dry wt unsaturated hydrocarbons. Nonsaponifiable lipids made up about 0.1% of the dry weight. The predominant unsaturated hydrocarbons (44–81%) were HEH (3,6,9,12,15,18-heneicosa hexene) and HEP (3,6,9,12,15-heneicosapentaene). Squalene (2,6,10,15,19,23-hexamethyl-2,6,-10,14,18,22-tetracosahexaene) made up 7.4% of the total hydrocarbons. The dominant alkane was n-pentadecane (nC-15) comprising 9–92% of the total n-alkanes and up to 15% of the total hydrocarbons. Also, n-heptadecane (nC-17) was prevalant. The alkanes, nC-18 through nC-22 and heavier than nC-32 were generally absent and there were usually more odd than even n-alkanes. Polynuclear aromatic hydrocarbons were absent except in polluted samples.

TABLE 5. Occurrence of Valuable Polysaccharides

Algae	Polysaccharide	Associated acid and derivatives
Brown algae		
Laminaria	laminarin	Alginic acid
	fucoidin	sodium alginate
	algin	ammonium alginate
Macrocystis	algin	calcium alginate
Sargassum	algin	chromium alginate
Fucus	fucoidin	
	algin	
Red Algae		
Gelidium	agar	Agarinic acid
Gracilaria	agar	sodium agarinate
Pterocladia	agar	potassium agarinate
Ahnfeltia	agar	calcium agarinate
		magnesium agarinate
Gigartina	carrageenan	Carageenic acid
Chondrus	carrageenan	Potassium carrageenate
		calcium carrageenate
Iridea	iridophycin	Iridophycinic acid
		sodium iridophycinate
Gloiopeltis	funorin	

Similar results were found by Payne *et al.*[18] for both *Macrocystis* and *Eisenia*. They found the nonsaponifiable fraction to be somewhat higher in concentration (0.5–1.5% dry weight). They also found polynuclear aromatics in what they believed to be unpolluted samples.

Extracellular products of marine algae are found in large enough quantities to be of potential importance. Fogg[19] found that extracellular organic matter can account for as much as 30% of the total organic matter synthesized in cultures of a marine species of *Chlamydomonas*. In some blue-green algae, Fogg found that up to 50% of the nitrogen fixed appeared in the culture medium in soluble form, possibly as polypeptides. Algae have been shown to produce large amounts of extracellular glycolic and oxalic acid as well as polysaccharides, up to 25% of the total organic matter produced. In all cases, the materials found in solution had been liberated by healthy cells and were not the products of autolysis.

5.3. Energy Content

Paine and Vadas[9] analyzed 74 species of marine algae from Washington state; over 278 separate determinations of calorific content were performed. Based on the means and variances for each major algae group given in Table

TABLE 6. Calorific Content of the Phyla of Marine Algae

Taxa	Dry weight (kcal/g)		Ash (% dry wt)		Ash-free dry weight (kcal/g)		Number of species
	$\bar{x}^a$	s^b	$\bar{x}$	s	$\bar{x}$	s	
Green algae	3.47	0.76	29.5	15.4	4.92	0.28	10
Red algae	3.19	1.04	20.5	21.2	4.58	0.47	39
Brown algae	3.06	0.56	32.4	9.9	4.49	0.33	25
Overall	3.17	0.84	25.7	16.7	4.59	0.40	74

$^a\bar{x}$ = mean.
^{b}s = standard deviation.

6, a one-way analysis of variance showed that each group was significantly different at a confidence level of 0.95. In descending order of calorific content, the green algae were the highest, the red algae next, and the brown algae the lowest. Table 7 gives the energy content of some important genera as well as their organic yields, energy yields, and estimated solar efficiency; this table will be referred to again in our discussion on potential energy yield.

6. Commercial Culture and Cultivation

Our discussion will be limited to the benthic algae since there has been no commercial culture or cultivation of marine phytoplankton or spermatophytes. Any discussion of algal culture or cultivation must, of necessity, be a discussion of oriental practices, for a large number of species are cultivated in the orient. No marine algae are commercially cultivated in the west, although *Macrocystis, Laminaria,* and *Rhodymenia* are extensively harvested from natural beds.

The most important genus cultivated in the orient is *Porphyra* in Japan. In all, five species are utilized: *P. tenera, P. yegoensis, P. pseudolinearis, P. augustus,* and *P. kuniedai. P. tenera* was first cultured during the seventeenth century by placing twigs in shallow water for the edible sporophyte (the large, asexually reproducing phase of the life cycle) to settle on. Twigs are still used in Korea. Recently, twigs were replaced by bamboo and then by palm fiber or synthetic net. Once the full life cycle of *Porphyra* was understood, tank culture of young sporophytes was developed. Presently, more than 70% of the young sporophytes are produced in prefectural and municipal laboratories.

Porphyra net culture has suffered from the settling of less desirable species. This problem was solved, however, by slightly drying and then freezing the net with the young sporophytes already attached. This procedure kills most competing species but does not damage *Porphyra.*

TABLE 7. Energy Conversion Potential

Taxa	Calorific, content (kcal/kg dry wt, $\times 10^3$)	Maximum yield (kg dry wt/m²-yr)	Calorific, yield (kcal/ m²-yr, $\times$ 10^3)	Solar conversion efficiency (percent)
Laminaria	2.74	24.1	66.0	3.47
Macrocystis	2.85	17.2	49.0	2.58
Chondrus	(3.19)[a]	16.4	(52.3)	2.75
Gracilaria	(3.19)	15.7	(50.1)	2.64
Fucus	3.43	15.3	53.9	2.84
Porphyra	4.20	7.7	32.3	1.70
Eucheuma	(3.19)	7.4	(23.6)	1.24
Gigartina	3.22	7.6	24.5	1.29
Iridea	2.92	7.6	22.2	1.17
Neoghardiella	(3.19)	5.6	(17.9)	0.94
Pelvetia	(3.06)	12.8	(39.2)	2.06
Hypnea	(3.19)	4.4	(14.0)	0.74
Sargassum	(3.06)	3.1	(9.5)	0.50
Thalassia	4.41	7.9	34.8	1.83
Ruppia	—	2.3	—	—
Natural phytoplankton population	—	6.9	—	—
Enteromorpha	3.01	(7.5)	(22.6)	1.19
Monostroma	4.56	(7.5)	(34.2)	1.80
Zostera	4.36	8.4	36.6	1.93

[a]Numbers in parentheses are average values for the phylum to which the genus belongs.

Open ocean cultivation of *Porphyra* was begun in Japan in 1967. Seeded and frozen nets are attached to floating rafts which are then anchored to the ocean floor. *P. pseudolinearis* and *P. yezoensis* are used because of their tolerance to high salinity.

In all, Japan produces about six billion sheets of dried *Porphyra*. These sheets weigh about three grams each and have a market value of over $230 million.

The large brown algae *Laminaria japonica* is a particularly important crop in China. Originally, rocks bearing young sporophytes were placed in the water, but now, young sporophytes are produced in greenhouses by methods similar to those used in Japan for *Porphyra*. The young sporophytes are attached to one of three types of rafts. The first is a basket raft consisting of cylindrical baskets with an unglazed pot containing fertilizer inside the basket; the plants are attached to the outside of the basket. The second is a single-line tube raft with round fertilizer pots slung along the length of the tube. The third is a double-line tube raft, shaped like a ladder, with elongate fertilizer pots slung from the cross pieces. All three types are constructed of bamboo. China has about 20,000 ha under cultivation with *Laminaria;* these produce approximately 150,000 tons dry wt of *Laminaria* annually.

A third important crop in the orient is the red alga *Eucheuma* in the Phillippines. Currently, over 10,000 tons dry weight is produced at a market value of $250–$700 per ton. *Eucheuma* is grown in coastal lagoons, other accessible shallow, protected coastal regions, and in pond polyculture with milkfish. The plants are propogated by cuttings from mature plants, and, while growing, are held in place by a variety of lines, nets, or pens.

Of lesser importance is the cultivation of *Undaria pinnatifida* in northern Japan. Initially, floats were anchored out in appropriate locations for young sporophytes to settle on. Now, strings of young sporophytes are produced in prefectural and municipal laboratories and then are set out on rafts or horizontal lines. One raft, 36.6 × 1.5 m, can produce up to 112.5 kg dry wt of *Undaria* per year. A similar form of *Undaria* cultivation has been started in northern China.

Gracilaria, Gelidium, and *Monostroma* are grown in brackish pond polyculture with crabs, shrimp, and milkfish in Taiwan, *Gracilaria* being the most important of the three. *Monostroma* is also grown and marketed with *Porphyra* in Japan.

Gloiopeltis is cultivated in Japan by placing boulders or concrete blocks in the water to provide hard substrate for sporophyte attachment. *Caulerpa* is grown in Japan in pond polyculture much as *Gracilaria* is in Taiwan.

In two remarkable experiments, the Russians have grown the Antarctic giant kelp *Phyllogigas* in pond culture with mussels while the Americans are investigating high technology offshore culture of *Macrocystis* in artificial upwelling systems.

7. Utilization of Marine Plants

The Chinese and Japanese have always been the most intensive users of marine algae as food. The Japanese in particular eat *Porphyra* (nori), *Laminaria* (kombu) and *Undaria* (wakame) literally as garden vegetables. The most diverse use of marine plants as food, however, was by the Polynesians, who ate over 75 different species. Besides China and Japan, other countries with current intensive use are the Phillippines, Malaysia, and Indonesia.

Some species have been used extensively as food in Europe; in particular, *Rhodymenia palmata* (known as "dulse" in Scotland, "dillish" in Ireland, and "so" in Iceland) and *Chondrus crispus* (Irish moss). The principal use of marine plants in Europe, however, is not as food but as fodder for livestock; marine algae are also widely used as fertilizers in Europe. In Great Britain, France, Scandinavia, and Iceland, *Rhodymenia palmata* and *Alaria esculents* are favored livestock feeds. In fact, European successes in using marine algae as fodder led to the development of the kelp industry on the U.S. west coast.

The medicinal applications of marine algae have long been recognized; they date from Chinese use of *Sargassum* and various Laminariales to treat goiter. The Chinese also used agar from *Gelidium, Pterocladia,* and *Gracilaria* to treat stomach and intestinal disorders; the Japanese later assumed the same practices. *Chondrus crispus* is still used in Europe for the treatment of stomach and intestinal disorders.

The importance of various polysaccharides extracted from marine algae has spurred the development of various industries in Great Britain, France, the United States, Japan, South Africa, Australia, New Zealand, and Russia. The polysaccharides on which these industries are based include algin, fucoidan, laminarin, agar, carrageenan, iridophycin, and funorin.

Algin is hydrophilic and therefore useful for thickening, suspending particles, stabilizing polymers and ink, and as an emulsifier in paints, foods, and cosmetics. In addition, it is useful in forming colloids and gels and in preventing drying.

Agar is probably the most useful algal derivative. It is used in gelatin, bakery products, dairy products, cosmetics, for water-proofing fabric, in photographic film, and as a lubricant. Besides the best known use of agar as a bacterial substrate medium, it is useful as a temporary preservative for meat.

Carrageenan is used in large quantities. Its gel-forming capacity is used to advantage in inks, nonedible jellies, cosmetics, and lotions. As a food additive, carrageenan is used in milk, ice cream, syrup, other desserts, and soups as a suspender and thickener.

The only other large-scale commercial use of marine algae is as a source of iodine. This represents probably the oldest pharmaceutical use of a marine plant.

8. Relationship of Primary Production and Chemical Composition to Marine Energy Crops

The effect of varying rates of production and chemistry on the utility of marine energy crops can be partitioned into three separate areas: preprocessing, processing by-products, and potential energy yield. Each of these areas is discussed separately.

8.1. Preprocessing

Preprocessing is the procedure by which a raw material is converted into a physical and chemical form suitable for an energy conversion process. For marine plants, the two most significant properties influencing preprocessing are water and ash content. The removal of both of these constituents is an energy-consuming process which lowers the efficiency of the overall energy conversion process. Table 8 gives the ash and water contents for a number of important genera of marine plants. The total range of values found in the literature is given, most of the intragenus variability being attributable to differences between species in the genus.

The red algae and some of the smaller brown algae appear to be the most desirable in terms of both ash and water content. The giant kelp, *Macrocystis*, appears to be the least desirable on both counts.

8.2. Processing By-products

Processing by-products can be significant to the overall operation of an energy conversion process. The relative concentrations of sulfur, bound nitrogen, and alkali metals are important considerations. Table 9 gives values from the literature for these chemical constituents. Some cautions, however, must be observed in interpreting these data. First, most of the trace elements, up to 80% of the initial alkali metals in the ash portion of the plant material are

TABLE 8. Preprocessing Factors

Taxa[a]	Ash content (% dry wt)	Water content (% wet wt)
Laminaria (b)	5–39	70–87 (I)
Macrocystis (b)	3–48	86–69 (I)
Chondrus (r)	9–22	75–80 (I)
Gracilaria (r)	—	72–87 (I)
Fucus (b)	18–24	68–82 (I)
Porphyra (r)	8–11	83–86 (I)
Eucheuma (r)	—	—
Gigartina (r)	22–30	65–68 (I)
Iridea (r)	25	72–81 (I)
Neoagardhiella (r)	—	—
Pelvetia (b)	—	—
Hypnea (r)	—	—
Sargassum (b)	37	62–84 (I)
Thalassia (s)	47	50–67 (I)
Ruppia (s)	—	—
Natural phytoplankton population	7–44	80–95 (I)
Enteromorpha (g)	36	71–86 (I)
Monostroma (g)	12	83 (I)

[a]Abbreviations used: b = brown algae; g = green algae; r = red algae; s = spermatophyte; I = water content value derived by an ignition technique.

TABLE 9. Processing Byproducts

Taxa[a]	Sulfur (% dry wt)	Bound N (% dry wt)	Alkali metals (% ash wt)
Laminaria (b)	0.54–1.04	0.25–4.24	11–41
Macrocystis (b)	0.77–1.20	1.07–1.58	36
Chondrus (r)	—	0.94–2.80	9–21
Gracilaria (r)	1.48–4.61	1.5	4–18
Fucus (b)	4.62	1.09–2.29	13–22
Porphyra) (r)	2.23	6.40	10–13
Eucheuma (r)	—	—	—
Gigartina (r)	—	3.54	22
Iridea (r)	8.16	1.00–3.10	25
Neoagardhiella (r)	6.27	1.40	—
Pelvetia (b)	—	1.49–2.83	—
Hypnea (r)	—	—	—
Sargassum (b)	0.70	0.80–1.20	5–40
Thalassia (s)	—	2.30	14
Ruppia (s)	—	—	—
Natural phytoplankton population	—	4.34–8.00	—
Enteromorpha (g)	0.50–3.40	3.78	14
Monostroma (g)	6.30	1.40	5

[a]Abbreviations used: b = brown algae; g = green algae; r = red algae; s = spermatophyte.

TABLE 10. Priority List[a]

Taxa	Maximum calorific yield (10^3 kcal/m²-yr)
1. *Laminaria*	66.0
2. *Chondrus*	52.3
3. *Fucus*	53.9
4. *Gracilaria*	50.1
5. *Macrocystis*	49.0
6. *Gigartina*	24.5
7. *Thalassia*	34.8
8. *Enteromorpha*	22.5
9. *Sargassum*	9.5
10. *Iridea*	22.2
11. *Porphyra*	32.3
12. *Monostroma*	34.2
13. *Neoagardhiella*	17.9

[a]Source: Science Applications, Inc.

TABLE 11. Summary of Important Characteristics of Selected Marine Plants

Taxa[a]	Priority ranking	Maximum calorific yield (10^3 kcal/m^2-yr)	Potential yield (kg dry wt/m^2-yr)	Water content range (% dry wt)	Ash content range (% dry wt)	Sulfur content range (% dry wt)	Nitrogen content range (% dry wt)	Alkali content (% ash wt)	Descriptive Characteristics
Laminaria (b)	1	66	24.1	70–89	5–39	0.54–1.04	0.25–4.24	11–41	Perennial or annual: single stipe, single leaf Generally 1–5 m, L. farlowwi = 5 m
Chondrus (r)	2	52.3	16.4	75–80	9–22	—	0.94–2.8	9–21	10–20 cm in carpet-like growth pattern Small, moss-like; single delicate leaf
Fucus (b)	3	53.9	15.3	68–82	18–24	~4.62	1.09–2.29	13–22	Perennial; branched stipe with large strong ribs 10–25 cm
Gracilaria (r)	4	50.1	15.7	72–87	—	1.48–4.61	~1.5	4–18	6–20 cm; G. sjoestedtii = 2 m, G. guerrucosa = 50 cm Attached to rocks with tough, cylindrical branches
Macrocystis (b)	5	49.0	17.2	86–89	3–48	0.77–1.2	1.07–1.58	~36	Perennial; stipe, 2–4 branches 6–48 m with most biomass in upper few meters
Gigartina (r)	6	24.5	7.6	65–68	22–30	—	~3.54	~22	15–50 cm, fronds thickly set with papillate outgrowth; G. corymbifera, 1 m × 30 cm wide)

elongated, cells in matrix of delicate interwoven fiber

No.	Organism							Description
7	Thalassia (s)	7.9	34.8	50–67	~47	—	~2.3	~14	Flowering plant, grown from rhizomes in dense beds. Tough fibrous stems with narrow leaves
8	Enteromorpha (g)	—	22.5	71–86	36	0.5–3.4	~3.78	~14	Less than 40 cm; E. intestinalis and E. linza reach 2 m in quiet waters
9	Sargassum (b)	3.1	9.5	62–48	~37	~0.70	0.80–1.20	5–40	Annuals and perennials 1–10m; tangled mass of leaves and stipes
10	Iridea (r)	7.6	22.2	72–81	~25	~8.16	1.0–3.1	25	20–40 cm; I. cordata ~ 1.2 m Long thin blade with short thick stipe at depths of 1–2 m
11	Porphyra (r)	7.7	32.3	83–86	8–11	~2.23	~6.4	10–13	Perennial 0.15–1 m; P. miniata ~ 6 m
12	Monostroma (g)	7.5	34.2	~83	~12	~6.3	1.4	~5	Annual; flat delicate blade. 4–12 cm; grows in calm water
13	Neoagardhiella (r)	5.6	17.9	—	—	~6.27	~1.4	—	10–41 cm Irregular branching with cells in matrix of longitudinal fibers

aAbbreviations used: b = brown algae; g = green algae; r = red algae, s = spermatophyte.

water-soluble and can be washed out during preprocessing. Second, as much as half of the sulfur is either water- or acid-soluble and could also be removed during preprocessing. Last, most of the nitrogen is in the form of water-soluble inorganic compounds or in the form of water-soluble protein, both of which could be removed during preprocessing.

8.3. Potential Energy Yield

All things considered, the most important single factor in selecting a candidate species for energy farming is the amount of energy bound in the tissues of the plant. Table 7 gives potential energy yields in kcal/kg ash free dry wt; missing data are estimated by using the average energy content for the major plant group to which the genus belongs. These energy values are used to estimate the potential calorific yield of the genus in question. Notice that the smaller brown algae consistently have the highest calorific yield, the red and green algae are next, and the large brown algae show relatively moderate to low calorific yields. The solar efficiency given in Table 7 is based on the total solar energy available in one year at 40¼ north latitude (ca. 1900 kcal/m²).

Show et al.[3] developed a ranking of genera of marine plants occurring in U.S. territorial waters based on their potential as marine bioenergy resources. This list is presented in Table 10 along with maximum calorific yields. The only general trend apparent in this list is that the first five genera (*Laminaria, Chondrus, Fucus, Gracilaria,* and *Macrocystis*) show a clear advantage in ranking score and in maximum potential calorific yield over the remainder of the list. Based on current knowledge, the 13 genera in Table 11 will probably form the backbone of any near future efforts to develop our marine biosolar resources. Table 11 characterizes each of these important genera physically and chemically.

References

1. J. H. Ryther, Cultivation of macroscopic marine algae and fresh water aquatic weeds, Progress Report 1 (May 1976–December 1976), under Contract EY-76-02-2948 to U.S. Department of Energy (1978).
2. M. R. Hart, D. de Fremery, G. K. Lyon, D. D. Duzmicky, and G. O. Kohler, Ocean Food and Energy Farm Kelp Pretreatment and Separation Processes, Western Regional Research Center, Agricultural Research Service, USDA, (1976).
3. I. T. Show, L. E. Piper, S. E. Lupton, and G. R. Stegen, A Comparative Assessment of Marine Biomass Materials, Electric Power Research Institute, AF-1169 (1979).
4. J. M. Radovich, P. G. Risser, T. G. Shannon, C. F. Pomeroy, S. S. Sofer, and C. M. Sliepcevich, Evaluation of the Potential for Producing Liquid Fuels from Biomaterials, EPRI AF-974, TPS77-716 Final Report, Palo Alto, California (1979).
5. E. Steeman-Nielsen, *J. Cons. Int. Explor. Mer.* **19**, 309–328 (1954).

6. O. J. Koblentz-Mishke, V. V. Volkovinsky, and J. G. Kabanova, in: *Scientific Exploration of the South Pacific* (W. S. Wooster, ed.), pp. 183–193, National Academy of Sciences Translations, Washington, D.C. (1970).

7. J. H. Ryther, in: *The Sea,* Vol. 2, (M. N. Hill, ed.), Wiley Interscience, New York (1963), pp. 347–380.

8. A. P. Vinogradov, Memoir No. 11, Sears Foundation for Marine Research, New Haven, Conn. (1953).

9. R. T. Paine and R. L. Vadas, *Mar. Biol.* **4**, 79–96 (1969).

10. V. I. Chapman, *Seaweeds and Their Uses,* 2nd edition, Methuen, London (1970).

11. E. Y. Dawson, *Marine Botany: An Introduction,* Holt, Reinhart, and Winston, New York (1966).

12. D. R. Hoagland, *J. Agric. Res.* **4**, 39–51 (1915).

13. H. G. Mautner, *Econ. Bot.* **8**, 174–192 (1954).

14. T. Platt and B. Irwin, *Limno. Oceanogr.* **18**, 306–310 (1973).

15. E. J. Lewis, *Proceedings of Seminar on Sea, Salt, and Plants,* Bavnager, India (1967), pp. 296–308.

16. N. B. Allen, *Proceedings of Seminar on Sea, Salt, and Plants,* Bavnager, India (1967), pp. 336–368.

17. S. S. Rossi, G. W. Rommel, and A. A. Benson, *Phytochemistry* **17**, 1431–1432, (1978).

18. J. R. Payne, B. de Lappe, and R. Risebrough, in: *Southern California Baseline Study, Intertidal,* Year II, Vol. III, Report 23, BLM/Science Applications, Inc. (1978).

5
Silvicultural Energy Farms

JEAN FRANCOIS HENRY

1. Introduction

Biomass in its various forms is an attractive alternative source of energy. Through photosynthesis, biomass collects and stores low-intensity solar energy which can then be harvested at will and released through direct combustion, thermochemical conversion, or biochemical conversion.

Most forms of plant biomass are suitable as energy feedstocks. Wood biomass, however, has received the most consideration because of its long and continuing precedent as a fuel and chemical feedstock. The basic technology for the conversion of wood to useful forms of energy or fuels is available. Wood, therefore, has the potential of making a significant contribution to the energy needs of many countries providing that reliable long-term supplies can be made available. In the near term, large amounts of wood fuel are potentially available in the form of noncommercial timber from existing forests and of residues from forest products industries' operations. About one-third of the seven to eight quads (quadrillion Btus) potentially available in the near term can probably be recovered at a cost competitive with that of fossil fuels. The projected increased demand for wood and wood residues by the forest products industries suggests that the amount of wood fuel potentially available for other users could be sharply reduced around the year 2005 if forest management is maintained at its 1970 level.[1-3] Also, by 2005, much of the resource potentially available for fuel will probably be proportionally more expensive than it is now because the inexpensive, easy to collect residues will be used mostly by the forest products industries.

JEAN FRANCOIS HENRY • Energy Planning and Design Corporation, 12864 Tewksbury Drive, Herndon, Virginia 22070.

Silvicultural energy farms, i.e., production entities devoted to the production of wood exclusively for its fuel or feedstock value, if implemented on a national scale, could make a large, reliable source of wood fuel/feedstock available for nonforest products users.[4]

Other approaches could be followed to increase the long-term supply of wood fuel. Increasing the level of forest management over present practice could probably double the output of the existing forests and therefore generate a surplus of biomass available for energy uses.[5,6] This approach, however, is a very long-term one (probably a 50-year program) requiring large investments over a long period of time and the availability of the resource could still be influenced by the demand for wood products.

2. The Silvicultural Energy Farm Concept

In conventional forestry, trees are grown in plantations to provide the raw material for a variety of products such as lumber, plywood, pulp and paper. In these plantations, the trees are generally widely spaced and grown to sizes large enough for the manufacture of wood products. Achieving these commercial sizes may require long growing periods or rotations. These may range from 30 to 80 years or more. Because of the size of the crop and the need to maintain the physical integrity required in the manufacture of wood products, conventional single tree harvesting and handling methods are generally used in forestry operations.

Tree crops, however, could be grown on much shorter rotations provided that tree size and form were not important considerations in the end use of the crop. Such is the case when wood chips used for fuel or feedstock are produced. Short rotation generally refers to rotations of 20 years or less. Short rotation tree farming is generally combined with close spacing of the trees in order to achieve full site utilization within the rotation period.

Short rotation silvicultural farming for wood fiber production has been proposed by a number of investigators.[7-12] Early research on short rotation tree crops indicated that biomass yields can far exceed those of conventional forestry. Mean annual biomass increments of 5–10 oven-dry tons (ODT) per acre were shown to be possible under short rotation management, well above the 1–3 ODT/acre generally recorded in conventional forestry.[13-17] Table 1 shows some representative productivities of candidate species for energy farming grown under short rotation conditions. Mean annual increments higher than those quoted in Table 1 have been recorded in some instances: about 9 ODT/acre-yr for hybrid poplars in Sweden,[55] 10–20 ODT/acre-yr for very dense red alder stands,[44] and about 10 ODT/acre-yr or more for eucalyptus.[56] It was, however, also noted that the high short-rotation yields require the application of intensive management to the plantation land. In

TABLE 1. Above Ground Biomass Productivities of Various Candidate Crops for Short
Rotation Energy Farming

Species	Spacing (ft²/tree)	Rotation (years)	Mean annual increment (ODT/acre-yr) [a]	References
American sycamore	1–24	2–4	2–6	8, 15, 16, 18,
(Platanus occidentalis)				19–23
Populus genus				
Aspen	3–22	2–15	0.3–2	24–27
Eastern cottonwood	6–72	4–10	3–7	28, 29, 30
Black cottonwood	1–16	2–4	1.2–6.3	31–33
Hybrid poplars	0.5–144	2–10	2–8.5	14, 34, 35, 36–49
Red alder	0.15–6	1–15	0.2–10	12, 32, 62, 44–47
(Alnus rubra)				
Eucalyptus	32–128	3–8	2–4	48, 49
(Eucalyptus spp)				
Loblolly pine	10–80	4–20	2–5	50–54
(Pinus taeda)				

[a]ODT/acre-yr: oven-dry ton per acre-yr (stem and branches).

many instances, the level of management is comparable to that required in the production of agricultural crops. The potential of short rotation silvicultural farming for fiber production prompted a number of investigators to advance the same concept as a possible source of biomass for energy production.[43,57–59]

As it is envisioned presently, the silvicultural energy farm is a woody biomass production entity relying on short rotations and intensive management to produce biomass exclusively for its fuel or feedstock value. In terms of wood biomass production, silvicultural energy farms offer a number of potential advantages over conventional forestry plantations: higher yield per unit area, lower land requirements for a given biomass output, earlier cash return on the investment, extensive mechanization similar to that practiced in agriculture, and ability of assimilating cultural and genetic advances quickly. Moreover, short rotation crops can be chosen among a variety of species which regenerate by coppicing, therefore reducing regeneration costs of the plantation. Silvicultural energy farms, however, have a number of disadvantages: establishment and management costs per unit area are generally higher than for conventional forest crops, only sites amenable to mechanized operations may be used and epidemic disease, and insect infestations may be more difficult to control than in conventional forest plantations. These advantageous and disadvantageous features of the silvicultural energy farm will have an important bearing on the design and operation of the farm and ultimately on the economics of biomass production.

3. Conceptual Design and Operation of the Silvicultural Energy Farm

Although many aspects of short rotation farming and its associated land management requirements have been and still are investigated, no full scale or even pilot silvicultural energy farm has been demonstrated. It is, therefore, necessary to use a conceptual design of a silvicultural energy farm to assess its economic and energy efficiency potential. A conceptual design including the features generally considered as essential for successful energy farming is presented below and used as a basis to estimate production costs of woody biomass.

3.1. Energy Farm Design Parameters and Layout

The energy farm model described here follows the design proposed by Inman et al.[60] However, other models have been proposed which include the same basic features.[34,61−63]

The design parameters adopted for the present discussion are summarized in Table 2. The farm has an annual production capacity of 250,000 ODT, enough to support an electric power plant of about 50 MW_e or a methanol conversion plant having a capacity of about 25 million gallons of methanol per year. The productivity is site dependent and can range between 5 and 12 ODT/acre-year. Correspondingly, the planted acreage will range between 50,000 acres and about 21,000 acres. Rotations are assumed to be six years and five crops (one first growth and four coppice crops) will be harvested during the lifetime of the farm. Land will be leased at 5% of its market value per year. The cost of preparation of the land for establishment of the plantation will vary as a function of the existing vegetative cover and condition of the land. The model assumes that bare root seedlings of fast growing hardwood species will be planted on a 4 ft × 4 ft pattern, i.e., about 2,725 trees per acre.

Irrigation will be supplied during the first half of each rotation to improve the establishment and development of the seedlings and to increase the development of shoots from the stumps during the coppice rotations. Nitrogen, phosphorus and potassium fertilizers will be applied annually to replace the nutrients mined by the crop and avoid depletion of soil nutrients. Weed competition for nutrients and moisture will be controlled by disking between the rows of trees during the first year of each rotation. It is assumed that canopy closure will be sufficient by the second year to eliminate competition by weeds. Harvesting will be performed by a self-propelled harvester presently under development. Conceptually, such a machine could be similar to a silage corn harvester cutting rows of trees, chipping the whole tree and blowing the chips in a wagon pulled by the harvester. Harvesting will be performed during the dormant season and the chips will be field stored. The chip will be transported by truck to the conversion plant located ideally in the center of the geographic area containing the energy farm. Work roads will be installed to facilitate

TABLE 2. Design Parameters for a Conceptual Silvicultural Energy Farm[a]

Items	Value
Annual production	250,000 oven-dry tons (ODT)
Productivity	5–12 ODT/acre-year
Planted acreage	Production/productivity, acres
Rotation	6 yr; 30-yr plantation lifetime
Land acquisition	Lease at 5% escalating market value per year
Land clearing/preparation	Cost based on existing vegetation cover
Planting	Bare-root seedlings planted in a 4 ft × 4 ft pattern
Irrigation	Automatic traveling sprinkler; 3 yr per rotation
Fertilization	Annual N, P, K applications; lime applied the first year of each rotation
Pest control	Mechanical weed control during the first year of each rotation
Harvesting	Conceptual self propelled harvester; harvest during dormant season; field storage of chipped biomass
Transportation	Trucks to on-site conversion plant
Work roads	450 mi of unpaved roads per site
Miscellaneous operations	Planning; supervision; field supplies

[a]As proposed by Inman *et al.*[60]

access to and within the plantation. Planning, supervision, supplies and maintenance of the equipment are also part of the farm operations.

An idealized layout for the energy farm is shown in Figure 1. The total planted area of the farm is divided in six annual modules. These modules are planted at one year intervals. After the first six years, the situation described in the figure is reached: the first module planted is in its sixth year of growth and ready to be harvested while module 6 is reaching the end of its first year of growth. The major field operations required by the various modules are indicated on the figure. From the seventh year on, the situation described on the figure is repeated with each module being switched by one position counterclockwise for each successive year. Thus, in year 7, module 1 assumes the state of module 6 on the figure, and so forth, and module 2 reaches its sixth year of growth and is ready to be harvested. During the first year of its second rotation (year 7), module 1 does not have to be replanted because regeneration will occur through coppicing. The other field operations indicated on the figure will, however, be performed during the second and subsequent rotations.

3.2. Energy Farm Operational and Cost Data

The operational and cost data used in estimating the biomass production costs are discussed in this section. Costs are quoted in 1978 dollars. Most operations and their associated costs will be site dependent and, therefore, ranges of costs are quoted when appropriate.

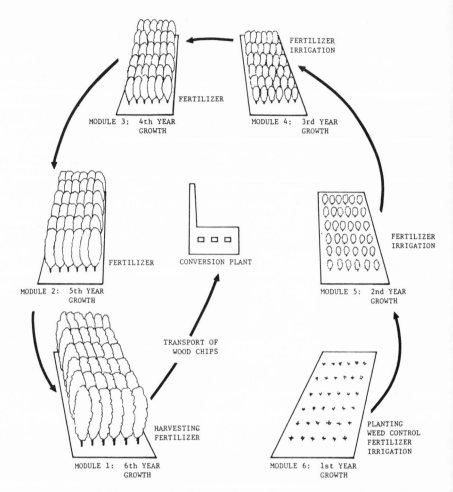

FIGURE 1. Idealized layout of a silvicultural energy farm operated on a 6-yr rotation.

3.2.1. Energy Farm Installation

The installation of the energy farm includes the following steps in approximate order: land acquisition, land preparation and improvement, access roads, installation of permanent equipment (such as irrigation mains, wells, and pumps), and planting.

3.2.1a. Land Acquisition. Land may be acquired either by direct purchase or long-term lease. The model developed by Inman et al.[60] assumes acquisition through long-term lease at an annual rate of 5% of the average escalating value of the land. The model also assumes that the total planted area (modules 1 to 6 in Figure 1) will be distributed within a geographic area ten times larger than the planted area. This assumption can vary from site to site and will have a bearing on the biomass transportation costs.

3.2.1b. Land Preparation. Extensive land preparation is a prerequisite to achieving good plantation establishment.[64-68] Land preparation may include clearing of existing vegetation, raking and burning of the slash, plowing, disking, subsoiling and application of herbicide as required by the site. Land improvement may include drainage and leveling as needed. The range of land preparation costs used in the production cost estimates are shown in Table 3. A recent analysis for a wooded site located in South Carolina confirmed the validity of these costs.[68-70]

3.2.1c. Work Roads. A network of improved work roads is provided for in the energy farm to minimize the equipment traffic over planted area. This network is made of parallel work roads separated by the distance that the harvester must travel to fill a wagon load of wood chips. The total mileage of roads for a 250,000 ODT/yr farm with a six year rotation is 454 miles. The road mileage per planted square mile will be a function of the productivity of the site. The construction cost of unpaved, 25-ft wide roads are reported in Table 3.

3.2.1d. Irrigation Systems. Several irrigation systems could be considered for the energy farm: flood, center-pivot, skid-tow, drip-trickle, or traveler. Some of these systems may be best adopted for some specific sites. For the purpose of the analysis, a self-propelled traveling sprinkler was adopted in the

TABLE 3. Estimated Installation Costs of a Silvicultural Energy Farm[a]

	Estimated cost (dollars/acre)		
Items	Low	Average	High
Land preparation and improvement			
Wooded land			
Mature forest	230	230	350
Farm wood lots	115	175	230
Open Land:			
Pasture	35	65	95
Cropland	5	12	17
Unpaved work roads	350	525	700
Irrigation System			
Permanent equipment (pump, well, mains)	—	115	—
Portable equipment (tractor, sprinkler, diesel motor, accessories)	—	170	—
Total	—	285	—
Planting, including planting stock	—	130	280

[a]As used by Inman *et al.*[60]

model. The traveling sprinkler has some limitations such as the need for open travel lanes, distortion of the spray pattern by the wind, and high working pressure. Conversely, the system has a number of features such as flexibility of operation, adaptability to irregular topography and irregular field shapes, which make it attractive for diverse situations. Basically, the traveling sprinkler is a mobile water cannon which sprays the crops as it travels along open irrigation lanes. Water is supplied from buried mains through hoses dragged behind the sprinkler during operation. Parallel 12 ft wide, 300 ft apart irrigation lanes were included in the conceptual design of the energy farm. The total area of the irrigation lanes amounts to about 4% of the planted area. Therefore, to maintain the output of the farm at its design value, about 4% more land than is needed to be planted will have to be acquired. Table 3 shows the costs required by a self-propelled traveling sprinkler system. It should be noted that the costs quoted may vary widely because of local conditions and should be used with caution and adjusted to account for site conditions.

    3.2.1e. Planting. The conceptual design assumes that purchased hardwood seedlings or cuttings will be planted mechanically on a 4 ft × 4 ft pattern (about 2,725 trees per acre). Planting costs include the purchase of the planting stock and the planting operations. Typical planting costs are reported in Table 3.

3.2.2. Energy Farm Operation

    Annual operations on the energy farm include irrigation of the modules in their first to third year of growth, fertilization of all modules, weed control in the module in its first year of growth, harvesting of one module, transportation of the wood chips, maintenance of the equipment and support activities. These activities are discussed in greater detail.

    3.2.2a. Irrigation. Well water will be applied to the modules in their first to third year of growth through the traveling sprinkler system discussed previously. The labor and fuel costs adopted in the analysis are shown in Table 4.

    3.2.2b. Fertilization. The amount of fertilizer to be applied yearly is determined on the basis of the nutrient analysis of the crop grown[11,16,71–73] assuming a 60% recovery factor. Formulated fertilizers (urea, 46% N; superphosphate, 46% P_2O_5; muriate of potassium, 60% K_2O) and lime will be applied by ground machinery during the first year of rotation. During each of the last five years of each rotation, the fertilizers will be applied by fixed-wing aircraft since the presence of the biomass crop would preclude the use of conventional ground equipment. Typical costs for fertilizer operations are presented in Table 4. These costs assume long-term supply contracts, include transportation and handling costs, and are averaged over a six-year rotation.

    3.2.2c. Weed Control. The control of weeds during at least the first year after planting has been shown to be essential for the establishment and

TABLE 4. Estimated Operating Costs for a Silvicultural Energy Farm[a]

Items	Units	Costs (1978 dollars)		
		Low	Average	High
Irrigation				
Labor	$/acre-yr	—	4	—
Fuel	$/acre-yr	—	43	—
Total	$/acre-yr	—	47	—
Fertilization				
Fertilizer and lime	$/acre-yr	20	27	45[b]
Application	$/acre-yr	5	7	12
Total	$/acre-yr	25	34	57
Weed control	$/acre	6	8	10
Harvesting	$/acre	17	24	30
Transportation	$/ODT	1.70	2.30	3.25
Planning	$/lifetime of farm	—	270,000	—
Supervision	$/yr	—	370,000	—
Support personnel	$/yr	—	50,000	—

[a] As used by Inman et al.[60]
[b] The low, average, and high costs refer to productivity levels of 5, 8, and 12 ODT/ acre-yr, respectively.

development of the seedlings by a number of investigations. Two weed control operations performed during the first growing season after planting and after each of the successive harvests have therefore been included in the model of the energy farm. These operations will be performed by disking or by an equivalent mechanical method. Estimated costs of these operations are given in Table 4.

3.2.2d. Harvesting. In the conceptual design, harvesting will be performed by a combine-type machine cutting the stems at ground level, chipping the biomass, stems and branches, and blowing the chips in a wagon pulled by the harvester. Chips will be stockpiled in reserved field storage areas. No such harvester exists at present. Field experiments conducted in Georgia have shown, however, that two-year-old strands of sycamore would be harvested with a silage corn harvester.[18] The characteristics of the conceptual harvester included in the energy farm model include a harvesting capacity of 60 ODT/ hr, an operating speed of 0.5–3.0 miles/hr, operation on slopes of up to 30%, a cutting width of 8 ft and a fuel consumption of 15 gallons/hr at peak load. The range of operating costs for a conceptual harvester of this type is shown in Table 4. The range of values quoted in the table reflect differences in terrain (slope), crop yields, and labor costs at various sites.

Harvesting will be performed during the dormant season both to take full advantage of the sixth year of growth and to avoid adversely affecting the regeneration of coppice during the following growing season. Tractors will pull the filled wagons to the storage areas where the chips will be stockpiled. Two major concerns are associated with wood chips storage: the loss of biomass with time through both chemical oxidation and microbial activity and the danger of spontaneous combustion. Some precautionary measures such as limiting the chip pile depth or mechanically turning the piles could reduce the material losses. Year round harvesting could be considered but would result in some losses in productivity. Storage of whole trees is another alternative, but this option would require another type of harvester and would increase biomass handling problems.

3.2.2e. Transportation. In the conceptual model described here, the system adopted for transporting chips from the field storage areas to the conversion facility consists of a fleet of semi-trailer trucks with hydraulic dump. The trucks will be loaded from the storage piles by front-end loaders. A range of values for transportation costs (excluding maintenance and depreciation) is shown in Table 4. The range of values reflects variation in transportation distance, labor rates, fuel rates and road speeds.

3.2.2f. Maintenance. The maintenance of the equipment owned by the energy farm operation is assumed to cost 10% of the equipment's purchase price per year. This equipment includes the harvesters, trucks, tractors and wagons, loaders, irrigation system, supervisory vehicles and field support vehicles. The maintenance of the work roads is assumed to be $115 per mile per year.

3.2.2g. Support Costs. Support costs include costs for planning, supervision and field support. These are estimated at about $270,000 for planning, $370,000 per year for supervisory personnel and $50,000 for field support personnel. Capital costs include supervisory and field support vehicles.

4. Biomass Production Economics

4.1. Biomass Production Costs

Biomass production costs were estimated for specific sites on the basis of the conceptual model proposed by Inman *et al.*[60] The economic model calculates, as appropriate, capital investment, schedules, depreciations, labor and fuel costs, tax, insurance, and repairs for the various components of the farm conceptual design. Together with financial data, such as debt to equity ratio, price escalation, and tax rates, these data are then used to estimate the return to equity and cash flow. In the next section, the financial model was operated so as to determine the biomass production cost required to yield a given cash flow rate of return. The results of the analysis are shown in Table 5. The two highest production costs are for the California and Illinois sites. The design assumes that high quality, expensive crop land will be used for energy farming

TABLE 5.   Estimated Biomass Production Costs[a]

| Site | Productivity (ODT/acre-yr)[b] | Biomass production costs | |
		$/ODT[c]	$/MMBtu[d]
Wisconsin	5.0	34.72	2.04
New England	5.0	36.95	2.17
Missouri	7.0	29.08	1.71
Illinois	8.0	48.08	2.83
Louisiana	12.0	23.44	1.38
Georgia	8.0	26.65	1.57
Mississippi	12.0	24.39	1.43
Florida	12.0	24.17	1.42
California	13.0	38.84	2.28
Washington	10.0	27.76	1.63

[a]Source: Reference 60.
[b]ODT/acre-yr: oven-dry ton/acre-yr
[c]$/ODT: 1978 dollars/oven-dry ton
[d]$/MMBtu: 1978 dollars/million Btu; 17 million Btu/oven-dry ton.

at these sites. This is not likely to occur, and the corresponding costs are probably not representative of those to be expected for energy farms. The production costs for the two northern sites, i.e., Wisconsin and New England, are higher than those for southern and southeastern sites, located in Louisiana, Georgia, Mississippi, and Florida. One of the major differences between these two groups of sites is the projected productivities, i.e., 5 ODT/acre-yr in the North versus 8–12 ODT/acre-yr in the South. Other site-specific factors such as land rental and preparation costs, transportation distance, etc., will also undoubtedly contribute to the difference in production costs. Comparing the production costs for the Louisiana, Georgia, and Mississippi sites located within the same general climatic area of the country also suggests that site productivity has an important impact on production costs.

Other estimates of the production cost of biomass on energy farms have been performed.[34,61−63,70,74−75] These estimates are generally consistent with those of Table 5 for sites having comparable productivities and comparable schedules of management despite slight differences in the financial assumptions and methods of calculation of the production costs used to derive these estimates. Table 6 summarizes the biomass production cost data of Table 5 and References 34, 61–63, 70, 74–75 by quoting means and standard deviations of production costs for various regions. The highest production costs are expected in the northern and northeastern parts of the country, i.e., a mean of about $30/ODT with a possible range reaching up to about $38/ODT. Intermediate production costs are expected in the Pacific and Appalachian regions, i.e., a mean of about $27/ODT with a possible high of about $30/ODT. The lowest costs are expected in the southern and southeastern part of the country, i.e., a mean of about \$24/ODT with a possible high of about \$27/ODT. The differences in cost between regions are probably mostly due to differences in productivities and major climatic conditions while the dispersion of the data within a region reflect the impact of local factors such as terrain, land availability, irrigation requirements, etc.

4.2. Production Cost Components

A cost breakdown by capital, operating, maintenance, and fixed costs for the data of Table 5 show that operating and maintenance costs constitute from 60–70% of the total costs, while capital and fixed costs constitute about 10% and 20–25% of the total costs, respectively.[60] (Fixed costs include interest, taxes, and return on the investment.) Similar conclusions were reached by others.[61] Biomass production therefore is not a capital-intensive operation. Efforts to reduce production costs should be directed towards reducing annual operating costs.

Table 7 presents a further costs breakdown by operations for two sample sites mentioned in Table 5. The major cost items are those associated with intensive management: fertilization and irrigation account for about 36% and 39% of the total production costs for the Wisconsin and Louisiana sites, respectively. Excluding fixed costs, transportation costs are the third highest operational costs, at about 14% of the total costs, followed by land leasing, at about 5–7%, and land preparation and harvesting, each at about 4% of the total cost.

TABLE 6. Means and Standard Deviation of Estimated Biomass Production Costs[a]

Region(s)	Mean ($/ODT)[b]	Standard deviation ($/ODT)[b]
Northern U.S. and Canada[c]	29.22	7.55
Appalachian[d]	26.54	3.30
Southern and Southeastern[e]	23.53	3.04
Pacific Coast[f]	27.48	1.91

[a]Data from Table 5 and references 34, 61–63, 70, 74, 75.
[b]1978 dollars/oven-dry ton.
[c]Includes Wisconsin and New England sites from Table 5 and Minnesota, Michigan, New York, and Canadian sites from references 34, 61, 75.
[d]Includes the Missouri site from Table 5 and the Kentucky, Kansas, and Tennessee sites.[61] The Illinois site of Table 5 located on good farm land is not included.
[e]Includes the Louisiana, Georgia, Mississippi, and Florida sites of Table 5 and the Virginia, Louisiana, Mississippi, South Carolina, Florida, and Texas sites of reference 61. The high cost estimates for South Carolina[70] are not included because the unfavorable conditions for irrigation would suggest the selection of a more attractive site.
[f]Includes the Washington site of Table 5 and the California sites of reference 74. The California site of Table 5 located on good farm land is not included.

TABLE 7. Production Cost Breakdown by Cost Category[a]

| | Production costs (1978 dollars) | | | |
| | Wisconsin | | Louisiana | |
Cost Category	$/ODT	Percent of total	$/ODT	Percent of total
Planning, supervision	1.99	5.73	1.99	8.46
Land lease	2.39	6.88	1.21	5.14
Land preparation	1.44	4.15	0.79	3.36
Roads	0.18	0.52	0.08	0.34
Planting	0.69	1.99	0.61	2.59
Irrigation	6.92	19.93	2.89	12.28
Fertilization	5.72	16.47	6.24	26.52
Weed control	0.19	0.55	0.11	0.47
Harvesting	1.57	4.52	1.06	4.50
Chip handling and transportation	4.85	13.96	3.19	13.56
Interests	1.32	3.80	0.79	3.36
Taxes	3.94	11.34	2.40	10.20
Return to investor	3.70	10.65	2.26	9.60
Salvage value	(0.17)	(0.49)	(0.09)	(0.38)
Total	34.73	100.00	23.53	100.00

[a]Source: Adapted from reference 60.

Analyses performed by others confirm that annual operating costs are the major elements in the biomass production costs, and that fertilization and irrigation costs are probably the highest or one of the highest items in the annual operating costs.[61,70]

5.   Energy Balance for Biomass Production

Silvicultural energy farms require intensive field management and are designed to achieve a high degree of mechanization. Both features require significant energy input. Therefore, in order to evaluate the silvicultural farm as an energy production operation, the energy produced in the form of biomass must be weighed against the energy consumed in the farming operations. In this comparison, energy forms gathered or manufactured by man are considered as inputs while solar energy, which is coincidental to man's purposes, is ignored. Neither the energy required to manufacture the field equipment nor the energy used in installing the energy farm are included in the energy balance. Both inputs have been shown to be negligible on an annual basis over the lifetime of the farm.[58,76]

Table 8 presents balances for the two sites analyzed in Table 7. The two

TABLE 8. Sample Energy Balances for Silvicultural Biomass Farms

Operation	Material[b]	Wisconsin[a]		Louisiana[a]	
		Energy $(10^9$ Btu/yr)	Percent of total (%)	Energy $(10^9$ Btu/yr)	Percent o total (%)
Supervision	Gasoline	1.67	0.42	1.67	0.60
Field supply[c]	Diesel/ gasoline	0.53	0.13	0.53	0.19
Harvesting	Diesel	9.43	2.36	7.87	2.84
Biomass handling[d]	Diesel	8.77	2.20	8.78	3.17
Biomass transportation	Diesel	28.83	7.22	17.32	6.25
Irrigation[e]	Diesel	224.93	56.34	112.48	40.60
Fertilizer manufacture	Urea, P_2O_5, K_2O	122.81	30.76	127.57	46.04
Fertilizer transportation	Diesel (rail)	0.98	0.24	0.64	0.23
Fertilizer application (air)	Gasoline	0.15	0.04	0.16	0.06
Ground operations (fertilizer application and cultivation)	Diesel	1.16	0.29	0.06	0.02
Total energy consumption		399.26	100.00	277.08	100.00
Total energy yield		4,250.00	—	4,250.00	—
Net energy yield		3,850.74	—	3,972.92	—
Net energy efficiency		0.91	—	0.94	—
Energy in/energy out		1/10.6	—	1/15.3	—

[a]Sites analyzed in Table 5 (adapted from reference 60).
[b]Energy contents from references 58, 78, 79.
[c]Includes delivery of equipment and supplies and moving of the personnel.
[d]Transportation to storage and loading of trucks.
[e]Includes water pumping and moving of equipment.

highest energy consuming operations are irrigation (pumping and application of the water) and manufacture of the fertilizers. Together these operations account for about 85% of the total energy input. The third highest energy input is that of transporting the harvested biomass from field storage to the conversion plant; this item accounts for 6–7% of the total energy input at both sites. Biomass handling and harvesting both account for 2–3% of the energy input. Other operations—cultivation, fertilizer application, supplies, etc.—require minimal energy input. The net energy efficiency is estimated by dividing the net energy yield (total biomass energy yield minus total energy input) by the total energy yield. Net energy efficiencies are of the order of 90–95%, resulting in input/output energy ratios of 1/10 to 1/15. It is therefore generally accepted that very favorable energy balances (energy in/energy out ratios higher than 1/10) can be expected for silvicultural energy farms.[58,60,61,70,76] In contrast, the energy balance for the production of agricultural energy feedstocks is much less favorable: energy in/energy out ratios in the range of 1/3 to 1/5 have been quoted for the production of corn grain and corn silage energy feedstocks.[77]

6.   The Potential of Silvicultural Energy Farming

If silvicultural energy farming is to make a significant contribution to national energy needs, large areas of land would have to be devoted exclusively to the production of biomass fuel. For example, about 40 million acres of land would have to be reserved for wood production to generate about 5 quads of wood fuel annually on the basis of an average mean annual productivity of 7 dry tons/acre. This area is only about 2% of the total land area of the United States but, also, it is about 10% of either cropland or commercial forest land in the country. Competition for adequate land could therefore occur particularly as hardwoods require fairly fertile land to achieve the productivities projected in energy farm designs.

The land potentially available for energy farming has been estimated by several authors. The criteria of suitability for energy farming adopted by these authors include a minimum of 25 inches of precipitation, arable land, and a slope of no more than 30% (17°).[80,81] Using the land classification system developed by the Soil Convervation Service, Salo et al.[80] estimated that about 320 million acres of pasture, forest, range, hayland, and open land formerly cropped could be diverted to energy farming. Assuming that about 10% of this area is devoted to energy farming, these authors predict a potential contribution of about 5 quads from silvicultural energy farming. Figure 2 shows the projected energy yields by USDA farm production regions. Intertechnology/ Solar estimated that 75–100 million acres could be devoted to energy farms resulting in a contribution of 10–15 quads of wood fuel annually. An independent analysis by Didercksen et al.,[82] suggests that about 110 million acres of noncropland have a good to medium potential for conversion to crop production. Although at least 40% of this land would be unusable for silvicultural farming because of low precipitation or steep slopes, this estimate confirms that large land areas could be available for energy farming.

7.   Conclusions

Large scale production of fiber from short rotation hardwood plantations is an industrial reality. About 10,000 acres of hardwoods grown on 10- to 15-yr rotations under intensive management are presently established in the coastal plain region of Virginia, and it is proposed to expand the production capacity to about 50,000 acres.[83] There is, therefore, little doubt that the concept of silvicultural farming is viable.

On the basis of presently available data, the analysis of conceptual designs suggest that woody biomass produced on energy farms could be competitive with fossil fuels or feedstocks in many areas of the country. The analyses also indicate that the energy balance for woody biomass production is very favorable when compared to the energy balance for production or recovery of other

FIGURE 2. Potential energy yield (quads) from silvicultural energy farms using 10% of land potentially available (Soil Conservation Classes I–IV, permanent pasture, forest, range, rotation hay and pasture, hayland, and open land formerly cropped). National total = 5 quads. From Salo et al.[80]

forms of biomass. Wood biomass also lends itself to many end-uses: direct combustion, conversion to synthetic fuels, or conversion to various chemicals or chemical feedstocks. Silvicultural energy farms must, therefore, be considered as one of the most attractive long-term renewable energy resources. Significant research and development efforts will, however, have to be expanded before this resource can realize its full potential.

Silvicultural energy farms have the potential of making a significant contribution towards reducing the U.S. consumption of fossil fuels provided sufficient land area is made available for energy farming. The amount of land devoted to energy farming will be determined by a number of often conflicting factors: economics, demand for food and fiber, expanding urbanization, environmental concerns, land ownership, social impacts and others. In view of the potential long-term benefit offered by silvicultural energy farming, it may well be worth defining policies and establishing incentives encouraging the development of this renewable energy resource.

References

1. U.S. Department of Agriculture, Forest Service, The Outlook for Timber in the United States, Forest Resource Report No. 20, Washington, D.C. (1974).

2. U.S. Department of Agriculture, Forest Service, Forest Statistics of the U.S., 1977, Review Draft, Washington, D.C. (1978).

3. K. Howlett and A. Gamache, Silvicultural Biomass Farms: Forest and Mill Residues as Potential Sources of Biomass, Vol. VI, MITRE Corporation/Metrek Division, Technical Report No. 7347, McLean, Va. (1977).

4. D. J. Salo and J. F. Henry, Wood based biomass resources in the United States: Near term and long term prospects, in: *Proceedings EPRI/GRI Workshop on Biomass Energy and Technology,* Santa Clara, Ca. (1978).

5. S. H. Spurr and H. J. Vaux, Timber: Biological and economic potential, *Science* **191**, 752 (1976).

6. J. B. Grantham, Anticipated competition for available wood fuels in the United States, paper presented at the American Chemical Society Meeting (1977).

7. A. M. Herrick and C. L. Brown, A new concept in cellulose production—Silage Sycamore, *Agri. Sci. Res.* **5**, 8 (1967).

8. G. F. Dutrow, *Economic Implications of Silage Sycamore,* U.S. Department of Agriculture, Forest Service Research Paper, Southern Forest Experiment Station New Orleans, La. (1971), pp. 50–66.

9. H. E. Young, *Wood Fiber Farming: An Ecologically Sound and Productive Use of Right-of-Ways,* University of Maine, School of Forest Resources (1972).

10. K. Steinbeck, J. T. May, and R. G. McAlpine, Silage cellulose—A new concept, in: *Forest Engineering Conference Proceedings,* American Society of Agricultural Engineers, Saint Joseph, Michigan (1968).

11. J. H. Ribe, Will short-rotation forestry supply future pulpwood needs?, *Pulp and Paper* December (1972).

12. J. H. G. Smith and D. S. DeBell, Opportunities for short rotation culture and complete utilization of seven Northwestern tree species, *Forestry Chronicle* **49**, 1 (1973).

13. P. R. Larson and J. C. Gordon, Photosynthesis and wood yield, *Agri. Sci. Rev.* **7**, 7–14 (1969).

14. E. J. Schreiner, *Mini-Rotation Forestry,* U.S. Department of Agriculture Forest Service, Res. Pap. NE-174, Northeast Forestry Experimental Station, Upper Darby, Pa. (1970).

15. K. Steinbeck and J. T. May, Productivity of very young *Platanus occidentalis* Plantings Grown at Various Spacings, in: *Forest Biomass Studies,* University of Maine Press, Orano, Maine (1971), pp. 153–162.

16. J. R. Saucier, A. Clark, and R. G. McAlpine, Above ground biomass yields of short rotation sycamore, *Wood Sci.* **5**, 1–6 (1972).

17. J. H. Ribe, Will short-rotation forestry supply future pulpwood needs?, *Pulp and Paper* December (1972).

18. R. P. Belanger and J. R. Saucier, Intensive culture of hardwoods in the South, *Iowa State J. Res.* **49** (3), Part 2, 339–344 (1975).

19. H. E. Kennedy, Jr., Influence of Cutting Cycle and Spacing on Coppice Sycamore Yield, U.S. Forest Service Research Note S.O.-193, Southern Forest Experimental Station, New Orleans (1975).

20. K. Steinbeck, Short-rotation hardwood forestry in the southeast, in: *Proceedings Second Annual Symposium on Fuels from Biomass* (W. W. Shuster, ed.), Rensselaer Polytechnic Institute, Troy, NY, June 20–22 (1978).

21. P. P. Kormanik, G. L. Tyre, and R. P. Belanger, A case history of two short rotation coppice plantations of sycamore on southern piedmont bottomlands, in: *IUFRO Biomass Studies,* College of Life Sciences and Agriculture, University of Maine, Orono, Maine (1973), pp. 351–360.

22. K. Steinbeck, R. G. McAlpine, and J. T. May, Short rotation culture of sycamore: A status report, *J. For.* **70**, 210–213 (1972).

23. J. Zavitkovski, *Biomass Farms for Energy Production: Biological Considerations,* SAF/CIF Convention, Proceedings, St. Louis, Missouri, October (1978).

24. A. R. Ek and J. P. Brodie, Preliminary analysis of short-rotation Aspen management, *Can. J. For. Res.* **5**, 245–258 (1974).
25. D. F. W. Pollard, Above ground dry matter production in three stands of trembling Aspen, *Can. J. For. Res.* **2** (27), (1972).
26. R. Doucet, Biomass d'un peuplement de peuplier faux tremble age de six ans, *Que. Minist. Terres For. Serv. Rech. Note* No. 7 (1977).
27. A. B. Berry, Production of dry matter from aspen stands harvested on short rotations, in: *IUFRD Biomass Studies* (H. E. Young, ed.) College of Life Science and Agriculture, University of Maine, Orono, Me. (1973).
28. Inter Group Consulting Economists, Ltd., Liquid Fuels from Renewable Resources: Feasibility Study, Volume C: Forest Studies, Report prepared for the Government of Canada, Interdepartmental Steering Committee on Canadian Liquid Fuels Program Options, Winnipeg, Manitoba, March (1978).
29. M. C. Carter and E. H. White, Dry weight and nutrient accumulation in young stands of cottonwood, Circular 190, Agr. Exp. Sta., Auburn University, Auburn, Alabama (1971).
30. R. M. Krinard and R. L. Johnson, Ten-year results in a cottonwood spacing study, USDA Forest Service Research Paper 50-106, Southeastern Forest Exp. Station (1975).
31. F. L. Schmidt and D. S. DeBell, Wood production and kraft pulping of short rotation hardwoods in the Pacific Northwest, in: *IUFRO Biomass Studies,* (H. E. Young, ed.), Coll. of Life Sci. and Agri., Univ. of Maine, Orono, Maine (1973).
32. D. S. DeBell, Short rotation culture of hardwoods in the Pacific Northwest, *Iowa State J. Res.* **49** (3), Part 2, 345 (1975).
33. P. E. Heilman, D. V. Peabody, D. S. DeBell, and R. F. Strand, A test of close-spaced short-rotation culture of black cottonwood, *Can. J. For. Res.* **2** (4), 456–459 (1972).
34. A. Musnier, Etude financiere et de gestion provisionelle des plantations et des fermes populicoles, Quebec Ministere des Terres et des Forets, Service de la Recherche, No. 31 (1976).
35. A. R. Ek and D. H. Dawson, Yields of intensively grown populus: Actual and projected, in: *Intensive Plantation Culture,* USDA Forest Service, General Technical Report NC-21, N. Central For. Exp. Sta., Rhinelander, Wisconsin (1976).
36. H. W. Anderson and L. Zsuffa, Yield and Wood Quality of Hybrid Cottonwood Grown in 2-Year Rotation, Forest Research Report No. 101, Ontario Ministry of Natural Resources (1975).
37. H. W. Anderson and L. Zsuffa, Farming Hybrid Poplar for Food and Fiber, Forest Research Report No. 103, Ontario Ministry of Natural Resources (1977).
38. J. F. Laundrie and J. G. Berbee, High yield of kraft pulp from rapid growth hybrid poplar trees, Research Paper FPL 186, USDA Forest Service, Wisconsin (1972).
39. D. H. Dawson, J. G. Isebrandts, and J. C. Gordon, Growth and dry weight yields and specific gravity, Research Paper NC-122, USDA Forest Service (1976).
40. United States Department of Agriculture Forest Service, Intensive Plantation Culture, General Technical Report No. 21, North Central Forest Experiment Station (1976).
41. T. W. Bowersox and W. W. Ward, Growth and yield of close-spaced young hybrid poplars, *For. Sci.* **22** (4), (1976) pp. 109–114.
42. J. B. Crist and D. H. Dawson, Anatomy and dry weight yields of two populus clones grown under intensive culture, USDA Forest Service Res. Paper NC-113, North Central Forest Exp. Sta., St. Paul, Minn. (1975).
43. R. S. Evans, Energy plantations: Should we grow trees for power plant fuel?, Can. For. Serv., Dept. Environ., Rep. VP-X-129, Vancouver, B.C. (1974).
44. J. H. G. Smith, Biomass of some young red alder stands, in: *IUFRO Biomass Studies* (H. E. Young, ed.), Coll. of Life Sci. and Agr., University of Maine, Orono, Maine (1973).
45. J. H. G. Smith and D. S. DeBell, Some effects of stand density on biomass of red alder, *Can. J. For. Res.* **4**, 335–340 (1974).

46. D. S. DeBell, Potential productivity of dense young thickets of red alder, *For. Res.* Note No. 2, Crown Zellerbach Corp., Central Research, Camas, Wash. (1972).
47. J. Zavitkovski and R. D. Stevens, Primary productivity of red alder ecosystems, *Ecology,* 53 (2) (1972).
48. E. C. Franklin, and G. Meskimen, Wood properties of some eucalypts for the southern United States, in: *Proceedings, 1975 National Convention, Society of American Foresters,* Washington, D.C. (1975).
49. United States Department of Agriculture, Forest Service, Cooperative Progress in Eucalypt Research, Southern Forest Exp. Station, Olustee, Florida (1975).
50. W. H. Smith, L. E. Nelson, and G. L. Switzer, Development of the shoot system of young loblolly pine, II. Dry matter and nutrient accumulation, *For. Sci.* **17**(1), 55 (1971).
51. G. L. Switzer and L. E. Nelson, Nutrient accumulation and nutrient cycling in loblolly pine plantation ecosystems: The first twenty years, *Soil Sci. Soc. Am. Proc.* **36**, 143–147 (1972).
52. J. C. Nemeth, Dry Matter Production and Site Factors in Young Loblolly Pine and Slash Pine Plantations, Ph.D. Diss., North Carolina State University, Department of Botany, Raleigh, N.C. (1972).
53. C. W. Ralston, Annual primary productivity in a loblolly pine plantation, in: *IUFRO Biomass Studies* (H. E. Young, ed.), College of Life Sciences and Agriculture, University of Maine, Orono, Maine (1973).
54. W. R. Harms and O. G. Langdon, Development of loblolly pine in dense stands, *For. Sci.* **22**, 331 (1976).
55. G. Siren and G. Sivertsson, Survival and dry matter production of some high yield clones of salix and populus selected for forest industry and energy production, Royal Coll. For., Dep. Reforestation, Res. Note 83 Stockholm, Sweden (1976).
56. Georgia-Pacific Corporation, Eucalyptus Study Intracompany memo, Fort Bragg, Ga. (1976).
57. G. C. Szego and C. C. Kemp, Energy forests and fuel plantations, *Chem Tech,* **3**(5): 275–284 (May 1973).
58. J. A. Alich, Jr. and R. E. Inman, Effective Utilization of Solar Energy to Produce Clean Fuel, NSF/RANN/SE/GI/38723 Contract, SRI Project No. 2643, Menlo Park, Ca. (1974).
59. C. L. Brown, Forests as energy sources in the year 2000: What man imagines, man can do, *J. For.* **74**, 7 (1976).
60. R. E. Inman, D. J. Salo, and B. J. McGurk, Silvicultural Biomass Farms, Vol. IV: Site-Specific Production Studies and Cost Analyses, MITRE Corporation/Metrek Division, MTR No. 7347 (1977).
61. Intertechnology/Solar Corporation, The Photosynthesis Energy Factory: Analysis, Synthesis and Demonstration, U.S. Department of Energy Contract No. EX-76-C-01-2548, Final Report, NTIS-HCP/T3548-01, Washington, D.C. (1978).
62. D. W. Rose, Cost of producing energy from wood in intensive culture, *J. Environ. Manage.* **5**, 1 (1976).
63. T. W. Bowersox and W. W. Ward, Economic analysis of a short rotation Fiber production system for hybrid poplar, *J. For.* **74**, 750 (1976).
64. R. E. Lohrey, Site preparation improves survival and growth of direct seeded pines, U.S. Forest Service Research Note 50-185, Southern Forest Exp. Station, New Orleans, La. (1974).
65. R. P. Schultz, Intensive culture of southern pines: Maximum yields or short rotations, *Iowa State J. Res.* **49**(3), 325 (1975).
66. J. S. McKnight, Planting cottonwood cuttings for timber production in the south, U.S.D.A. Forest Service Res. Paper 50–60 (1970).
67. D. S. deBell and J. C. Harms, Identification of cost factors associated with intensive culture of forest crops, *Iowa State J. Res.* **50**(3), 295 (1976).

68. R. Hunt, Effects of Site Preparation on Planted Sweetgum, Sycamore and Loblolly Pine on Upland Sites, Third Year Measurement Report, International Paper Company, Southlands Experiment Forest, Bainbridge, Ga. (1975).

69. D. J. Salo, J. F. Henry, and R. E. Inman, Design of a Pilot Silvicultural Biomass Farm at the Savannah River Plant, ERDA Contract No. EG-77-C-01-4101, MITRE Corporation/ Metrek Division, MTR-7960 (1979).

70. D. J. Salo, J. F. Henry, and A. W. DeAgazio, Pilot Silvicultural Biomass Farm Layout and Design—Comparative Energetic and Cost Assessment of Irrigation Alternatives at The Savannah River Plant, ERDA Contract No. EG-77-C-01-4101, MITRE Corporation/ Metrek Division, MTR-79W00102 (1979).

71. Hansen, E. A., in: Intensive Plantation Culture (Five Years Research), USDA Forest Service General Technical Report NC-21, North Central Forest Exp. Station (1976).

72. J. R. Boyle, J. J. Phillips, and A. R. Ek, Whole tree harvesting: Nutrient budget evaluation, J. For. **71**, 760 (1973).

73. J. R. Boyle, Nutrients in relation to intensive culture of forest crops, *Iowa State J. Res.* **49**(3), pt2, 297 (1975).

74. J. F. Henry, M. D. Fraser, W. B. Scholten, and C. W. Vail, Economics of energy crops on specific Northern California marginal crop lands, paper presented at the Fuels from Biomass Sympsoium, California Energy Commission, Sacramento, Ca. (August 3, 1977).

75. Inter Group Consulting Economists Ltd. Liquid Fuels from Renewable Resources: Feasibility Study, Vol. C: Forest Studies, report prepared for the Government of Canada, Interdepartmental Steering Committee on Canadian Renewable Liquid Fuels, Winnipeg, Manitoba (1978).

76. P. R. Blankenhorn, W. K. Murphey, and T. W. Bowersox, Energy Expended to Obtain Potentially Recoverable Energy from the Forests, Tappi Conference Papers, Forest Biology Wood Chemistry Conference, Madison, Wisconsin (June 20–22, 1977).

77. E. S. Lipinsky, T. A. McClure, J. L. Otis, D. A. Scantland, and W. J. Sheppard, Systems Study of Fuels from Sugarcane, Sweet Sorghum, Sugar Beets and Corn, Vol. IV: Corn Agriculture, final Report, ERDA Contract No. W-7405-ENG-92, BMI-19574 A (Vol. IV) (March 31, 1977).

78. D. Pimentel, Agricultural production: Resource needs and limitations, in: *Transactions of the 40th North American Wildlife and Natural Resources Conference 1975,* Wildlife Management Institute, Washington, D.C. (1975).

79. Federal Energy Administration, Project Independence Blueprint (1974).

80. D. J. Salo, R. E. Inman, B. J. McGurk, and J. Verhoeff, Silvicultural Biomass Farms, Volume III: Land Suitability and Availability, MITRE Corporation/Metrek Division, MTR 7347, McLean, Va. (1977).

81. InterTechnology/Solar Corporation, Solar SNG: The Estimated Availability of Resources for Large Scale Production of SNG by Anaerobic Digestion of Specially Grown Plant Matter, Report No. 011075, American Gas Association, Project No. IU 114-1, Warrenton, Va. (1975).

82. R. Didericksen, A. Hidlebaugh, and K. Schmede, Potential Cropland Study, Soil Conservation Service, U.S. Department of Agriculture, Washington, D.C. (1977).

83. B. F. Malac, and R. D. Heeren, Hardwood plantation management, Southern *J. Appl. For.* **3**(1), 3 (1979).

PART II
CONVERSION PROCESSES

Section A
Direct Combustion Processes

6
Basic Principles of Direct Combustion

FRED SHAFIZADEH

1. Introduction

The combustion process may be defined as an interaction between fuel, energy, and the environment rather than merely a chemical reaction of fuel with oxygen to release energy, or a physical reaction of the fuel with the environment involving heat and mass transfer. The process is enormously complex even with relatively simple or homogenous fuels. A quantitative description of this process requires the combined capabilities of a physical chemist specializing in chemical kinetics, a physicist specializing in heat and mass transfer, and an engineer specializing in fluid dynamic and unit operations communicating with the skills of an applied mathematician.

Such a description has not been attempted partly because until recently biomass fuel has been out of vogue for being unsophisticated, inconvenient, or uneconomical. A major problem in scientific understanding or quantitative description of biomass fuel is the heterogeneity and complexity of the substrate. Although the title of this chapter denotes direct combustion, it should be noted that biomass fuels, which are normally solid, organic materials and thermally degradable, may be delivered directly to the combustion furnace but are not combusted directly. These materials, in a hot furnace or under the influence of a sufficiently strong source of energy, are converted to a mixture of volatiles and a carbonaceous char which burn with entirely different combustion characteristics. The volatile products mix with air and burn with flames (at the gas phase within the furnace); the carbonaceous char burns with glowing ignition

FRED SHAFIZADEH • Wood Chemistry Laboratory, Department of Chemistry, University of Montana, Missoula, Montana 59812.

(at the solid phase on the grate). These two modes of combustion have entirely different chemical mechanisms and kinetics, which in turn are affected by both the chemical and physical composition of the substrate and characteristics of the combustion system which control the heat and mass transfer.

The heat of combustion (ΔH) for various compounds and fuels, calculated on the basis of complete combustion under ideal adiabatic conditions at 25°C could also be misleading. It should be at least corrected for the change in entropy (ΔS) or the energy lost in converting the solid fuel into gaseous combustion products as shown by Equation (1),

$$\Delta G = \Delta H - T\Delta S \qquad (1)$$

where G is the free energy, H is enthalpy, T is the absolute temperature, and S is the entropy. The combustion products carry out and dissipate some of the heat of combustion which should be taken into consideration. In addition, incomplete combustion and various deviations from the ideal conditions lower the efficiency of the combustion process and the net usable energy that could be obtained.

The combustion efficiency also relates to the problems of environmental quality. Complete combustion of organic materials provides carbon dioxide and water, which are harmless components of the natural environment. Incomplete combustion, however, provides carbonaceous residues (fly ash), smoke (soot and tar aerosol), and odorous and noxious gases (containing carbonyl derivatives, unsaturated compounds, and carbon monoxide) besides wasting considerable fuel.

In view of these considerations, the following sections provide a discussion of the fuel composition, the pyrolytic transformations, and the combustibility and heat of combustion of the pyrolysis products in relation to the physical aspects which control the problems of heat and mass transfer.

2. Composition

Biomass fuels which originate from plants generally contain absorbed and condensed moisture and some inorganic material (as the original component or contaminant) in addition to various types of organic compounds. These components could vary greatly according to the state and history of the fuel and the parts and species of the plant from which they originate.

Plant cells are generally distinguished from animal cells by the presence of cellulosic cell walls and large vacuoles. The living cells, such as those in the green leaves, contain some proteins in their protoplasm and considerable water in vacuoles which maintains their turgidity. The woody tissues contain some living (parenchyma) cells, but are largely composed of dead (prosenchyma)

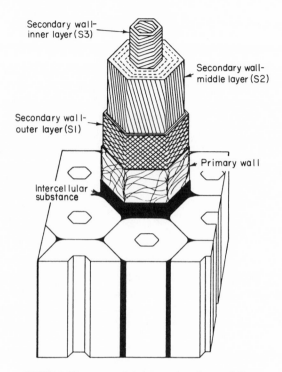

Secondary wall-
inner layer (S3)

Secondary wall-
middle layer (S2)

Secondary wall-
outer layer (SI)

Primary wall

Intercellular
substance

FIGURE 1. Diagrammatic presentation of a wood fiber.

cells which consist of the cell walls and intercellular materials (middle lamella). Figure 1 shows a diagrammatic presentation of a wood fiber (tracheid).

2.1. Moisture Content

The cavity within the dead cells (lumen) could contain considerable water, which is lost on drying. In addition to the condensed water within the cell cavities, the cell walls also contain some absorbed moisture which equilibrates with the ambient relative humidity. Therefore, there could be a wide range of moisture content (MC) depending on fuel condition. Biomass fuels such as wood, wood residues, and bagasses, as received without drying, could have about 50% MC that could be reduced to about 20% by air drying. However, some agricultural residues that readily dry, such as straw, may have about 10–12% equilibrium MC. The moisture content not only acts as a heat sink and lowers the combustion efficiency but also has a limiting effect on the economics and transportation range of the fuel.

2.2. Ash Content

Mineral components of plant materials vary substantially according to the species, locality, and soil contamination. Wood contains about 0.5% ash mainly composed of alkali and earth alkali cations present as carbonates, carboxylic salts, and some silica crystals. The latter compound is especially abundant in cereal straws that may have an ash content of 18% or more. Bark contains more ash than wood. Contamination could substantially increase the inorganic content of the municipal and agricultural wastes. The silica and other insoluble inorganic compounds act as a heat sink, but the soluble ionic compounds could have a catalytic effect on the gasification and combustion of the biomass fuel as discussed later.

2.3. Organic Content

Cellulose microfibrils embedded in a matrix of hemicelluloses and lignin form the plant cell walls and main components of biomass. In addition to these materials, there are some lipids and hydrocarbons (terpenes) which are soluble in ether and various types of phenolic compounds, carbohydrates and proteins, which may be soluble in benzene, alcohol, or water. The soluble components are collectively called extractives. The leaves and bark generally contain more extractives and less cell wall materials than wood and woody tissues. Table 1 shows the approximate analysis of several types of biomass.

The cellulose component, which is composed of D-glucopyranose units linked linearly with β-$(1 \rightarrow 4)$ glycosidic links is the same in all types of biomass except for the degree of polymerization. However, the nature of the hemicelluloses and lignin could vary.

Acetyl-4-O-methylglucuronoxylan (xylan) forms the main hemicellulose of the hardwoods, and glucomannan (mannan) forms the principal hemicellulose of the softwoods. Furthermore, the softwood lignin contains guaiacyl propane units (phenolic groups having one methoxyl group). Hardwood lignin, in addition to this, contains syringyl propane units (with two methoxyl groups).

TABLE 1. Approximate Analysis of Some Biomass Species

Species	Total ash (%)	Solvent soluble (%)	Water soluble (%)	Lignin (%)	Hemicellulose (%)	Cellulo (%)
Softwood	0.4	2.0	—	27.8	24.0	41.0
Hardwood	0.3	3.1	—	19.5	35.0	39.0
Wheat straw	6.6	3.7	7.4	16.7	28.2	39.9
Rice straw	16.1	4.6	13.3	11.9	24.5	30.2
Bagasse	2.3	8.4	10.0	18.5	29.0	33.6

The higher content of acetyl and methoxyl groups in hardwoods explains why this material has been used in destructive distillation processes to obtain acetic acid and methanol.

As discussed later, the carbohydrates, on heating, readily break down at the glycosidic link to monomeric materials and have a lower heat of combustion because they contain several molecules of water in their chemical composition that could be presented as $[C_6(H_2O)_5]_n$ or $[C_5(H_2O)_4]_n$.

3. Pyrolysis and Heat of Combustion of Biomass and Its Components

Since the combustion of biomass involves the thermal degradation of the fuel and subsequent oxidation of the products, it is important to know how these products are formed, what their heat content is and how they burn.

As the biomass is heated in a combustion chamber or otherwise, it breaks down and evaporates, leaving a carbonaceous residue containing the mineral components. The volatile degradation product consist of: a gaseous fraction containing CO, CO_2, some hydrocarbons, and H_2; a condensable fraction, containing water and low molecular weight organic compounds such as aldehydes, acids, ketones, and alcohols; and finally, a tar fraction containing higher molecular weight sugar residues, furan derivatives, and phenolic compounds. Fine, airborne particles of tar and charred material form the smoke.

The proportions of the volatiles formed, the residue which is left, and the rate of evaporation or weight loss can be determined by thermogravimetry (TG) and its derivative (DTG), and the changes in enthalpy (ΔH) can be measured by differential thermal analysis (DTA) or differential scanning calorimetry (DSC), which are collectively called thermal analysis. Thermal analysis of cottonwood and its components, shown in Figures 2 and 3, indicates that pyrolysis of the cellulose and hemicellulose (carbohydrate) components gives mainly volatile products, whereas lignin gives mainly char. Furthermore, the pyrolysis of wood reflects the thermal properties of its components.

These methods can be used to measure the energy which is consumed at different temperatures for drying, distillation, pyrolysis, and heating of the biomass before ignition (the heat of preignition). The energy that is generated after complete combustion of the degradation products can be measured in a calorimeter. Alternatively, the released energy can be measured during the course of the pyrolysis and combustion as a function of time or temperature by thermal evolution analysis (TEA). Figure 4 shows the heat of combustion of the volatiles produced by gasification of Douglas-fir needles at different temperatures before and after removal of the extractives. This figure and similar data obtained for various forest fuels dramatically demonstrate the contribution of the extractives to the combustibility of forest fuels. The balance between the heat of combustion and the heat of preignition at different temperatures or

Fred Shafizadeh

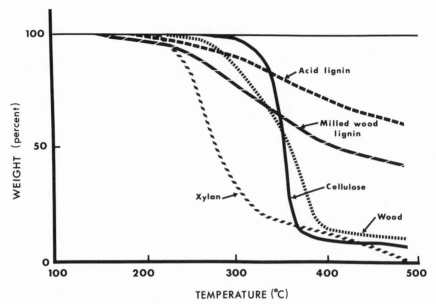

FIGURE 2. Thermogravimetry of a hardwood (cottonwood) and its components.

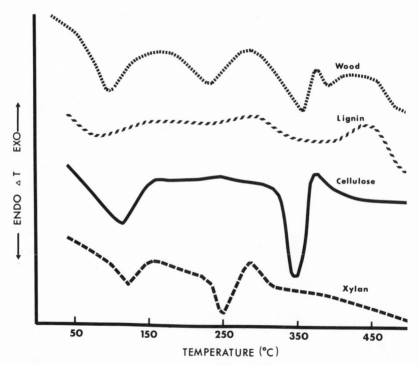

FIGURE 3. Differential thermal analysis of a hardwood (cottonwood) and its components.

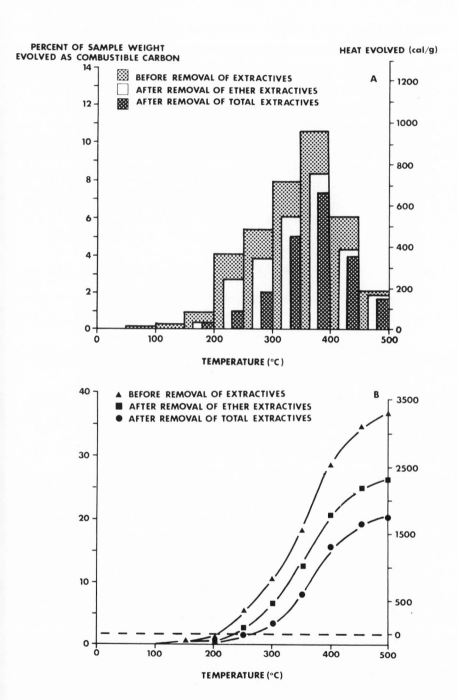

FIGURE 4. Evolution of the combustible volatiles and the corresponding heat of combustion from Douglas fir foliage. (A) in temperature intervals and (B) cumulative.

given conditions determine the combustibility of the fuel and the net energy released.

The heat contents of different types of forest fuels and their components are given in Table 2. This table also shows the amount of char left at 400°C on heating at the rate of 200°C/min and the heat of combustion of the gaseous and carbonaceous products. Table 3 shows the distribution of the heat of combustion between the char and the volatile products.

The heat of combustion for different types of biomass closely relates to the amount of oxygen required. As shown in Figure 5, each gram of oxygen liberates 3,349 calories. Thus, the heat of combustion relates to the oxidation state of the fuel. Within the biomass category, the heat of combustion of a fuel and its pyrolysis products also correlate to the carbon content. Pure cellulose, which is composed entirely of sugar units with the elemental analysis of $C_6(H_2O)_5$, has a relatively low heat content ($\Delta H^{25°} = -4{,}143$ cal/g) because of the high level of oxidation. However, 85% of its heat content is released to the volatiles because of the thermal cleavage of the sugar units (discussed later).

Lignin, which is composed of coniferyl units (and some related syringly

TABLE 2. The Heat of Combustion of Natural Fuels and Their Pyrolysis Products as Char and Combustible Volatiles

Fuel			Char		Combustible volatiles	
Source	type	$\Delta H^{25°}_{comb}$ (cal/g)	yield %[a]	$\Delta H^{25°}_{comb}$ (cal/g)	yield %[a]	$\Delta H^{25°}_{comb}$ (cal/g)
Cellulose	Filter paper	−4143	14.9	−7052	85.1	−3634
Douglas-fir lignin	Klason	−6371	59.0	−7416	41.0	−4867
Poplar wood						
Populus ssp.	Excelsior	−4618	21.7	−7124	78.3	−3923
Larch wood						
Larix occidentalis	Heart wood	−4650	26.7	−7169	73.3	−3732
Decomposed Douglas fir						
Pseudotsuga menzeisii	Punky wood	−5120	41.8	−7044	58.2	−3738
Ponderosa pine						
Pinus ponderosa	Needles	−5145	37.0	−6588	63.0	−4298
Aspen						
Populus tremuloides	Foliage	−5034	37.8	−6344	62.6	−4238
Douglas fir bark						
Pseudotsuga menzeisii	Outer (dead)	−5122	52.8	−5798	47.2	−4366
Douglas fir bark						
Pseudotsuga menzeisii	Whole	−5708	47.1	−6406	52.9	−5087

[a] Heating rate 200° C/min to 400° C and held for 10 min.

TABLE 3. Distribution of the Heat of Combustion of Forest Fuels

| Fuel | | Char (cal/ | Gas (cal/g | Total |
source	type	g fuel)	fuel)	(cal/g)
Cellulose	Filter paper	−1050	−3093	−4143
Douglas-fir lignin	Klason	−4375	−1995	−6370
Poplar wood				
Populus ssp.	Excelsior	−1546	−3072	−4618
Larch wood				
Larix occidentalis	Heart wood	−1914	−2736	−4650
Decomposed douglas fir				
Pseudotsuga menzeisii	Punky wood	−2944	−2176	−5120
Ponderosa pine				
Pinus ponderosa	Needles	−2438	−2708	−5146
Aspen				
Populus tremuloides	Foliage	−2398	−2636	−5034
Douglas-fir bark				
Pseudotsuga menzeisii	Outer (dead)	−3061	−2061	−5122
Douglas-fir bark				
Pseudotsuga menzeisii	Whole	−3017	−2691	−5708

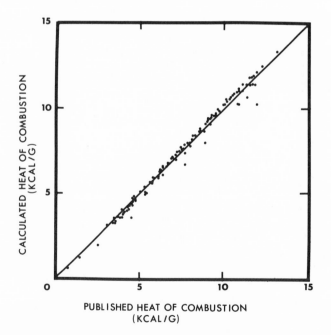

FIGURE 5. Correlation between the published heats of combustion and calculated values based on 3,349 cal/g oxygen consumed.

units in hardwoods), with the approximate elemental analysis of $C_{10}H_{11}O_2$, has a lower degree of oxidation and a considerably higher heat of combustion ($\Delta H^{25°} = -6,371$ cal/g). On pyrolysis, it forms mainly char because it is not readily cleaved to low molecular weight fragments. Samples of wood which are mainly composed of lignin and carbohydrates show intermediate heats of combustion and volatilization characteristics. The ether extractives (terpenoid hydrocarbons and lipids) have a still lower oxygen and higher heat content ($\Delta H^{25°} = -7,700$ to $-8,500$ cal/g) and affect the heat of combustion of ponderosa pine and aspen foliage, which have a high extractive content.

Figure 6 shows the correlation between the heat of combustion of various biomass fuels and their pyrolysis products (chars and volatiles) with their respective carbon content. The least squares line through the individual points in this figure fits Equation (2):

$$\Delta H^{25°}_{comb} \text{ (cal/g)} = 94.19 \text{ (\%C)} + 55.01 \tag{2}$$

This correlation may be expected because the oxygen is consumed mainly for oxidation of the carbon atoms.

These correlations could be used for determining the heat of combustion or the rate of heat release of different fuels under different conditions. Figure 7 shows the rate and the amount of oxygen consumed for combustion of the

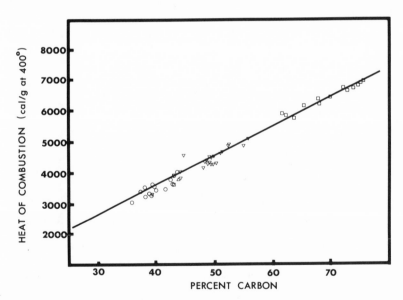

FIGURE 6. Heat of combustion at 400°C vs. percent carbon: (∇) Fuels, ($\square$) char, (O) volatiles.

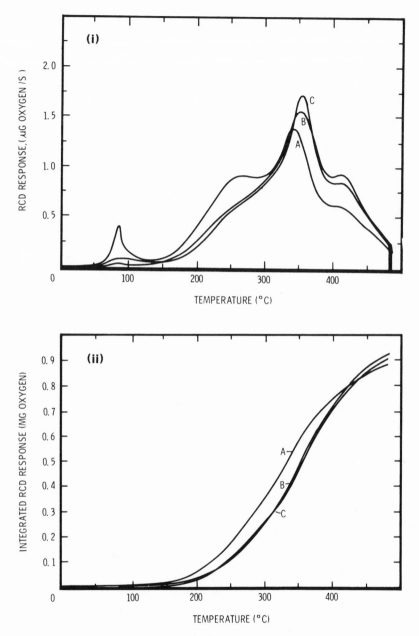

FIGURE 7. The rates (i) and amounts (ii) of oxygen consumed for combustion of volatile pyrolytic products of: (A) new Douglas-fir needles, (B) old Douglas-fir needles, and (C) old ponderosa pine needles.

volatile extractives and pyrolysis products from different foliage (needles) determined with a special oxygen detector (Reaction Coulometer).

Thermal degradation of cellulose and hemicelluloses to flammable volatile products and chars involves a series of highly complex reactions and a variety of products which have been extensively investigated in the author's laboratory. These reactions, which take place both concurrently and consecutively may be classified in the following categories:

1. At temperatures below 300°C, pyrolysis of cellulose in air or inert atmosphere proceeds through a series of reactions, which include free radical initiation, elimination (of water), depolymerization, formation of carbonyl and carboxyl groups and evolution of CO and CO_2, producing mainly a charred residue. Figure 8 shows some of the reactions involved in the autoxidation of cellulose in air.

2. At the temperature range of 300–450°C, the glycosidic linkage of the polysaccharide breaks by substitution involving one of the free hydroxyl groups (transglycosylation) to provide a mixture of levoglucosan, other derivatives of the glucose unit, and oligosaccharides, as shown in Figure 9 and Table 4 for cellulose. This mixture is generally referred to as the tar fraction.

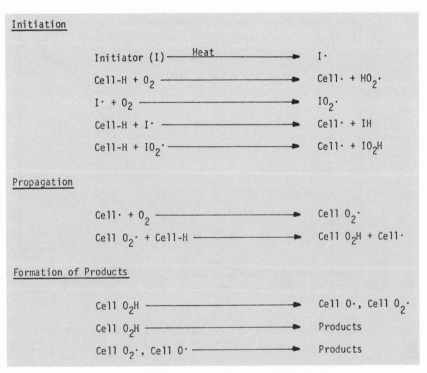

FIGURE 8. The thermal autooxidation of cellulose in air.

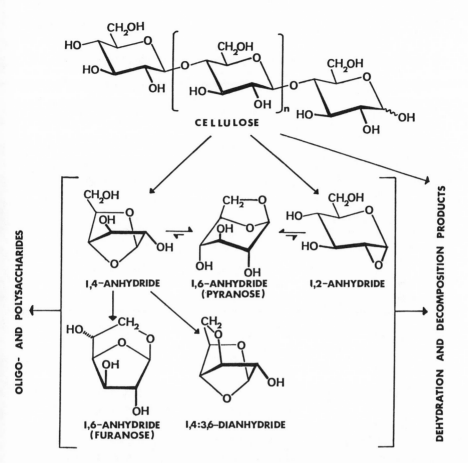

FIGURE 9. Pyrolysis of cellulose to glucose derivatives (anhydrides), oligosaccharides and decomposition products found in tar.

TABLE 4. Analysis of the Pyrolysis Products of Cellulose at 300°C under Nitrogen

	Experimental condition		
Product	Atm. pressure (%)	1.5 mm Hg (%)	1.5 mm Hg, 5% SbCl₃ (%)
Char	34.2	17.8	25.8
Tar	19.1	55.8	32.5
levoglucosan	3.57	28.1	6.68
1,6-anhydro-β-D-glucofuranose	0.38	5.6	0.91
D-glucose	trace	trace	2.68
hydrolyzable materials	6.08	20.9	11.8

Side labels on figure:

OLIGO- AND POLYSACCHARIDES

DEHYDRATION AND DECOMPOSITION PRODUCTS

CELLULOSE

1,4-ANHYDRIDE   1,6-ANHYDRIDE (PYRANOSE)   1,2-ANHYDRIDE

1,6-ANHYDRIDE (FURANOSE)   1,4:3,6-DIANHYDRIDE

TABLE 5. Pyrolytic Products of Cellulose and Treated Cellulose at 550°C

Product	Neat[a]	+5% H_3PO_4	+5% $(NH_4)_2HPO_4$	+5% $ZnCl_2$
Acetaldehyde	1.5[b]	0.9	0.4	1.0
Furan	0.7	0.7	0.5	3.2
Propenal	0.8	0.4	0.2	T[c]
Methanol	1.1	0.7	0.9	0.5
2-Methylfuran	T	0.5	0.5	2.1
2,3-Butanedione	2.0	2.0	1.6	1.2
1-Hydroxy-2-propanone } Glyoxal	2.8	0.2	T	0.4
Acetic acid	1.0	1.0	0.9	0.8
2-Furaldehyde	1.3	1.3	1.3	2.1
5-Methyl-2-furaldehyde	0.5	1.1	1.0	0.3
Carbon dioxide	6	5	6	3
Water	11	21	26	23
Char	5	24	35	31
Tar	66	16	7	31

[a] The values in this column are for the respective products before the inclusion of additives.
[b] Percentage yield based on the weight of the sample.
[c] T = trace amounts.

3. Concurrently, or at somewhat higher temperatures, dehydration, rearrangement, and fission of sugar units provides a variety of carbonyl compounds, such as acetaldehyde, glyoxal and acrolein, which readily evaporate.

4. Condensation of the unsaturated products and cleavage of the side chains through a free radical mechanism leaves a highly reactive carbonaceous residue containing trapped free radicals.

Heating of the cellulosic material at or above 500°C provides a mixture of all of these products as shown in Table 5 for cellulose. Addition of an acidic catalyst or slow heating promotes the dehydration and charring reactions. Therefore, higher temperature and smaller particle size, which result in faster heating rates, promote the gasification process, whereas, lower temperatures, larger particle size, and the presence of moisture and inorganics favor the production of char, water, and CO_2.

4. Combustion Process

As noted before, the pyrolysis and ensuing combustion of biomass fuels proceed by two alternative pathways. In the first pathway, which operates at higher temperatures, pyrolysis or thermal decomposition of the biomass provides a mixture of combustible gases. These gases mix with air to fuel the flaming combustion that could rapidly spread in the gas phase. In the second pathway, which dominates at lower temperatures, pyrolysis gives mainly carbonaceous char and a gas mixture containing water and carbon dioxide that is not very flammable. Oxidation of the resulting active char then provides glowing or smoldering combustion. This type of combustion proceeds as a front in the solid phase at a slower rate.

As graphically shown in Figure 10, during the flaming combustion, a relatively high rate of heat release and heat flux generated by the combustion of volatile products provide the energy required for gasification of the substrate and the propagation of an intensive fire. When the temperature or intensity of the heat flux falls below certain levels, oxidation of the char could result in smoldering combustion, an incomplete combustion in the solid phase accompanied by smoking or emission of unoxidized pyrolysis products. This process is often observed with low density fibrous or porous materials. In these materials, the char is slowly oxidized by the in-diffusing air and the slow rate of heat release in the absence of substantial heat loss by radiation, convection or conduction provides the low heat flux required for further charring and prop-

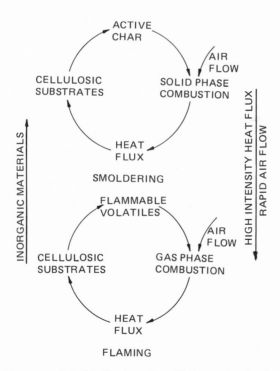

FIGURE 10. Graphic presentation of the flaming and smoldering combustion showing the respective roles of combustible volatiles and active char produced by pyrolysis under heat flux at different conditions.

agation of the smoldering combustion. In a furnace, the smoldering or smoking is observed as a transitory stage before the fire is built up or when it dies down. Assuming complete combustion, for a uniform fuel burning at a steady state, the combustion intensity may be expressed by Equation (3),

$$I_R = -\Delta H \frac{dw}{dt} \tag{3}$$

where I_R is the reaction intensity, ΔH is the heat of combustion, and dw/dt equals the rate of mass loss.

For biomass fuel this equation gives only the approximate rate of heat release at a steady rate of combustion. This is because the composition and the ΔH of the pyrolysis products could change and partial combustion of the products affect the fire intensity before the steady state is reached or after it has been terminated, irrespective of the rate of fuel consumption (dw/dt). In order to obtain the rate of heat release and net energy generated, the moisture content of the fuel and the inorganics present should also be taken into consideration.

The heat of combustion, $\Delta H_{comb}^{25°}$, is calculated by assuming that combustion of the fuel, shown in Equation (4), takes place at 25°C:

$$\text{Fuel} + O_2 \rightarrow CO_2 + H_2O \tag{4}$$

Since combustion takes place at a higher temperature ($x°$), $\Delta H_{comb}^{25°}$ should be corrected as follows:

$$\Delta H_{comb}^{x°} = \Delta H_{comb}^{25°} + \int_{25°}^{x°} Cp(CO_2)\ dT + \int_{25°}^{x°} Cp(H_2O)\ dT$$
$$+ \Delta H_{vap}^{25°}(H_2O) - \int_{25°}^{x°} Cp(\text{fuel})\ dT - \int_{25°}^{x°} Cp(O_2)\ dT \tag{5}$$

These corrections include the heat capacities required to raise the temperature of the combustion products from 25°C to $x°$C and the heat of evaporation of water. The heat capacity values can easily be obtained for CO_2, H_2O, and O_2 by using Spencer's integrated heat capacity equations, which are a function of temperature alone. These calculations give 89.24, 174.8, and 86.99 cal/g for raising the temperature of CO_2, H_2O, and O_2, respectively, from 25°C to 400°C. A value of 582.3 cal/g is used for the heat of vaporization of H_2O at 25°C. The quantity of each substance needed in the combustion of 1 g of dry fuel is obtained from the carbon–hydrogen analysis, assuming the balance is oxygen or produces the same effect in heat capacity. The only term in the expression which must be experimentally determined is the heat capacity of the fuel, which is determined by DSC. Experimentally determined $\Delta H^{25°}_{comb}$ and the corresponding calculated corrections and $\Delta H^{400°}_{comb}$ for several fuels are given in Table 6. In these calculations the correction for temperature is small compared to the total heat of combustion, although certainly not negligible. The dominant factor in the calculation is the heat of vaporization of the water produced by combustion. This accounts for the larger correction necessary for cellulose, which has a relatively high hydrogen content, and indicates that the correction will decrease for fuels such as lignin, which has a high degree of unsaturation in its molecular structure. Since the contribution of the heat capacity of product gases is small compared to the heat of vaporization term in Equation (5), the heat of combustion correction is not expected to vary significantly with temperature.

The above corrections are on the dry weight basis and do not take into consideration the nitrogen content of the air. For a fuel of $y\%$ moisture content, further corrections will be required as follows:

$$\Delta H^{x°}_{moist} = \Delta H^{x°}_{dry} \left(\frac{100 - y}{100} \right) - \frac{y}{100} 582.3 - \frac{y}{100} 0.435 \, (x - 25) \quad (6)$$

This equation takes into consideration the reduced amount of the combustible matter as well as the energy required for the evaporation and heating of the moisture content.

A similar correction will be needed for the inorganic content, particularly with fuels such as municipal and agricultural waste which have a high ash content. The combustion system is usually designed to minimize the losses due to the moisture content of the fuel and heat content of the combustion products, by passing the flue gases over the fresh fuel. Also, to increase the efficiency, attempts are made to carry the combustion at higher temperatures and recover the heat content of the combustion gases as much as possible. Therefore, to calculate the energy lost in the flue gases or the efficiency of the system, the temperature of the flue gases could be replaced for the temperature of the furnace.

TABLE 6. Heats of Combustion of Natural Fuels at 25° and 400°C

Sample	$\Delta H^{25°}_{comb}$ (cal/g)	$\Delta H^{400°}_{comb}$ (cal/g)	Correction 25° − 400° (cal/g)
Cellulose	−4143	−3853	290
Wood (*Populus* ssp.)	−4616	−4341	277
Punky wood (Douglas fir)	−5120	−4878	242
Ponderosa pine needles	−5145	−4904	241

The efficiency of the combustion process and the system as a whole also depends on the intake and temperature of air which contains ca. 21% oxygen and 78% nitrogen. Oxygen deficiency gives incomplete combustion and excess air cools the system.

Theoretical considerations of the rate of burning dw/dt or the fire dynamics are much more complex than the evaluation of the heat of combustion discussed so far. However, generally speaking, the rate of burning is a function of the heat and mass transfer in the whole system as well as the fuel characteristics such as the composition and the particle size. At lower temperatures the rate of burning is controlled by the kinetics of pyrolysis and combustion, but, at higher temperatures when these reactions take place very rapidly, the heat and material transfer become the controlling factor. Fuels with smaller particles having proportionally larger surface areas exposed to the heat or radiation flux readily dry, pyrolyze, and burn. Whereas, those with larger dimensions, such as logs, burn more slowly because of the limitation in the rates of heat and mass transfer. The principles involved in this phenomenon have been theoretically established for model substrates under idealized conditions. Although these principles serve as a guideline, quantitative applications of the results are hampered by the complexity of the system.

For instance, kinetic studies with cotton cellulose have shown that within the temperature range of 259–341 °C, the substrate is activated by intermediate, physical and chemical changes such as glass transition and break down to smaller molecules of about 200 glucose units. The activated molecules are subsequently pyrolyzed to volatiles or char by the competing pathways discussed before. Thus, the kinetics for pyrolysis of cellulose could be expressed by the following model:

$$
\begin{array}{ccc}
& & k_v \nearrow \text{Volatiles} \\
\text{Cellulose} \xrightarrow{k_i} \text{``Active cellulose''} & & W_v \\
W_{\text{cell}} \qquad\qquad W_A & k_c \searrow & \\
& & \text{Char} + \text{Gases} \\
& & W_c \qquad W_g
\end{array}
\tag{7}
$$

where

$$
-\frac{d(W_{\text{cell}})}{dt} = k_i[W_{\text{cell}}]
\tag{8}
$$

$$
\frac{d(W_A)}{dt} = k_i[W_{\text{cell}}] - (k_v + k_c)[W_A]
\tag{9}
$$

$$
\frac{d(W_c)}{dt} = 0.35 k_c[W_A]
\tag{10}
$$

For pyrolysis of pure cellulose under vacuum, the rate constants k_i, k_v, and k_c were found to correspond with $k_i = 1.7 \times 10^{21} e^{-(58,000/RT)}$ min^{-1}, $k_v = 1.9 \times 10^{16} e^{-(47,300/RT)}$ min^{-1} and $k_c = 7.9 \times 10^{11} e^{-(36,000/RT)}$ min^{-1}, respectively.

In this study, vacuum has been used to decouple or minimize the problem of mass transfer by removing the volatile products and preventing their secondary reactions in the heated zone. Although small samples have been used to minimize the problems of heat transfer, the experimental range is still limited because at temperatures above 400°C the substrate is cooled by rapid evolution of the pyrolsis products and the reaction is controlled by the rate of heat transfer rather than the kinetics of pyrolysis.

Conceptually, the combustion of fuel particles starts with pyrolysis of the surface materials, which on exposure to heat flux emit combustible volatiles and convert to a carbonaceous layer. This layer becomes thicker as the pyrolysis front or the heated zone moves inside the particle and the evolving volatile products move out. The temperature profile within the cross section of the particle is elevated by the heat flux and lowered by the (heat of) evaporation of the pyrolysis product. Since the outcoming pyrolysis products could react with the remaining carbonaceous layer, as shown below, this system could not be treated purely as a physical model. More basic data than thermal diffusivity will be required for determination of the temperature profile.

$$H_2O + C \rightarrow CO + H_2 \tag{11}$$

The rate of combustion of the volatile products are highly dependent on the aerodynamics of the system. Burning of the volatile products in a rapid and turbulent flow may be analyzed by the theory of diffusion flames. According to this theory, the fuel reacts with active free radical species notably •OH which diffuse into the fuel zone forming intermediate free radicals. These free radicals then diffuse into the combustion zone for oxidation. Thus, there is very little direct contact between the original fuel and the oxidant.

A variety of free radical reactions are involved in the initiation, propagation, branching, oxidation, and termination of this process. A discussion of these reactions that have been investigated for relatively simple fuels is beyond the scope of this article. However, it should be noted here that the intermediate transformations of the pyrolysis products are by no means limited to these free radical reactions. The transformations include further fission, dehydration, and disproportionation of the sugar units discussed previously. In these reactions formation of unsaturated intermediates, subsequent polymerization and further pyrolysis could provide carbonaceous particles or soot in the gas phase. The soot formation may also involve the following reaction at temperatures of ca. 700°C:

$$2 CO \rightleftharpoons CO_2 + C \tag{12}$$

The charred residue which is left after emission of the volatile pyrolysis product is highly reactive and porous. It is characterized by a large surface area that increases the adsorptive capacity of the product. The surface area and the physico–chemical properties of the char are related to the pyrolysis condition. These properties have been investigated for cellulose chars. It has been shown that chars produced at 550°C (a temperature that is easily reached within the pyrolysis front) are most reactive and have the largest surface area. As shown in Figure 11, there is also a high concentration of trapped carbon free radicals ($C^\bullet$) which peak for char produced at 550°C.

Although adsorption of inert gases (N_2 or CO_2) on the char surface is a rapid and reversible physical phenomenon, the chemical interaction of the active sites with oxygen is a highly exothermic chemical process that proceeds at a finite rate increasing with temperature. The chemisorption of oxygen on the char surface corresponds with the formation of surface oxides ($C-O$ and $C=O$) that can be detected by infrared spectroscopy. The initial heat of chemisorption is very high (ca. 110 kcal/mole of O_2), but as the more active sites are occupied, it falls off and levels (at ca. 75 kcal/mole of O_2). Conversely, the energy of activation which can be calculated from the Elovich equation increases linearly (from 13–25 kcal/mole with the oxygen uptake of 0–2.5 mole O_2/g of char).

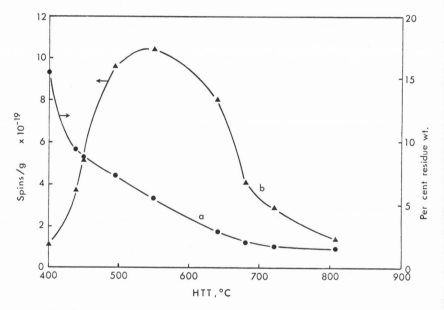

FIGURE 11. The weight of residual char (a) and the concentration of free radicals (b) at different heat treatment temperatures (HTT).

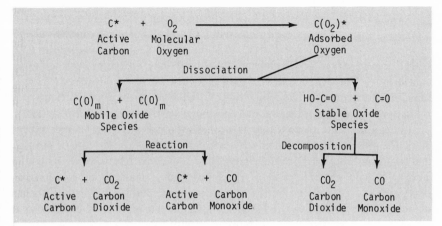

FIGURE 12. Oxygen adsorption, chemisorption, and oxidation of char.

Further studies have shown that nascent cellulose char is highly pyrophoric and on exposure to oxygen could ignite spontaneously at relatively low temperatures. The mechanism of the oxidation reactions involved are not clear. However, it is believed that oxygen molecules are originally adsorbed on the active sites (C^*) which may or may not be free radicals. As shown in Figure 12, the adsorbed oxygen subsequently reacts to produce CO_2, CO, and new active sites. At 500°C these complex reactions could be summarized in the following equations:

$$O_2 + C \rightarrow CO_2 \qquad \Delta H = -88.5 \text{ kcal/mole} \qquad (13)$$
$$O_2 + 2C \rightarrow 2CO \qquad \Delta H = -22.9 \text{ kcal/mole} \qquad (14)$$

These equations show that incomplete oxidation of the char to CO produces about one-quarter of the total energy based on the available carbon, and one-half of the total energy based on the available oxygen. Since the oxidation reactions are activated and exothermic, when sufficient oxygen is available the rate of reaction and the corresponding heat release could rapidly increase. This will raise the surface temperature to the level of glowing or incandescence, where much of the energy is lost to the surrounding environment by radiation.

The ratio of CO to CO_2 is influenced by various anions and cations. The former compounds such as phosphates and borates increase the ratio of CO to CO_2 and are known as glow retardants or smolder inhibitors. The latter, such as sodium and potassium ions, reduce the ratio of CO to CO_2 and promote the smoldering combustion. These compounds should not be confused with flame retardants which reduce the production of flammable volatiles by promoting the char production in the pyrolysis pathways discussed before, although, ammonia phosphates and borax also function as flame retardants.

The glowing combustion is highly dependent on diffusion of oxygen to the char and the counter-current emission of CO and CO_2. In this counter-current flow, depending on the prevailing condition, the CO may be oxidized to CO_2 within the envelope of the glowing char, or it may escape as in smoldering combustion. The CO_2 could partly diffuse back to the char surface where it may be reduced to CO. The remaining CO in the combustion products not only reduces the heat of combustion but also increases the toxicity and pollution.

It is interesting to note that when houses are heated with charcoal braziers, the glowing charcoal is covered with ashes to reduce the rate of burning and to prevent the escape of toxic CO to the confined atmosphere of the house. This is a good example of how the knowledge of biomass combustion has developed and propagated by the experience and traditions to the extent that its science and technology is taken for granted. Utilization of biomass as a major source of fuel and energy, however, requires extensive modernization of this technology based on scientific information and design. This presentation indicates the enormous complexity of the subject and the leads and guidelines that have been established through the scientific investigation of the related but opposite problem of fire prevention. In view of these considerations and the expected role of biomass as a major source of fuel and energy, the need for development of precise knowledge on combustion of these materials and the scientific task ahead seems rather obvious.

ACKNOWLEDGMENT

The author is pleased to acknowledge the support and interest of the Center for Fire Research, National Bureau of Standards, for investigations on combustion of cellulosic materials discussed in this article.

References

1. F. Shafizadeh, *Adv. Carbohydr. Chem.* **23**, 419 (1968).
2. F. Shafizadeh, K. V. Sarkanen, and D. A. Tillman, *Thermal Uses and Properties of Carbohydrates and Lignins,* Academic Press, New York (1976).
3. R. A. Susott, W. F. DeGroot, and F. Shafizadeh, *J. Fire Flammability* **6**, 311 (1975).
4. F. Shafizadeh, *Appl. Polym. Symp.* **28**, 153 (1975).
5. A. G. W. Bradbury, Y. Sakai, and F. Shafizadeh, *J. Appl. Polym. Sci.* **23**, 3271 (1979).
6. F. Shafizadeh and A. G. W. Bradbury, *J. Appl. Polym. Sci.* **23**, 1431 (1979).
7. R. A. Susott, F. Shafizadeh, and T. W. Aanerud, *J. Fire Flammability* **10**, 94 (1979).
8. A. M. Kanury, *Introduction to Combustion Phenomena,* Gordon and Breach Science Publishers, New York (1977).
9. J. N. Bradley, *Flame and Combustion Phenomena,* Chapman and Hall Ltd., London (1972).
10. A. G. W. Bradbury and F. Shafizadeh, *Combustion and Flame* **37**, 85 (1980).
11. A. G. W. Bradbury and F. Shafizadeh, *Carbon* **18**, 109 (1980).

7
The Andco-Torrax System

STANLEY D. MARK, JR.

1. Introduction

Other chapters in this book have discussed municipal solid waste (MSW) as a biomass material. MSW is without question one of the most difficult fuel materials to handle in any conversion process for producing energy. It would be expected, therefore, that processes which are successful with MSW could be applied to other biomass.

The technology for processing MSW developed rather slowly from the early part of this century up to the mid 1960s. Combustion processes were usually based on adaptations of coal-burning stoker equipment. These processes advanced most rapidly in Europe and in Japan during that time. In order to accelerate the development of MSW processing technology, the U.S. Federal Government enacted The Solid Waste Act of 1965, which provided a program of demonstration grants for the development and evaluation of new solid waste conversion technology. The demonstration grants program resulted in the pilot-scale testing of several new systems in the U.S. for the conversion of MSW, including a slagging process called the Andco-Torrax System (ATS). This chapter provides information on the ATS.

There was a coincident interest in the U.S. and elsewhere beginning in the 1960s in applying the principles of destructive distillation or pyrolysis to MSW processing. Laboratory scale work on the pyrolysis of waste materials was conducted by the U.S. Bureau of Mines[1] as well as by many other agencies and companies. The ATS and several other processes[2] are based on these principles.

STANLEY D. MARK, JR. • Andco Incorporated, 25 Anderson Road, Buffalo, New York, NY 14225.

2. Description of Process

The ATS has been described in the European[3-6] and North American[7,8] technical literature. It is used mainly to convert MSW into usable energy forms by the pyrolysis and the primary and secondary combustion of its combustible materials. The noncombustible materials in MSW are melted to form a glassy slag which usually flows readily at 1200°C (2190°F) and higher. MSW is processed without sorting or pretreatment, except for shearing or crushing bulky items to a maximum dimension of about one meter. The process is often referred to as slagging pryolysis.

Pyrolysis processes have been used for years by industry. Typical examples are found in the production of charcoal and methanol from wood and coal gasification. The pyrolysis process requires heating the fuel (in the case of the ATS, the organic materials contained in solid waste) to a temperature at which the volatile matter will distill (thermal decomposition) leaving carbonaceous char and inert materials behind. Carbon and volatiles do not burn immediately in the pyrolysis process owing to an intentional deficiency of oxygen. In the ATS, the carbon from the pyrolyzed material is subsequently burned to a mixture of carbon monoxide and carbon dioxide using preheated air, thus releasing heat energy at sufficiently high temperatures to convert all noncombustibles contained in the MSW to a molten slag and to pyrolyze the incoming waste.

2.1. Gasifier

The principal component of any ATS is the gasifier shown in Figure 1. Refuse is charged into the inlet hopper of the gasifier without prior preparation with the exception that large pieces of refuse such as appliances are sheared or crushed before charging. The top of the refuse bed is maintained at a preset level by adding fresh material as needed. The MSW descends by gravity through the three process zones of the gasifier: drying, pyrolysis, and primary combustion.

The function of the drying zone is to evaporate the moisture in the MSW and to act as a plug to restrict the in-flow of air during charging. The drying zone extends downward below the bottom of the lantern section of the gasifier. The lantern is the plenum volume wherein the combustible gas and vapor mixture from the pyrolysis process is collected and then flows to the next step in the process.

Below the drying zone is the pyrolysis zone where the dried refuse is thermally decomposed to a residual mixture of carbon char and inerts, thus producing the mixture of combustible gases and vapors.

The heat for pyrolyzing and drying the refuse is supplied by the partial combustion of the carbon char using air preheated to about 1000°C (1830°F) supplied at the base of the gasifier in the primary combustion zone. The heat

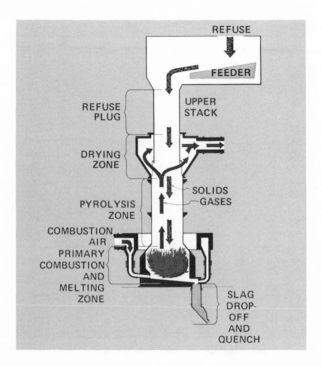

FIGURE 1. Andco-Torrax gasifier.

generated by this combustion process also transforms the noncombustible materials to a molten slag. The molten slag is drained continuously through a sealed slag tap into a water quench tank to produce a black glassy aggregate which contains no carbon or putrescible material.

The chemical and physical properties of the slag are shown in Tables 1–3. The slag described was produced from MSW having the composition shown in Table 4. This quality of MSW will produce slag which is 3–5% of the volume and 15–20% of the weight of the original charge of MSW.

2.2. Fuel Gas System

At least 90% of the energy content of MSW is contained in the gas stream which leaves the gasifier. This energy is in the form of combustible gases, vapors, and entrained particles and as sensible and latent heat. The complete combustion of this gas stream produces about the same volume of products of combustion per unit of heat released as would be the case with other gaseous fuels.

This combustible gas stream, with or without cleaning to remove

TABLE 1. Chemical Analysis of Slag[a]

Constituent	Average % (by weight)	Range %
SiO_2	45	32.00–58.00
Al_2O_3	10	5.50–11.00
TiO_2	0.8	0.48–1.30
Fe_2O_3	10	0.50–22.00
FeO	15	11.00–21.00
MgO	2	1.80–3.30
CaO	8	4.80–12.10
MnO	0.6	0.20–1.00
Na_2O	6	4.00–8.60
K_2O	0.7	0.36–1.10
Cr_2O_3	0.5	0.11–1.70
CuO	0.2	0.11–0.28
ZnO	0.1	0.02–0.26

[a]Typical sample analysis from demonstration plant in the United States. (Analysis will change significantly with refuse composition.)

entrained material, has potential application as an energy source in power boilers, utility boilers, and cement kilns. Such applications become increasingly feasible where close coupling of the gasifier with the recipient combustion device is possible.

The composition and properties of the combustible gas stream are dependent, of course, on the refuse mix. The ingredients in the stream are primarily carbon monoxide, carbon dioxide, hydrocarbon vapors and gases, hydrogen, nitrogen, water vapor, and particulate materials. The heating value of the gas will normally be in the range of 1130–1600 kcal/Nm³ (120–170 Btu/ft³).

2.3. Integrated Complete Combustion System

In most cases, a complete MSW conversion system is required which can operate independently of other facilities, except possibly to provide steam, hot water, or electricity to nearby users. The preferred ATS system for accomplishing this objective is described next.

TABLE 2. Physical Analysis of Slag[a]

Dry bulk density	1.40 g/cm³
True residue density	2.80 g/cm³

[a]Average density of several samples taken from the demonstration plant in the United States.

TABLE 3. Screen Analysis of Residue[a]

U.S. standard screen	Opening size (mm)	Percent residue on screen
$3\frac{1}{2}$	5.66	4
4	4.75	4
7	2.82	15
10	2.00	30
12	1.68	10
20	0.84	30
30	0.59	5
40	0.42	1
Fines	—	1

[a]Approximate distribution of particle sizes from several samples from U.S. demonstration plant.

The equipment consists of five major subsystems, namely:

Subsystem	Function
Gasifier	Solid waste pyrolysis, primary combustion, and slagging.
Secondary combustion chamber	Complete oxidative release of heat energy from volatile and entrained materials resulting from the pyrolysis of solid waste.
Regenerative towers	Primary combustion air preheating.
Waste heat boiler	Conversion of heat energy to steam or hot water.
Gas cleaning system	Removal of particulate material.

The relationship and arrangement of these subsystems can be understood by referring to Figures 2 and 3 which are schematic representations of the system.

TABLE 4. Composition of Average North American Municipal Solid Waste

Component	Average % (by weight)
Mixed paper	45
Wood	4
Textiles	1
Plastics and rubber	5
Food and yard waste	25
Fines and ash	7
Metal	6
Glass	7

Lower heating value = 2414 kcal/kg (4345 Btu/lb)

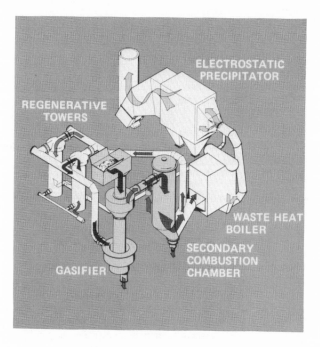

FIGURE 2. Andco-Torrax System.

2.3.1. Gasifier

The functions of the gasifier were described previously; it produces a combustible off gas and liquid slag.

2.3.2. Secondary Combustion Chamber

The combustible gas–vapor mixture from the gasifier is burned to completion in the secondary combustion chamber as shown in Figure 4. The fuel–gas mixture flows from the gasifier lantern section at a temperature of 450–550°C (800–1000°F). The fuel gas is mixed with ambient combustion air in a high energy mixing burner at the inlet to the secondary combustion chamber. The thorough mixing and the long residence time in the refractory-lined chamber assures complete combustion with minimum excess air.

Temperatures are maintained in the secondary combustion chamber at 1150–1300°C (2100–2370°F). Much of the inert particulate material which enters the secondary combustion chamber is captured in the liquid slag which forms on its inner surfaces and then drains off at the base. This slag is water-

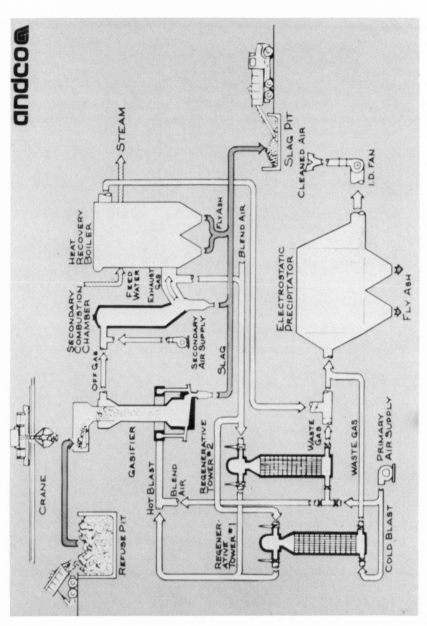

FIGURE 3. Andco-Torrax Process flow diagram.

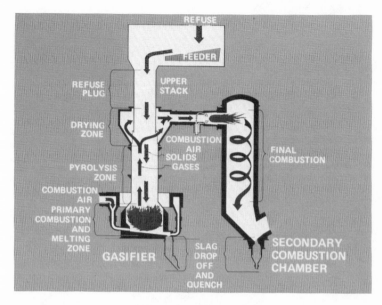

FIGURE 4. Andco-Torrax unit.

quenched and comprises approximately 10% of the total slag produced in the process.

2.3.3. Regenerative Towers

Regenerative towers or stoves are the preferred means for heating primary combustion air. The towers are heated by the hot products of combustion consisting of nitrogen, carbon dioxide, oxygen, and water vapor from the secondary combustion chamber. Approximately 15% by volume of these products of combustion is passed through the towers. Two towers are usually used in cyclical operation as shown in Figure 5. Refractory lined regenerators of this type have long been used in iron, steel, and glass making.

During the heating or "gas" cycle, products of combustion from the secondary combustion chamber are introduced into the top of one of the regenerative towers. The heat from these gases is transferred to the checkerwork bringing it up to temperature levels of approximately 1150°C (2100°F) at the top and 260°C (500°F) at the base. The waste gas exiting the checkerwork is returned to a duct at the inlet of the gas cleaning system.

During the "blast" cycle, the combustion products from the secondary combustion chamber are diverted to the second tower to heat its checkerwork. Ambient process air is introduced at the base of the fully heated first tower and passes up through the checkerwork absorbing the stored heat. The exit temperature of the air from the checkerwork ranges from 980°C (1800°F) to 1100°C (2010°F). A constant process temperature may be maintained by blending the heated air with ambient air before introduction into the gasifier.

2.3.4. Waste Heat Boiler

The major portion (about 85%) of the volume of combustion products leaving the secondary combustion chamber passes through the waste heat boiler. As much as three pounds of saturated steam is produced for every pound of MSW, depending on its heating value. The steam can be generated at pressures up to 60 atmospheres without incurring severe boiler tube corrosion. The process gas leaves the waste heat boiler at approximately 260°C (500°F) and combines with the flow from the regenerative towers before entering the gas cleaning system.

2.3.5. Gas Cleaning System

The waste gases leaving the regenerative towers and the waste heat boiler are combined before entering the gas cleaning system which usually consists of an electrostatic precipitator sized to handle the maximum gas flow from the secondary combustion chamber. Wet scrubbers or other devices may be preferred for installation economy and to remove objectionable gaseous constitu-

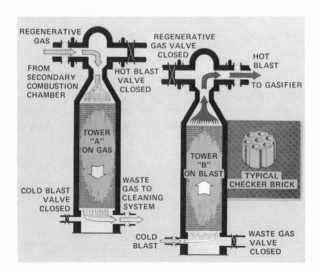

FIGURE 5. Regenerative towers.

ents as well as particulate material. The normal gaseous emissions will contain about 70% nitrogen, 10% carbon dioxide, 5% oxygen, and 15% water vapor by volume.

3. Review of Plant Operations

By mid-1979, there had been five plants built for the ATS. The first plant was for the demonstration of the technology; the other four plants were for commercial use.

3.1. Demonstration Plant

The first plant was located in Orchard Park which is adjacent to Buffalo, New York. The plant was designed to handle about 68 metric tons per day of MSW consisting mainly of household and commercial wastes with admixtures from time to time of industrial solid waste. Initial operation of the plant began during the second quarter of 1971. The first year of operation consisted mainly of developing process, design, and operating improvements.

From the summer of 1972 until the end of 1976, the plant operated as an engineering development facility to evaluate design features and to test the process on MSW alone and on a number of refuse materials admixed to MSW. During this period, about 150 test runs were made at refuse rates ranging from 2 to 5.5 tonnes/hr. Extensive data were collected on equipment design and on process heat and mass balances. The plant was shut down in 1977 because its use for development and demonstration purposes was no longer needed.

3.2. Commercial Plants

Table 5 contains information on the four commercial ATS plants which have been built through mid-1979.

In the case of the Frankfurt plant, it was necessary to position all the main process equipment, except for the gasifier, outside an existing incineration

TABLE 5. Commercial Plant Data

Location	Date operations began	Number of units	Unit design (capacity in mt/hr)	Type of by-product energy
Luxembourg	9/76	1	3.0–8.3	Electricity
Grasse, France	10/77	1	3.8–7.0	Process steam
Frankfurt, W. Germany	7/78	1	4.0–8.0	Electricity
Creteil, France	11/79	2	5.4–8.3	Electricity

building. Figure 6 is a photograph taken during the construction of the Frankfurt plant showing from right-to-left the regenerative towers, secondary combustion chamber, boiler, and electrostatic precipitator.

4. Heat and Mass Balances

Data from the operation of the Orchard Park plant were used to prepare a computer model for the calculation of heat and mass balances for the ATS. Other computer programs have been developed to calculate the proximate and ultimate analyses of MSW and other feedstock materials, if the ingredients in the feedstock are known.

FIGURE 6. Andco-Torrax unit at Frankfurt, West Germany.

RATE   : 2OO. TONS/DAY
REFUSE : LHV = 2414. KCAL/KG, HHV = 2752. KCAL/KG

REFUSE ANALYSIS		PROXIMATE ANALYSIS		ULTIMATE ANALYSIS	
% COMBUSTIBLES	58.2O	% FIXED CARBON	7.37	% CARBON	29.21
% WATER	22.O3	% VOLATILES	50.83	% HYDROGEN	3.94
% ASH-INERTS	19.77	% WATER	22.O3	% SULFUR	O.13
		% INERTS	19.77	% OXYGEN	23.64
				% CHLORINE	O.91
				% NITROGEN	O.37

GASIFIER
- - - - - - -

HOT BLAST TEMP	1O37.	DEG C
HOT BLAST AIR FLOW	3817.	NM3/H
NAT. GAS TO TUYERES	O.	NM3/H
NAT. GAS TO SLAG TAP	3O.1	NM3/H
SLAG TAP AIR FLOW	283.	NM3/H
SLAG TAP AIR TEMP	3O.	DEG C

OFFGAS DATA:
FLOW	11918. NM3/H
TEMPERATURE	462. DEG C

HEATING VALUE:

SENSIBLE	187. KCAL/NM3
LATENT	94. KCAL/NM3
CHEM + PART	1496. KCAL/NM3
- - - - - - - - - - - - - - -	
TOTAL	1777. KCAL/NM3

SECONDARY COMBUSTION CHAMBER
- -
PROCESS AIR FLOW	3O877. NM3/H
PROCESS AIR TEMP	25. DEG C
EXCESS AIR	54.1 %
PROCESS NAT. GAS	O. NM3/H
S. TAP/PILOT NAT. GAS	36.1 NM3/H
S. TAP/PILOT AIR FLOW	34O. NM3/H
S. TAP/PILOT AIR TEMP	2O. DEG C

EFFLUENT CONSTITUENTS (BY VOLUME)

% O_2	5.49
% N_2	68.67
% CO_2	11.1O
% H_2O	14.61

EFFLUENT FLOW	415O7. NM3/H
EFFLUENT TEMP	126O. DEG C

REGENERATIVE TOWER
- - - - - - - - - - - - - - - - -
EFFICIENCY 97 %

GAS FLOW	3675. NM3/H
BLEND AIR FLOW	455. NM3/H
GAS TEMP IN	115O. DEG C
GAS TEMP OUT	26O. DEG C
HOT BLAST FLOW	3817. NM3/H
BLAST TEMP IN	4O. DEG C
BLAST TEMP OUT	1O37. DEG C

WASTE HEAT BOILER
- - - - - - - - - - - - - - -
EFFICIENCY 97 %

GAS FLOW	38O78. NM3/H
GAS TEMP IN	126O. DEG C
GAS TEMP OUT	29O. DEG C

GAS CLEANING EQUIPMENT
- - - - - - - - - - - - - - - - - - - -
EFFICIENCY 98 %

GAS FLOW	41935. NM3/H
PARTICULATE	126. KG/H
GAS TEMP IN	287. DEG C
GAS TEMP OUT	282. DEG C

ENERGY RECOVERY
- - - - - - - - - - - - -
ENERGY	17637738. KCAL/H
STEAM	23153. KG/H

KG STEAM/KG REFUSE	2.78
SYSTEM EFFICIENCY	69.57 %

FIGURE 7. Typical heat and mass balances for Andco-Torrax System, Part 1.

HEAT AND MASS BALANCE

INFLUX	KG/H	KCAL/H	OUTFLUX	KG/H	KCAL/H

GASIFIER

INFLUX	KG/H	KCAL/H	OUTFLUX	KG/H	KCAL/H
REFUSE	8333.	20116667.	OFFGAS	12816.	20065171.
HOT BLAST	4933.	1329817.	SLAG	1530.	519991.
S. TAP AIR	366.	1281.	LOSSES		1115015.
TOTAL FUEL	23.	250000.			
TOP LEAKAGE	690.	2413.			
TOTAL	14346.	21700177.	TOTAL	14346.	21700177.

SECONDARY COMBUSTION CHAMBER

INFLUX	KG/H	KCAL/H	OUTFLUX	KG/H	KCAL/H
OFFGAS	12816.	20065171.	WASTE GAS	53019.	20156830.
PROCESS AIR	39906.	91230.	SLAG	170.	57777.
S. TAP AIR	439.	472.	LOSSES		242266.
TOTAL FUEL	28.	300000.			
TOTAL	53189.	20456873.	TOTAL	53189.	20456873.

WASTE HEAT BOILER

INFLUX	KG/H	KCAL/H	OUTFLUX	KG/H	KCAL/H
WASTE GAS	48637.	18490967.	WASTE GAS	48637.	3558403.
FEEDWATER	23153.	3259902.	STEAM	23153.	17637738.
			LOSSES		554729.
TOTAL	71790.	21750870.	TOTAL	71790.	21750869.

REGENERATIVE TOWERS

INFLUX	KG/H	KCAL/H	OUTFLUX	KG/H	KCAL/H
WASTE GAS	4382.	1665863.	WASTE GAS	4931.	317141.
BLEED AIR	549.	1920.	HOT BLAST	4933.	1329817.
COLD BLAST	4933.	29209.	LOSSES		50033.
TOTAL	9864.	1696992.	TOTAL	9864.	1696992.

GAS CLEANING SYSTEM

INFLUX	KG/H	KCAL/H	OUTFLUX	KG/H	KCAL/H
WASTE GAS	53567.	3875544.	WASTE GAS	53442.	3790237.
			PARTICULATE	126.	7796.
			LOSSES		77511.
TOTAL	53567.	3875544.	TOTAL	53567.	3875544.

FIGURE 8. Typical heat and mass balances for Andco-Torrax System, Part 2.

The ingredient composition of what is considered to be typical American MSW was used as the basis for calculating heat and mass balances. The ATS unit size was arbitrarily set at 200 tonnes/day. The results are shown in Figures 7, 8, and 9 which are typeset reproductions of the actual computer print outs. These results are also shown diagramatically in Figures 10 and 11. Note that the overall thermal efficiency calculations exclude the electrical energy consumed by the ATS unit. A conservatively high estimate for the electrical energy is 80 kwh/tonne MSW processed.

ASSUMPTIONS:
- - - - - - - - - - -

   1) SUPPLEMENTAL FUEL TYPE IS NATURAL GAS:
   2) SUPPLEMENTAL FUEL DATA:
      COMPOSITION: 90.00 % METHANE   O. % CARBON MONOXIDE
      (BY VOLUME)   5.00 % ETHANE   O. % CARBON DIOXIDE
                  O.   % PROPANE   5.00 % NITROGEN
                  O.   % BUTANE   O. % WATER VAPOR
     SPECIFIC GRAVITY: O.600 (AIR = 1.0)
     LOWER HEATING VALUE: 8300. KCAL/NM3
   3) BOILER DATA:
     STEAM PRESSURE: 34.0 ATMOSPHERES
     STEAM TEMPERATURE: 385.0 DEGREES CELSIUS
     STEAM ENTHALPY: 761.8 KCAL/KG
     FEEDWATER ENTHALPY: 140.8 KCAL/KG

NOTES:
- - - - - -

   1) AMBIENT TEMPERATURE IS 15.56 DEGREES CELSIUS.
   2) AMBIENT PRESSURE IS 1.0 ATMOSPHERE (760 MM HG).
   3) GASES AND SOLIDS HAVE ZERO ENTHALPY AT 15.56 DEGREES CELSIUS.
   4) WATER AND WATER VAPOR HAVE ZERO ENTHALPY AT 15.56 DEGREES CELSIUS.
   5) PROCESS AND COMBUSTION AIR IS DRY.
   6) ONE NORMAL CUBIC METER OF GAS IS AT 0.0 DEGREES CELSIUS AND 1.0 ATMOSPHERE.
   7) ALL SUPPLEMENTAL FUEL IS COMBUSTED TO CARBON DIOXIDE AND WATER VAPOR WITH 100 % HEAT RELEASE.
   8) ANCILLARY EQUIPMENT HEAT LOSSES ARE DETERMINED AS FOLLOWS:
     LOSS = (1 − N) * HOT FLUID SENSIBLE HEAT IN
     WHERE N IS THE EQUIPMENT ASSUMED EFFICIENCY.
   9) OVERALL SYSTEM EFFICIENCY IS DETERMINED AS FOLLOWS:

$$\text{EFFICIENCY} = \frac{(\text{STEAM HEAT OUT} - \text{FEEDWATER HEAT IN})}{(\text{REFUSE HEAT IN} + \text{TOTAL FUEL HEAT IN})}$$

   WHERE REFUSE AND FUEL HEATS ARE DETERMINED FROM THEIR LOWER HEATING VALUES.

FIGURE 9. Typical heat and mass balances for Andco-Torrax System, Part 3.

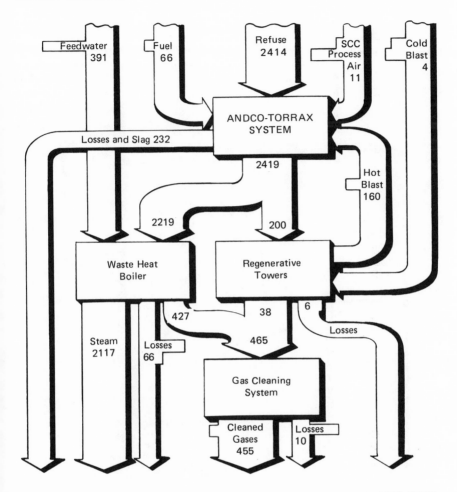

FIGURE 10. Typical heat balance for a 200-MTPD ATS unit processing 2414 Kcal/KG (LHV) MSW (Kcal/KG of MSW).

5. Some Applications of the Process

The ATS was developed specifically for MSW, but testing has been done with admixtures of other wastes. In principle, the ATS can handle any material which has the following properties:

1. Adequate heating value to sustain required temperatures in the gasifier and secondary combustion chamber.
2. Sufficient permeability to permit flow of gases when charged in the gasifier.

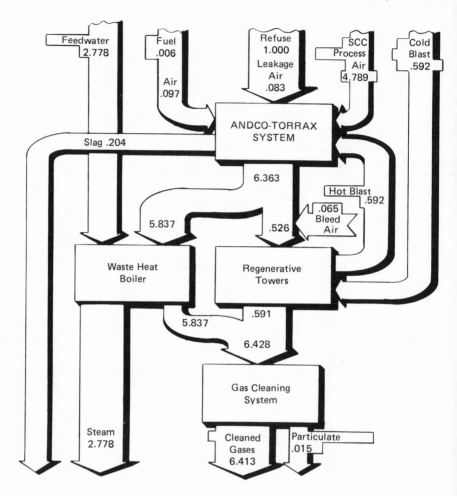

FIGURE 11. Typical mass balance for a 200-MTPD ATS unit processing 2414 Kcal/KG (LHV) MSW (Kcal/KG of MSW).

Based on plants tests and on the above properties criteria, the ATS can successfully process admixtures of the following materials and MSW:

- Waste oils and sludges
- Plastic waste
- Hospital wastes
- Toxic wastes such as DDT, PCB, etc.

- Fly ash
- Incinerator residues
- Chemical sludges
- Biomass materials in general

Further plant testing is required to determine the practical range of admixture of each of these materials.

5.1. Tires, Waste Rubber, and MSW

It has been demonstrated conclusively that the ATS works well with admixtures of up to 10% of unshredded tires and waste rubber to MSW. Higher amounts of admixed rubber may be acceptable but have not yet been tested. The general effects of rubber admixture is to increase temperatures at the gasifier hearth and in the secondary combustion chamber. An increase occurs also in the steam production rate.

The ATS unit in the Luxembourg plant began being used in 1979 for regular disposal of tires and rubber. During codisposal operations with these materials, the admixed materials range from 1% to 5% of the MSW.

5.2. Sewage Sludge and MSW

Undigested sewage sludge (78% water content) in admixed amounts up to about 30% has been satisfactorily processed with MSW. Temperatures within the gasifier and secondary combustion chamber remained within normal process limits. Steam production rates decreased due to the high water content of the charge material.

The normal methods for disposal of sewage sludge are land-filling or incineration in multiple hearth furnaces. The possible presence of heavy metals and pathogens raise questions as to the advisability of land-filling, while the energy costs in operating conventional sludge incinerators are becoming prohibitive. There are several new processes, including ATS, which may permit safe combustion of sewage sludge without excessive fuel costs.

References

1. W. S. Sanner, C. Ortuglio, J. G. Walters, and D. E. Wolfson, "Conversion of Municipal and Industrial Refuse Into Useful Materials By Pyrolysis", U.S. Bureau Mines, R1 7428 (1970).
2. J. L. Jones and S. B. Radding (eds.), *Advanced Thermal Processes for Conversion of Solid Wastes and Residues,* ACS Symposium Series 76, American Chemical Society, Washington, D.C. (1978).
3. E. Legille, F. A. Berczynski, and K. G. Heiss, A slagging pyrolysis conversion system, CRE-Montreux, Reprint IEEE Cat. No. 75CH1008-2 CRE, 232–237 (1975).
4. C. Melan, Eigenheiten technischer stand und geschatzte Kosten der hochtemperatur-pyrolyse System Andco-Torrax, *Mull und Abfall,* 363–368 (1978).
5. D. Bohn, Das Andco-Torrax Verfahren, *Abfallwirtschaft in Forschung und Praxis-Bielefeld* (1977).
6. D. Bohn, Das Andco-Torrax Verfahren zur Vergasung von Haushaltsabfaellen, *Abfallwirtschaft an der Technische Universitat Berlin BD2* (1978).
7. P. E. Davidson, Andco-Torrax: A slagging pyrolysis solid waste conversion system, *Bulletin Canadian Mining and Metallurgy,* (1977).
8. P. E. Davidson and T. W. Lucas, The Andco-Torrax high-temperature slagging pyrolysis system, in: J. L. Jones and S. B. Radding (eds.), *Advanced Thermal Processes for Conversion of Solid Wastes and Residues,* ACS Symposium Series 76, American Chemical Society, Washington, D.C. (1978), pp. 47–62.

Section B
Thermochemical Conversion Processes

8

Basic Principles of Thermochemical Conversion

STEPHEN M. KOHAN

1. Introduction

This chapter addresses some of the basic principles of the thermochemical conversion of biomass to other, more useful products. Thermochemical conversion usually denotes such concepts as combustion (discussed in Chapter 6), gasification, pyrolysis, and liquefaction, all of which involve the high-temperature (and occasionally high-pressure) processing of biomass.

Starting with a general lignocellulosic biomass of composition $(CH_2O)_n$, Figure 1 schematically presents alternative thermochemical processing routes. As suggested by the illustrative formulas of the compounds, the ratio of H/C (atoms) doubles when cellulosic biomass is converted to methane or methanol, and may remain approximately unchanged when the biomass is converted to liquids via the catalytic liquefaction technology. The H/C ratio may be increased either by adding hydrogen (in the form of steam) or by removing carbon in the form of char by pyrolysis. In catalytic liquefaction, alkali metal catalysts promote the removal of oxygen from the biomass by carbon monoxide. Many forms of biomass contain small amounts of sulfur, nitrogen, and ash, and these materials will be found in varying proportions in the liquid and solid products from the pyrolysis and liquefaction technologies.

The remaining sections of this chapter consider the basic steps involved in the gasification, pyrolysis, and liquefaction of biomass. This field is rapidly

STEPHEN M. KOHAN • Electric Power Research Institute, 3412 Hillview Avenue, Palo Alto, California 94303.

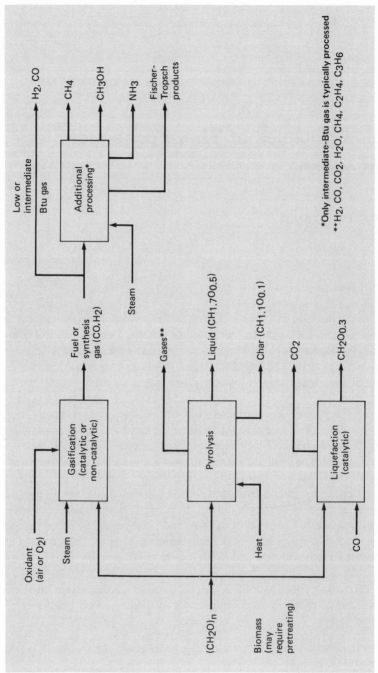

FIGURE 1. Generalized block flow diagram for the thermochemical conversion of biomass.

evolving under the sponsorship of the Department of Energy, Electric Power Research Institute, Gas Research Institute, and others.

2. Kinetics and Thermodynamics Framework

This section considers principally gasification and ideal gas concepts. When considering the conversion of a specific feedstock to a desired product using a certain type of technology, the preferred sources of information on which to base predictions of yield structures, product properties, optimum operating conditions, and other similar concepts are, in descending order of preference*: (a) commercial plant data; (b) semi-commercial-scale plant data; (c) large pilot plant data; (d) pilot plant data; (e) bench scale data; and (f) limited or no data.

The basic problem is to determine how the feedstock components (fixed carbon, oxygen, etc.) become distributed among the possible products of the reaction. This section presents a framework which can be used to estimate this distribution subject to selected kinetic, thermodynamic, heat balance, and material balance constraints. Many corporations have computer programs and extensive equilibrium, solubility, and physical property data bases to facilitate this type of calculation.

2.1. Heating Values

Heats of combustion are widely reported for coals, woody biomass, and other potential feedstocks. Heats of combustion may in some instances be estimated from data on heats of formation, following well-known procedures. Typically, heats of combustion are reported at reference temperatures of 25°C (77°F) or 15°C (60°F), and are additionally reported as gross or higher heating value, in which the water of reaction is a liquid at the reference temperature; and net or lower heating value, in which the water of reaction is a vapor at the reference temperature. For the purposes of this chapter, the heat of combustion of biomass, when mentioned, is the higher heating value and refers to 25°C (77°F). Table 1 lists the heat of combustion of several tree species. Radovich et al.[1] and Chereminisoff and Morresi[2] list heating values and chemical compositions of other biomass species.

2.2. Standard States

Figure 2 represents the general "system" under consideration. Feedstock and oxidant (if needed) and steam (if needed) cross the thermodynamic system

---

*Subject to the usual data-related qualifications of heat and material balance closure, etc.

TABLE 1. Heats of Combustion of Typical
New England Woods[a]

Tree species	kJ/kg	Btu/lb
White ash	20,730	8920
Fir balsam bark	21,150	9100
Yellow birch	20,100	8650
Yellow birch bark	22,940	9870
White cedar	19,520	8400
Elm bark	17,660	7600
Eastern hemlock	20,030	8620
Eastern hemlock bark	20,660	8890
Red maple	19,940	8580
Red maple bark	19,030	8190
Oak (white)	20,470	8810
Pine (yellow)	22,330	9610
Pine (white)	20,920	9000
Pine (white) bark	20,750	8930
Poplar	20,730	8920
Poplar bark	20,470	8810

[a]Source: Potential of Wood for an Energy Source in
New England, New England Federal Regional
Council (September, 1977).

boundary at a temperature T_r. Products (of combustion, gasification, or liquefaction) leave the system at T_r. The general biomass feedstock contains carbon, hydrogen, oxygen, nitrogen, and sulfur and can be represented (on a dry, ash free basis) as $C_A H_B O_C N_D S_E$.

A convenient definition of the standard state is developed as follows. For the generalized combustion reaction of the biomass feedstock, Equation (1) gives the reaction stoichiometry:

$$C_A H_B O_C N_D S_E(s) + nO_2\,(g) \overset{\Delta}{\to} ACO_2\,(g)$$
$$+ \tfrac{1}{2}B\,H_2O(l) + \tfrac{1}{2}DN_2\,(g) + ESO_2\,(g) \quad (1)$$

where (s), (l), and (g) denote solid, liquid, and gaseous states at T_r and the coefficient n is determined by the reaction stoichiometry. The relevant enthalpies are defined as follows:

h_i = enthalpy* of component i (J/kg-mol) at T_r

h_f = enthalpy* of feed material (J/kg-mol) at T_r

*Enthalpy base is not yet specified.

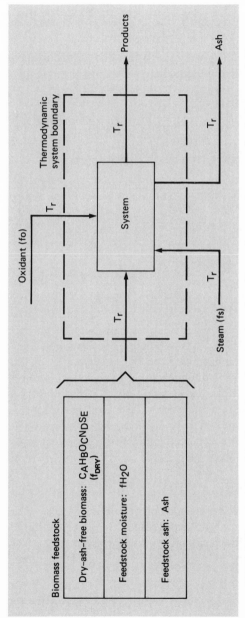

FIGURE 2. Representation of the thermodynamic system boundary. T_r = reference temperature; f = feed rate, mols/time.

Then for 1 mol of feed material, following Equation (1),

$$\text{heat into system} = 1 \times h_f + nh_{O_2}$$
$$\text{heat out of system} = Ah_{CO_2} + \tfrac{1}{2}Bh_{H_2O(l)} + \tfrac{1}{2}Dh_{N_2} + Eh_{SO_2}$$

Ignoring heat losses and changes of potential and kinetic energies, the heat of reaction, ΔH_R, can be represented as

$$\Delta H_{R,T_r} = [\Sigma\, H_{\text{products}} - \Sigma\, H_{\text{reactants}}]\,|_{T_r} \tag{2a}$$
$$\Delta H_{R,T_r} = Ah_{CO_2} + \tfrac{1}{2}Bh_{H_2O(l)} + \tfrac{1}{2}Dh_{N_2} + Eh_{SO_2} - h_f - nh_{O_2} \tag{2b}$$

(where T_r denotes the reference temperature)

Define the following component enthalpies to be zero at temperature T_r:

$$h_{CO_2} = h_{H_2O(l)} = h_{N_2} = h_{SO_2} = h_{O_2} = 0 \tag{3}$$

Then equation (2b) becomes

$$\Delta H_{R,T_r} = -h_f \tag{4}$$

where, by convention, a negative sign means heat is evolved. Let us assume that $T_r = 25°C$ (77°F), a common reference temperature for measuring and reporting heats of combustion. Then the interpretation of Equation (3) and Equation (4) is that by suitably choosing an enthalpy datum, the fuel or feedstock (solid or liquid) at a temperature of 25°C enters the system thermodynamically at its heat of combustion. The fuel or feedstock may be solid (biomass, municipal waste, coal, or lignite) or liquid (petroleum distillate or residual fractions). The next section presents a series of tables for component enthalpies based on these concepts.

2.3. Enthalpy Tables

The temperature dependence of heat capacities or enthalpies of many compounds (e.g., H_2, CO_2) is expressable as a cubic, quadratic, or other mathematical function of the temperature with the assumption of ideal gas behavior (e.g., system pressure = 0 kPa abs). These expressions may be written (for species i) as follows:

$$C_{p,i}(T) = a_i + b_iT + c_iT^2 + d_iT^3 \quad \text{for } T_1 \le T \le T_2 \tag{5a}$$
$$h_i(T) = e_i + f_iT + g_iT^2 + j_iT^3 \quad \text{for } T_3 \le T \le T_4 \tag{5b}$$

where

$$C_{p,i} = \text{heat capacity of species } i$$
$$h_i = \text{enthalpy of species } i \text{ above an arbitrary temperature}$$
$$a_i, \ldots, j_i = \text{experimentally determined coefficients}$$
$$T_1, T_2, T_3, T_4 = \text{limits of temperature ranges for equation applicability.}$$

To incorporate the datum conditions from Equation (3) into Equation (5b), the values of the component enthalpies are constrained to have certain values at the datum temperature 25°C by the following method:

- enthalpy of species i at temperature T is

$$h_i(T) \tag{5c}$$

- modified enthalpy $h_i'(T)$ of species i at temperature T constrained to have numerical value V_i at $T = 25°C$:

$$h_i'(T) = h_i(T) - h_i(25°C) + V_i \tag{6}$$

Table 2 lists nine compounds and the associated values of V_i. The values of V_i for O_2, N_2, CO_2, and SO_2 reflect the datum conditions of Equation (3). The

TABLE 2. Values of V_i for Different Compounds[a]

Species	V at 25°C	
	kJ/kg	Btu/lb
O_2	0	0
H_2	141,666	60,957.7
$H_2O(v)$	2,440	1,050
H_2S	16,507	7,102.8
N_2	0	0
CO	44,954	19,343.5
CO_2	0	0
SO_2	0	0
CH_4	55,453	23,861.0

[a]Source: J. H. Perry, in: *Chemical Engineers' Handbook*, Third Edition (J. H. Perry, ed.), McGraw-Hill Book Company, New York (1950), pp. 244-246.

value of V_i for $H_2O(v)$ is its heat of vaporization at $25\,^\circ C$. The values for V_i for H_2, H_2S, CO, and CH_4 are the higher heating values at $25\,^\circ C$.

Table 3 presents the component enthalpies, h_i' reflecting the procedures outlined for Equation (6). These enthalpies reference a modified $25\,^\circ C$ datum and ideal gas behavior. The environments for the products of most combustion and gasification reactions are high temperatures and low pressures, in which the ideal gas assumption used to calculate these enthalpies may be valid.

The following sections discuss principles of biomass gasification and liquefaction. Kohan and Barkhordar[3] and Jones, et al.[4] have presented additional details of the applications of these principles to a specific woody biomass feedstock.

TABLE 3. Enthalpies of Ideal Gases

Temperature ($^\circ C$)	Component enthalpies (kJ/kg-mol)								
	O_2	H_2	H_2O	H_2S	N_2	CO	CO_2	SO_2	CH_4
50	721	286544	44819	563871	709	283675	958	1015	89122
100	2199	288000	46501	565615	2151	285119	2947	3118	89324
150	3708	289456	48218	567409	3611	286584	5013	5294	89540
200	5245	290912	49966	569249	5086	288068	7154	7539	89767
250	6808	292370	51743	571134	6575	289569	9366	9849	90006
300	8393	293830	53550	573061	8079	291087	11647	12220	90258
350	10001	295294	55384	575030	9596	292620	13993	14648	90521
400	11629	296761	57247	577040	11127	294169	16400	17129	90795
450	13277	298233	59138	579090	12672	295734	18865	19660	91081
500	14943	299710	61056	581180	14230	297313	21383	22235	91378
550	16626	301193	63002	583309	15801	298906	23952	24852	91687
600	18325	302682	64975	585476	17386	300515	26567	27506	92006
650	20039	304178	66976	587682	18985	302137	29224	30193	92337
700	21768	305681	69004	589926	20598	303774	31919	32909	92679
750	23509	307192	71060	592206	22224	305425	34650	35651	93032
800	25263	308711	73143	594524	23865	307089	37411	38414	93395
850	27028	310239	75254	596879	25520	308768	40199	41195	93770
900	28804	311776	77393	599269	27189	310460	43009	43989	94155
950	30589	313323	79559	601696	28872	312166	45839	46792	94551
1000	32383	314880	81754	604158	30570	313886	48684	49601	94958
1050	34185	316448	83976	606655	32283	315620	51541	52412	95375
1100	35993	318027	86227	609187	34011	317367	54404	55221	95802
1150	37807	319618	88506	611753	35754	319127	57272	58023	96240
1200	39626	321222	90813	614354	37512	320901	60138	60815	96689
1250	41449	322838	93148	616988	39286	322688	63001	63592	97148
1300	43276	324467	95512	619656	41075	324488	65855	66352	97616

3. Gasification of Biomass

3.1. Types of Gasification Technologies

Gasification technologies may conveniently be classified according to the method of contacting the biomass particles with the gas phase:

- fixed bed* (one or more stages)
- fluidized bed
- other (e.g., entrained bed or molten bath)

Table 4 lists gasifier types currently being developed by the Department of Energy. Jones et al.[5] have presented numerous additional categories of thermal gasifiers for solid wastes and residues. Shelton[6] has presented a convenient triangular-diagram representation of materials amenable to self-sustaining thermal processing. At limits of 1–2% ash in as-received wood, for example, Shelton's data suggest that thermal processing of wood will not be practical without the addition of supplemental fuel when wood moisture contents exceed 70%; statements to this effect can also be found in the literature. Additionally, kelp with 82% water, or peat containing 90% water, would not be suggested for thermal processing on an as-received basis; instead, mechanical dewatering or field drying may be used to reduce moisture contents to acceptable levels. Shelton states that the autothermic region for thermal processing for general combustible feedstocks has been determined by experiment to lie near 21,000 kJ/kg of combustibles (9000 Btu/lb).

In fixed bed gasification, countercurrent or cocurrent flow of gas and solids creates distinct temperature zones in the gasifier (or in one stage of a multizone gasifier) that assist in the conversion of fixed carbon to gas. Zones frequently found in countercurrent, downflow fixed-bed gasification include drying/devolatilizing at the top (low temperature); the steam decomposition and steam–carbon reducing reaction zone; the high temperature carbon oxidation zone where hot oxidizing gases are generated by carbon combustion; and the ash collection and discharge zone (the ash may be dry or slagged depending on the technology under consideration). Fixed bed gasifiers are sensitive to fines in the feed material since these may cause bed plugging and channeling. Tars and liquid products in addition to gaseous hydrocarbons heavier than methane can be produced in fixed bed gasifiers.

Numerous additional combinations of fixed-type bed reactors are possible, such as the tumbling solids kiln being studied at the University of Arkansas, and the moving staged stirred bed being developed by Garrett Energy Research and Engineering. The type and condition of the feed, its moisture

---

*Includes multiple-hearth furnaces, rotary kilns, and pyrolysis vessels.

TABLE 4. Biomass Gasification Experimental Units Sponsored by the U.S. Department of Energy[a]

Contractor/Developer	Technology type	Nominal capacity			Representative biomass feedstocks	Primary energy products	Comments
		ST/D of as-received biomass feedstock	Dry ST/D[b]				
			50% moisture feedstock	10% moisture feedstock			
Battelle Columbus Division	Multi solid fluidized bed[c]	—	—	0.2	Wood	IBG[d]	Catalytic gasification; gasification with steam and recirculating hot solid
Battelle Pacific Northwest Division	Agiated fluidized bed	1.2	0.6	1.1	Wood	IBG	Catalytic gasification; gasification with steam, air, oxygen, and/or CO_2
Garrett Energy Research and Engineering	Multiple hearth furnace (Herreshoff type)	4.0	2.0	3.6	Manure, sawdust, cotton gin trash	IBG	Distinct hearth zones include direct contact drying; pyrolysis; combustion; ash cooling

Gilbert/Commonwealth Companies; Environmental Energy Engineering	Multiple operating modes	3.5–6	1.7–3	3.1–5.4	Wood, corn stover, cotton gin trash, bagasse	Gas, liquids, char (depends on operating mode)	Equipment may be operated as fluid bed, entrained bed, packed bed, falling particle bed
Texas Tech University	Variable velocity fluidized bed	0.5	0.25	0.4	Manure, wood, corn stover, mesquite cotton gin trash, wheat straw	LBG[e], IBG	Gasification with steam, air, or oxygen (future)
University of Arkansas	Rotary pyrolytic kiln	40	20	36	Wood waste	LBG, charcoal	Technology licensed by A&P Coop; charcoal is desired product.
University of Missouri, Rolla	Fluid bed (top feed)	2.5–3[f] 24[g]	1.2–1.5[f] 12[g]	2.2–2.7[f] 21.6	Wood	LBG, IBG	A. Coors gasifier; gasification with steam, air, oxygen (future), catalytic gasification (future)
Wright-Malta	Pressurized indirectly heated rotary kiln	6	3	5.4	Wood, peat, cornstalks	IBG	Gasification with catalyst and steam

[a]Source: Sixth Biomass Thermochemical Conversion Contractors' Meeting, Tucson, Arizona, January 16–17, 1979.
[b]ST/D = Short tons per day (1 ST = 907.2 kg).
[c]Unit normally operates with 10% moisture feedstock.
[d]IBG = Intermediate Btu gas.
[e]LBG = Low-Btu gas.
[f]With sleeve.
[g]No sleeve.

content and ash composition (e.g., high or low silica; high or low alkali metals) frequently determine the preferred contacting alternative.

Fluidized bed gasifiers accept a wide size distribution of feed solids and achieve a reasonably uniform bed temperature distribution through rapid fluid–solids mixing. Ash and unconverted carbon exit the bed in the product gas stream or are withdrawn at the gasifier base. Temperatures at the gasifier base control carbon loss. In first-generation single stage gasifiers, these temperatures are limited by the requirement of operating below the ash softening point to minimize clinker formation in the bed. In advanced (multiple stage or "sticky ash") gasifiers, deliberate operation in the ash agglomeration temperature range may reduce carbon loss (e.g., see Jequier *et al.*[7] for a description of this concept as applied to coal gasification).

Entrained gasifiers are commercially used for coal gasification (Koppers-Totzek); advanced pressurized entrained gasifiers are being developed by the Electric Power Research Institute (the Texaco gasifier) and others. A general requirement is for ground or pulverized coal to increase reaction rates in these short contact time vessels. It may not prove practical or desirable to grind or pulverize biomass because of the inherent moisture content and soft and fibrous nature of many types of biomass. For example, currently few of the DOE-sponsored biomass experimental units (see Table 4) are of the entrained type.

3.2. Governing Equations

Numerous reactions are possible in biomass gasification, many of which are coupled. Von Fredersdorff and Elliott[8] present an extensive kinetic framework for carbon reactions in coal gasification; many of these concepts should apply equally well to biomass gasification, particularly the gas-phase kinetics. Examples of endothermic carbon reactions include:

- devolatilization $C + heat \rightarrow CH_4 +$ condensible hydrocarbons
                                                            $+ char$
- steam–carbon $C + H_2O + heat \rightarrow CO + H_2$
- reverse Boudouard $C + CO_2 + heat \rightarrow 2CO$

Examples of exothermic reactions include:

- oxidation $C + O_2 \rightarrow CO_2 + heat$
- hydrogasification $C + 2H_2 \rightarrow CH_4 + heat$
- water gas shift $H_2O + CO \rightarrow CO_2 + H_2 + heat$
- methanation $3H_2 + CO \rightarrow CH_4 + H_2O + heat$
                      $4H_2 + CO_2 \rightarrow CH_4 + 2H_2O + heat$

By suitably arranging the method of gas–solids contacting and by employing multiple stages or other devices, a specific type of gasifier will attempt to exploit certain of the above reactions in preference to other possible reactions.

One possible approach to estimating the products of biomass gasification in single-staged reactors is presented below:

1. Define a basis for the calculation (e.g., 1000 kg)

2a. Select the biomass feedstock

2b. Express the feedstock to the gasifier on a molar basis, with feed moisture and ash streams shown separately (Figure 2)

3. Select gasifier type and gasification conditions (temperature, pressure) based on considerations such as feed moisture level, size, etc., and also with regard to the final use to which the gas will be put (e.g., low-pressure fuel gas; high-pressure synthesis gas, etc.)

4. Set some peripheral conditions:

(a) *Liquid products.* Certain types of gasifiers (e.g., fixed beds) produce liquid hydrocarbons which condense when the products of gasification are cooled to 50–100°C for gas cleanup. Other types of gasifiers (e.g., entrained; high-temperature fluid beds) which operate at high temperatures (1000–1300°C) produce essentially no "condensibles" because the potential liquid products are cracked to gases in these temperature ranges. For a selected biomass feedstock, the quantity and composition of the liquid product is best estimated by using experimental data. The carbon and hydrogen in the condensible products then become unavailable for hydrogen or CO production unless recycled to the gasifier. Condensible products also affect the gasifier heat balance since these products exit the gasifier as a vapor (using the criteria described in Section 2, the heat in the condensible products at the gasifier exit temperature would be the sum of the heat of combustion (25°C), latent heat of vaporization, and vapor sensible heat from 25°C to the gasifier exit temperature).

(b) *Nitrogen.* Since the gasification reactions produce a reducing atmosphere, a fraction of the feedstock nitrogen will be converted to ammonia (which subsequently may have to be removed). By analogy with coal gasification, the largest fractions of feedstock nitrogen (e.g., 50–100%) may be converted to ammonia in the downflow fixed bed type of reactors.

(c) *Sulfur.* Possible products of sulfur conversion include H_2S, COS, CS_2, mercaptans, etc. The first two compounds have received the most attention from the viewpoint of downstream product purification. Many biomass feedstocks are low in sulfur, suggesting that minimal sulfur-removal facilities may be needed when fuel gas is the desired product. By analogies with coal gasification, the molar ratio of H_2S to COS in the product gas may be about 20 for fixed beds; about 15 for entrained beds; and about 6 for fluid beds.

(d) *Carbon loss.* Complete conversion of the carbon in the feedstock to useful products is theoretically possible but of no practical significance since infinite solids residence times would be required. For each gasifier type or gasification configuration, a practical balance exists between the amount of carbon which is rejected (usually as a char) and the difficulty of further conversion of the remaining carbon particles.* As gasification reactions progress, the remaining carbon becomes less reactive as it becomes depleted of hydrogen. Consequently, a small fraction (2–5%) of the incoming feed carbon may generally be expected to be rejected as char (e.g., 95 wt%C; 5 wt%H) along with the gasifier ash reject.

(e) *Methane and other non condensible hydrocarbon gas yields.* High gasification pressures and low temperatures favor high equilibrium yields of methane and higher hydrocarbons. In coal gasification, von Fredersdorff and Elliott[8] point out that reactive (coal) fuels in fluid beds yield methane concentrations substantially in excess of that predicted when using β-graphite/hydrogen equilibrium calculations. Thus gasifier methane yields may be difficult to predict solely from equilibrium considerations and may depend on the reactivity of the feedstock, the type of gasifier, and the prior thermal history of the feedstock. Very small gasifier methane yields from any feedstock would be expected for high-temperature entrained and fluid beds. Significant gasifier methane yields are observed for pressurized downflow fixed bed coal gasifiers because of the top low-temperature devolatilizing zone (e.g., 5–10% of the feed carbon may be converted to methane); this may likewise be directionally true for biomass feedstocks in downflow fixed-bed gasifiers (e.g., see Mudge and Rohrmann's[10] data for atmospheric-pressure fixed bed wood gasification), and it is expected that gasifier methane yields for biomass feedstocks should be in proportion the fraction of volatile matter in the feedstock.

Table 5 lists the elemental balances thus far developed for the "peripheral conditions." The carbon, hydrogen, and oxygen remaining for further reaction with air or oxygen and steam are denoted by A′,B′,C′, respectively.

5. Set gasification conditions. The remaining unknowns are six in number: the products of reaction (H_2, H_2O, CO, and CO_2) and the steam (fs) and oxygen (fo) feed rates. The number of equations involving these unknowns are generally five:† (a) three equations involving elemental balances of C, H, and O; (b) one equation of shift equilibrium

---

*This "balance point" will not be the same for biomass and coal gasification when considering a specific gasifier. Coal ash is generally inert towards gasification reactions and tends to block or shield remaining active carbon sites by its presence as the gasification reaction progresses. In contrast, certain types of biomass (e.g., wood) have high proportions of alkali metals (Na,K) in the ash which catalyze the gasification reactions. For example, Feldmann[9] has experimentally shown that wood ash is a superior catalyst for wood gasification.
†An equation involving steam decomposition may be considered to be the sixth equation.

TABLE 5. Elemental Balance: "Peripheral Conditions"

Elements		C	H	O	N	S
Moles in (DAF[a]) feed to gasifier		A	B	C	D	E

Product compounds	Total moles of compounds	Moles of elements in compounds				
moles as condensible liquid	(—)	Al	Bl	Cl	Dl	El
moles as NH_3	(Da)		3Da		Da	
moles as H_2S	(Eh)		2Eh			Eh
moles as COS	(Ec)	Ec		Ec		Ec
carbon loss[b]	(—)	Ao	0.6Ao			(if desired)[c]
moles as CH_4	(Am)	Am	4Am			
moles as C_2H_6	(Ae)	2Ae	6Ae			
other compounds[d]	(—)	—————— as needed ——————				

Carbon remaining
for conversion
to CO and CO_2 $A - Al - Ec - Ao - Am - 2Ae = A'$

Hydrogen remaining
for conversion
to H_2 and H_2O $B - Bl - 3Da - 2Eh - 0.6Ao - 4Am - 6Ae = B'$

Oxygen remaining
for CO, CO_2,
and H_2O $C - Cl - Ec = C'$

Nitrogen remaining for N_2 $D - Dl - Da = D'$

Sulfur balance $E = El + Eh + Ec$

[a]DAF = dry, ash-free.
[b]C–H ratios shown for a char of 95 wt%C, 5 wt%H.
[c]Assuming no sulfur in the char will result in a conservative design for downstream gas purification systems.
[d]Excluding H_2, H_2O, CO, and CO_2; could include CS_2 or other species suggested by available data.

$$K_{eq} = \frac{(H_2)\,(CO_2)}{(CO)\,(H_2O)} \tag{7}$$

and (c) one equation for the overall heat balance.

In downflow fixed bed reactors a common assumption is that the feedstock moisture (fH_2O) is evaporated by the hot rising product gases. The feedstock moisture is therefore not available for the production of H_2 and CO in this type of reactor and would not be involved in the calculations for equations (a) and (b) mentioned previously. These equations are discussed below:

(a) *Elemental balances*. These equations are easily formulated from the inputs (A', B', C', fs, fo, and possibly fH_2O) and the outputs (H_2, H_2O, CO, and CO_2). As a rule, it is generally thought to be desirable to minimize the molar ratio of steam to oxidant feeds (fs/fo) to minimize costs.

(b) *Shift equilibrium.* The shift reaction is considered to be a heterogeneous phenomenon which occurs on the fuel surface.[8] At elevated temperatures typical of gasification systems, the shift reaction is generally considered to be at thermodynamic equilibrium. These temperatures (used for equilibrium calculations) vary with the type of gasifier: (1) for dry ash fixed bed gasifiers, the temperature is kept below the ash deformation temperature to avoid slagging; steam injection is also used for temperature moderation; (2) for slagging fixed bed gasifiers, the temperature is that needed to produce a fluid slag (which must be contained by water-cooled refractory material); (3) for dry ash fluid bed gasifiers, the temperature is generally set 25–50°C below the initial ash deformation temperature in order to maintain reasonable reaction rates and to avoid clinker formation in the bed; (4) for agglomerating ash ("sticky ash") fluid bed gasifiers, the temperature is deliberately kept within the ash softening range to reduce carbon loss; special ash-withdrawal and steam and oxidant injection geometries may be required to accomplish this; and (5) for short residence time slagging entrained gasifiers, the temperature is generally reduced at the gasifier exit to several hundreds of degrees below the flamezone temperature because of heat losses and endothermic reactions (special water-cooled refractories are needed for this gasifier).

(c) *Heat balance.* The heat balance is performed around the gasifier as a whole.

Inputs include:

• Feedstock at its heat of combustion plus any sensible heat effects if the feedstock enters the gasifier at a temperature above 25°C (e.g., from a drying operation)

• Steam at or above gasifier pressure and an appropriate temperature

• Oxidant (air or oxygen of 95–98% purity) at or above gasifier pressure and an appropriate temperature

• Any solid (e.g., char) or liquid (e.g., tar) recycled to the gasifier

Outputs include:

• The gas stream: (1) Product noncondensible gases at the gasifier exit temperature.* Table 3 may be conveniently used at this point. Nitrogen entering with the oxidant as well as unconverted steam (discussed later) both exit with the product gases. (2) Product condensible materials at the gasifier exit temperature* (heat of combustion (25°C), latent heat vaporization at 25°C and sensible heat effects); for reference, the heats of vaporization of many petroleum fractions lie in the range of 460–700 kJ/kg (200–300 Btu/lb), and (3) Some ash and unconverted carbon (char); ash heat effects may be conve-

---

*Not necessarily the shift equilibrium temperature.

niently estimated by using heat capacities of 1.0–1.7 kJ/kg°C (0.25–0.40 Btu/ lb°F); unconverted carbon heat effects include its heat of combustion at 25°C (32,740 kJ/kg or 14,087 Btu/lb) plus sensible heat effects

• Heat loss, which may be estimated as 1–2% of the heat in the incoming feedstock

• Bottom ash and char: (1) if nonslagged ash is generated, use the same guidelines as item (3) of the gas stream discussion (above); (2) if slagged ash is generated, add an additional term to the ash heat effects to account for the heat of fusion of the ash (generally 700–1150 kJ/kg or 300–500 Btu/lb).

In general the steam temperature, oxidant temperature, and gasifier exit temperature may be varied to obtain heat-balance closure.

Steam decompositions (i.e., the percent of feed steam which is decomposed to H_2 and O_2) generally vary from 25–85% in coal gasification.[8] Practically 100% steam decomposition may be obtained in the BGC slagging fixed-bed-coal gasifier.[11] Decomposition is a function of gasifier temperatures and the geometry of injection. High steam decompositions are desirable since less oxidant would be required and smaller quantities of (unreacted) steam would be condensed downstream and have to be treated.

3.3. Downstream Processing

3.3.1. Gas Purification

Depending on the end use of the gas, some or most of the sulfur compounds may have to be removed. Numerous commercial processes are available to accomplish this removal; solvents such as carbonates or amines may be used when the gas is cooled to 130–150°C (260–300°F). In cooling the raw gas from the gasifier exit temperature to 130–160°C:

(a) Sensible heat may be recovered if minor amounts of tars or oils are contained in the gas; if large amounts of condensibles are contained in the gas, the gas should simply be quenched to the desired temperature because the condensible material will foul heat recovery equipment;

(b) the condensate produced by cooling the raw gas should be treated to remove organics, ammonia, etc;

(c) high-pressure saturated (and possibly superheated) steam can be produced in the waste heat boiler; a heat balance and 30°C or greater pinch points* will determine the steam quantities involved, and the tube metallurgy

---

*Minimum temperatures separating the temperature of the gas stream from the temperature of the water or steam at any point inside of the waste heat boiler.

will limit the maximum gas temperature (usually $1100°C$) which can be used to generate steam.

The choice of chemical or physical systems for the removal of H_2S from the gas depends on the desired purity level, gas pressure, and many other factors.[12] The chemical route (carbonates or amine solvents) involves the reversible chemical reaction of H_2S and CO_2 with the polar group of the solvent; the solvent is regenerated using closed steam stripping at a lower system pressure. Diisopropyl amine and methylethyl amine systems are widely used in this service in petroleum hydroprocessing. Chow et al.[13] present details of a carbonate system. Generally, up to 99% removal of H_2S can be achieved; and the amount of COS and CO_2 which can be removed is a function of the H_2S/CO_2 partial pressures, solvent, and other factors.

Physical organic solvents absorb acid gases (H_2S and CO_2) at high pressures; the solvent is regenerated by flashing at lower pressures. Because of solubility differences between H_2S and CO_2, many of these systems can be designed for selective absorption of H_2S. Chandra et al.[14] present details of a physical absorption system.

After the H_2S is removed from the gas, it must be recovered in sulfur recovery units for environmental reasons. Since many biomass materials are low in sulfur, it is not likely that sulfur recovery units can be justified solely on the basis of the revenue generated from the sale of sulfur.

If the feed to the sulfur recovery units contains above 10–15% H_2S (molar), conventional Claus plants can be used.[15] If the H_2S concentration is below about 10%, the more expensive Stretford[16] technology has to be used. Claus plants generally can recover 95–97% of the sulfur in the acid gas stream. The recovery is a function of the H_2S concentration in the feed stream, the number of stages selected, and other factors. Beavon et al.[17] list the reactions as:

$$H_2S + \tfrac{3}{2}O_2 \rightarrow H_2O + SO_2 \tag{8a}$$

$$2H_2S + SO_2 \rightarrow (\tfrac{3}{2}x)\,Sx + 2H_2O \tag{8b}$$

$$\text{Overall:}\quad 2H_2S + O_2 \rightarrow 2S + 2H_2O \tag{8c}$$

One-third of the incoming H_2S is burned to SO_2, which reacts with the remaining H_2S from elemental sulfur. Equation (8b) is reversible and limits the conversion to sulfur to 95–97%. The Beavon sulfur removal process[17] or Shell's SCOT process[18] are used to remove additional sulfur compounds from the Claus plant tail gas for environmental reasons.

The Stretford process is a wet oxidative extraction process for the removal of H_2S down to a level of about 10 parts per million volume (ppmv) in the treated gas stream:

$$H_2S + Na_2CO_3 \rightarrow NaHS + NaHCO_3 \tag{9}$$

$$NaHS + NaHCO_3 + 2NaVO_3 \rightarrow S + Na_2V_2O_5 + Na_2CO_3 + H_2O \quad (10)$$
$$Na_2V_2O_5 + \tfrac{1}{2}O_2 \text{ (air)} \rightarrow 2NaVO_3 \quad (11)$$

Hydrogen sulfide is removed from the gas stream by reaction with sodium carbonate [Equation (9)]. Subsequent oxidation and reduction steps produce elemental sulfur using vanadium-based chemistry [Equations (10) and (11)]. Low temperatures ($25-55°C$) promote high sulfur recoveries. The CO_2 partial pressure in the feed gas should be kept below a few atmospheres to prevent interaction with the bicarbonate chemistry shown. A Bechtel report to the city of Seattle presents additional details of the Stretford system.[19]

3.3.2. Shift Conversion

The H_2/CO ratio in the gas may require adjusting prior to synthesis. The shift reaction

$$H_2O + CO \rightarrow H_2 + CO_2$$

is generally used for this purpose, with the addition of steam to drive the reaction to the right. The H_2/CO ratio achieved in the conversion is a function of the catalysts used, system temperature, and steam-to-carbon ratio. If complete conversion to H_2 is not desired, a high-temperature sulfur insensitive catalyst can be used. Temperatures are generally limited to a maximum of $455-480°C$ ($850-900°F$) at the bed outlet for metallurgy and catalyst-stability reasons.

If hydrogen is the desired product, an additional lower temperature shift step must be added because low temperatures ($200-400°C$) favor hydrogen production. Many commercial low-temperature shift catalysts are copper-based and are deactivated by sulfur; as a consequence, an acid gas removal step must preceed this shift. Recently, Haldor Topsoe A/S and Exxon Research and Engineering[20] have marketed a sulfur resistant low temperature shift catalyst, type SSK; a certain minimum sulfur content in the process gas is required to maintain the catalyst in its active, sulfided state. Whether or not the SSK catalyst will find application in biomass gasification should be determined based on economics and other factors, including whether or not the sulfur content of the feedstock produces a gas which exceeds the minimum H_2S requirement referred to previously. In any event, an acid gas removal step will probably be required after a low temperature shift using sulfur-tolerant catalyst. The steam and power requirements of these technologies have to be determined for any integrated-plant study. Process licensors should be contacted for this information for gas purification and sulfur recovery technologies.

Tables of shift equilibrium constants or nomographs for biomolecular reactions are used together with enthalpy information (Table 3) and the equation for shift equilibrium [Equation (7)] in calculating shift conversion heat and material balances.

3.3.3. Additional Processing

A synthesis gas consisting principally of H_2 and CO (which generally requires oxygen-blown gasification) can be catalytically upgraded to numerous products. Among these are methane (SNG), methanol, hydrogen, and ammonia. The literature on the chemistry of these upgrading steps, the hardware which is employed commercially, and commercial operating experience is voluminous. Only highlights are mentioned here.

Synthesizing SNG, methanol, or ammonia from syngas generally involves fixed catalyst bed technology employing recycle streams and cold gas quench, heat exchange or other options to remove the exothermic heats of reaction. Fresh synthesis gas is mixed with recycled synthesis gas and reintroduced over the catalyst bed to achieve the desired per-pass conversion of reactants to product (catalyst activity and selectivity determine the achievable per pass conversions). Frequently, the maximum temperature rise across a catalyst bed is set by considerations of catalyst stability (sintering) and desired catalyst lifetime. For the general recycle process, heat and material balances are used to determine the performance of one or more reactors. Heat inputs include feedstock, steam (if needed), and compressor work (makeup and recycle). Heat outputs include product streams, purge streams (to prevent buildup of inert material in the loop), heat recovery (steam, hot water), and heat rejection (cooling). In the following paragraphs, individual products are briefly discussed.

3.3.3a. Substitute Natural Gas (SNG). The preferred methanation reaction is

$$3H_2 + CO \rightarrow CH_4 + H_2O \qquad (12)$$

Methane formation is favored by high pressures and low temperatures. Input H_2/CO mole ratios are usually in the range of 3.1/1 to 3.15/1. The maximum temperature in any reactor is usually limited to 480°C (900°F).

The equilibrium constant for the reaction given by Equation (12) is:

$$K_{eq} = \frac{(CH_4)(H_2O)}{(CO)(H_2)^3} \left[\frac{\text{total moles}}{\text{absolute pressure}} \right]^2 \qquad (13)$$

Unlike the shift reaction equilibrium [Equation (7)], the methanation equilibrium shown by Equation (13) is pressure-sensitive because three moles of hydrogen are involved in the reaction. Examples of successive recycle calculations for methanation can be found in an ERDA report[21] prepared by C. F.

Braun & Co. (shift and methanation equilibria are satisfied at the outlet of each converter). Commercial methanation catalysts are sulfur-sensitive and H_2S must be removed from the gas before synthesis.

An alternative to separate shift and methanation steps is combined shift/ methanation, which may be conceptualized as occuring in adiabatic packed-tube reactors. High-pressure superheated steam may be generated from such a design. Carbon formation may occur by the Boudouard reaction, and sufficient steam should be added to avoid carbon-forming regions (triangular C—H—O diagrams may be used for this purpose).[22]

3.3.3b. Methanol. Synthesis gas is catalytically converted to methanol by the following reactions:

$$2H_2 + CO \rightarrow CH_3OH$$
$$3H_2 + CO_2 \rightarrow CH_3OH + H_2O$$

Low temperatures and high pressures favor the formation of methanol. Conventionally, excess hydrogen is used in the makeup syngas according to formulas such as:

$$\frac{H_2}{2CO + 3CO_2} = 1.1 \text{ to } 1.2$$

The development of active, copper-based catalysts (which are sulfur-sensitive) has permitted the newer fixed-bed commercial methanol plants to be designed for 50–150 atm, with bed temperatures in the range of 250–380°C. Considerable attention has been given to the optimization of the heat recovery and energy useage in the methanol synthesis loop by process licensors.[23]

The Electric Power Research Institute is sponsoring the development of a liquid-phase methanol synthesis concept at Chem Systems, Inc.[24] Fresh plus recycle gases are fed to a ebullated catalyst bed which is fluidized by an inert, nonmiscible hydrocarbon liquid. High per pass conversions to methanol are possible because the methanol is absorbed in the solvent. Excellent temperature control is over the reaction is also possible. Chow *et al.*[13] present preliminary heat and material balances for this concept. The mole ratio of H_2 to CO is adjusted to 2/1 prior to synthesis.

Catalytica Associates, Inc., have evaluated the potential for the development of a sulfur-tolerant methanol synthesis catalyst for EPRI.[25] No such catalysts are presently known. Little economic incentive was found for the development of this type of catalyst.

3.3.3c. Ammonia. The ammonia synthesis reaction can be represented by

$$3H_2 + N_2 \rightarrow 2NH_3$$

Synthesis occurs at 17,200 kPa (2500 PSIG) to 24,100 kPa (3500 PSIG). Refrigeration for the loop must also be provided if a liquid product is desired. A Bechtel report for the City of Seattle[19] presents process details. No future improvements in ammonia synthesis technology are anticipated.

4. Liquefaction of Biomass

This section discusses two technologies which can be used to directly produce liquid fuels from biomass: pyrolysis and catalytic liquefaction. Technologies which can be used to produce liquids indirectly from biomass (e.g., methanol) have been discussed in the section on biomass gasification.

4.1. Types of Liquefaction Technologies

Pyrolysis technologies can produce acidic, oxygenated liquids from biomass materials by heating the biomass in the absence of oxygen. These low-pressure processes produce large amounts of char, which is generally low in sulfur and ash. The economics of producing liquid fuels from biomass by pyrolysis is strongly dependent on the value assigned to the coproduct char. In some instances, regional markets may exist for the char as an activated carbon material for such uses as water purification, or as a compliance fuel for boilers (utility or industrial). Pyrolysis technologies have been developed by Tech Air,[26] Energy Resources Company, Inc.,[27] and Occidental Petroleum.[28] Roberts et al.[29] have discussed the kinetics of the pyrolysis of cellulosic material.

Carbonaceous materials such as biomass or coal can be liquefied (or gasified) with the use of alkaline metal catalysts (e.g., Na_2CO_3). A 3-ton/day pilot plant for waste wood liquefaction, under DOE sponsorship, is in operation in Albany, Oregon to test the feasibility of this concept.[30] The original work motivating these current efforts was performed on cellulosic material (e.g., municipal waste, wood residue) and manures at the Pittsburgh Energy Research Center of DOE (and its predecessor ERDA).[31] In biomass liquefaction, the alkaline catalyst promotes the removal of oxygen from the biomass material in a slurry–recycle type of operation. Recent experiments suggest that mild acid hydrolysis of the wood feedstock before liquefaction significantly improves the conversion of wood to oil.

Other biomass liquefaction options (e.g., the extractive processing of *Euphorbia* plants) have been discussed by Jones, et al.[4]

4.2. Governing Equations

For pyrolysis, the calculation procedure follows much the same format as discussed for biomass gasification. However, about half of the carbon in the feedstock may be present in the char. It is difficult to predict the elemental composition and heating value of the char, and use should be made of existing experimental data concerning the feedstock and pyrolysis technology under consideration. The char should be high in carbon and low in moisture content, and generally contain the majority of the ash present in the feedstock. To predict the composition and heating value of the oil product, reference should again be made to experimental data. Pyrolytic gaseous production of methane, ethane, propane, unsaturated compounds, and the like should be determined when possible by reference to experimental data.

Water is a product of the pyrolysis reaction. Kohan and Dickenson[32] have presented an equation which suggests a method to determine the desirable moisture content of the biomass feedstock sent to the pyrolysis unit:

$$
\begin{array}{l}
\text{water} \\
\text{entering} \\
\text{with} \\
\text{biomass}
\end{array}
+
\begin{array}{l}
\text{water} \\
\text{of} \\
\text{reaction}
\end{array}
=
\begin{array}{l}
\text{water (of} \\
\text{saturation) in} \\
\text{pyrolysis gas} \\
\text{product}
\end{array}
+
\begin{array}{l}
\text{water content of oil} \\
\text{product (for viscosity} \\
\text{control, product} \\
\text{stability, or other reasons)}
\end{array}
\qquad (14)
$$

In Equation (14), water inputs include the feedstock moisture and the water of reaction. Water exits the technology principally in the liquid oil product and the gas. The amount of biomass drying can be determined by using the above equation so that no water will be condensed (and treated for the removal of organic matter) in downstream reaction-mix quenching steps.

For catalytic liquefaction, the basic chemistry concepts are being developed by DOE contractors and Lawrence Berkeley Laboratory personnel. One version of the idealized chemistry is suggested by Equation (15):

$$
\underset{\text{(feed)}}{C_A H_B O_C} + Y \text{ "O"} \xrightarrow{Na_2CO_3} \underset{\text{(product)}}{C_A H_B O_{C-Y}} + Y O_2 \qquad (15)
$$

Carbon monoxide has been investigated as the reducing agent, "O." The stoichiometric coefficient, Y, is set based on the desired oxygen content of the product. The amount of oxygen in the product has been observed to be as low as 14–15 wt% for certain cellulosic wastes.[31] Bechtel has used 23 wt% oxygen in a preliminary design study of wood liquefaction.[33]

Wood liquefaction can be assumed to proceed in a relatively straightforward recycle system as shown on Figure 3. The governing equations have been presented in many standard textbooks and will be repeated here for reference (an ash-free basis will be used for convenience).

Let the dry ash-free biomass $(C_A H_B O_C N_D S_E)$ be represented by f_{DAF} (mass/hr). Then

$$
F_0 = f_{DAF} + f_{H_2O}
$$

and (1) overall mass balance:

$$
F_1 + F_2 = f_{DAF} + f_{H_2O} \qquad (16a)
$$

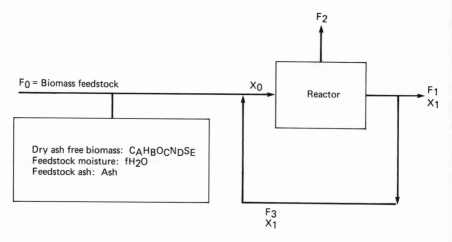

FIGURE 3. Diagram showing the conversion of biomass by recycling of the unconverted feed-stock. F = flowrate, mass/time; F_1 = product; F_3 = recycle; X = weight fraction of biomass; X_1 = unconverted biomass.

(2) specified weight fraction, X_0, of dry ash-free biomass in slurry fed to reactor:

$$X_0 = (f_{DAF} + F_3 X_1)/(F_0 + F_3) \qquad (16b)$$

X_0 is set by limitations such as slurry viscosity or feed pumpability and may range from 0.2 to 0.5.

(3) weight fraction of unconverted biomass, X_1:

$$X_1 = (f_{DAF} + X_1 F_3)(1 - PPC)/(F_0 + F_3 - F_2) \qquad (16c)$$

PPC represents the desired per pass conversion (as a fraction) of biomass to product. In petroleum hydrocracking, values of per pass conversion range from 50% to 80% depending on the feedstock type, catalyst type, desired product slate, and other factors. These concepts are all at early stages of development for biomass liquefaction.

(4) secondary stream F_2 (principally gases):

$$F_2 = (f_{DAF} + X_1 F_3)(PPC)\left(\frac{16Y}{f_{DAF}}\right) + f_{H_2O} \qquad (16d)$$

where Y is the coefficient of "O" in Equation (15). Equation (16d) suggests that water (f_{H_2O}) entering the reactor leaves as a vapor in stream F_2.

Equations (16a–d) may be solved for the four unknowns F_1, F_2, F_3, and X_1. The quantities f_{DAF}, f_{H_2O}, and X_0 are assumed to be specified for the problem.

Additional points include:

(1) Specification of partial pressures. If CO is the active species in Equation (15), then a minimum partial pressure of CO at the reactor inlet is needed to achieve reasonable reaction kinetics. This value is currently being experimentally determined by DOE, LBL, and others. Bechtel has used a value of 13,800 kPa (2000 PSIA) in a conceptual design.[33]

(2) Excess reactant. Expressions such as Equation (15) can suggest the theoretical chemical comsumption of CO or other reactants to produce a desired product slate. In practice, 20–80% excess reactant may be utilized to improve reaction kinetics; account for the loss of gaseous reactant to both liquid-phase solubility and system leaks; and similar effects. Typical tradeoffs involved in the specification of excess reactant include reduced reactor sizes because of improved kinetics versus increased capital and operating costs for recycle compressors, among and other factors. Bechtel has used a figure of 40% excess CO in a conceptual study.[33]

(3) In addition to feedstock moisture, stream F_2 will contain the CO_2 of reaction, unreacted CO, and any inerts (e.g., CH_4, H_2, etc.) entering with the reaction gas mixture.

(4) Temperatures in the reactor should lie in the range of 315–350°C (600–650°F).

4.3. Parallel or Downstream Processing

4.3.1. Gasification and Recycle

In the catalytic liquefaction technology, oil, unconverted biomass, ash, and other material exit the reactor. By a series of staged separations, gaseous and liquid–solid material can be separated. The liquid–solid streams can be separated (by filtration, flashing, or other means) into a low-ash liquid product stream and a recycle stream containing ash, unconverted biomass, and some oil. This liquid recycle stream can be gasified (along with supplemental biomass feedstock if needed) to produce a portion of the required reaction gases.

The gaseous product from the reactor can be combined with the gasification products, sent to acid gas removal, compressed, and returned to the reactor.

4.3.2. Solids Removal

In coal liquefaction (e.g., H-Coal or Solvent-Refined Coal), the separation of the coal ash from the liquid or solid fuel products has proven to be problematical. Concepts such as rotary pressurized precoat filters, solvent deashing, and vacuum flashing have met with varying degrees of success. It is possible that the structure of biomass-derived liquids may be somewhat simpler than the multibenzene-ring asphaltenic structure attributed to many coal-derived liquids. Consequently, biomass-derived liquids (which tend to be oxygenated) may prove to be less difficult to process for solids removal and have lower viscosities than coal-derived liquids.

4.3.3. Liquid Recovery/Drying

Direct contact of hot gases with wet solids (e.g., biomass feedstocks) is possible when the solids contain no volatile solvents. Fluid-bed and rotary-drum drying are common options. For example, to dry green wood, about 3490–3950 kJ/kg of water evaporated (1500–1700 Btu/lb) must be supplied; required gas temperatures should be in the 315–480°C (600–900°F) range; and the exhaust gases should exit the dryer at 20–50°C (50-100°F) above the dewpoint to achieve reasonable drying rates.

If solvent recovery from solids is desired, indirect heat transfer equipment should be used. An example would be an indirectly heated rotary calciner. In this case, the flue gases should exit the dryer at a temperature of 120–200°C (250–400°F) above that of the vaporized fluid to achieve reasonable equipment sizes. Typically, for every 100 kJ of thermal energy furnished to the dryer, one-third is transferred through the calciner walls and two-thirds end up in the high-temperature flue gas (which may then be used for steam raising, direct contact drying, or similar applications).

In performing dryer calculations, the concepts discussed in Section 2 of this chapter may be used, together with the enthalpy information in Table 3.

References

1. J. M. Radovich, P. G. Risser, T. G. Shannon, C. F. Pomeroy, S. S. Sofer, and C. M. Sliepcevich, Evaluation of the Potential for Producing Liquid Fuels from Biomaterials, Electric Power Research Institute, Final Report AF-974 (January, 1979), pp. 3–1 to 3–7.
2. P. N. Chereminisoff and A. C. Morresi, *Energy from Solid Wastes*, Marcel Dekker, Inc., New York (1976), Chapter 15, pp. 363–411.
3. S. M. Kohan and P. M. Barkhordar, Mission Analysis for the Federal Fuels from Biomass Program, IV: Thermochemical Conversion of Biomass to Fuels and Chemicals, Final Report prepared by SRI International for the U.S. Department of Energy under Contract EY-76-C-03-0115 PA131 (January, 1979), NTIS SAN-0115-T3.
4. J. L. Jones, S. M. Kohan, and K. T. Semrau, Mission Analysis for the Federal Fuels from Biomass Program, Volume VI: Mission Addendum, Final Report prepared by SRI International under Contract EY-76-C-0115 PA131 (January, 1979), pp. 51–90, NTIS SAN-0115-T4.
5. J. L. Jones, R. C. Phillips, S. Takaoka, and F. M. Lewis, *Proceedings of the ASME Eighth Biennial National Waste Processing Conference*, American Society of Mechanical Engineers, New York, (May, 1978).

6. R. D. Shelton, in: *Advanced Thermal Processes for Conversion of Solid Wastes and Residues* (J. L. Jones and S. B. Radding, eds.), American Chemical Society, Washington, D.C. (1978), pp. 165–190.

7. L. Jequier, L. Longchambon, and A. van der Putte, The Gasification of Coal Fires, in: *J. Inst. Fuel*, **33**, 584–591 (1960).

8. C. G. von Fredersdorff and M. A. Elliott, in: *Chemistry of Coal Utilization, Supplementary Volume* (H. H. Lowry, ed.), John Wiley & Sons, New York (1963), pp. 892–1022.

9. H. F. Feldmann, Conversion of forest residues to a methane-rich gas, Paper presented at the Sixth Biomass Thermochemical Conversion Contractors' Meeting, Tucson, Arizona (January 16–17, 1979).

10. L. K. Mudge and C. A. Rohrmann, in: *Advanced Thermal Processes for Conversion of Solid Wastes and Residues* (J. L. Jones and S. B. Radding, eds.), American Chemical Society, Washington, D.C. (1978) pp. 126–141.

11. D. Hebden, J. A. Lacey, and W. A. Hursler, paper presened at the 30th Autumn Research Meeting of the Institution of Gas Engineers, London, (November, 1964).

12. R. P. Schaaf and R. N. Tennyson, *Oil Gas J.* **75**(2), 78–86 (1977).

13. T. K. Chow, D. W. Stanbridge, and G. A. White, Screening Evaluation: Synthetic Liquid Fuels Manufacture, Electric Power Research Institute Final Report AF-523 (August, 1977).

14. K. Chandra, B. McElmurry, E. W. Neben, and G. E. Pack, Economic Studies of Coal Gasification Combined Cycle Systems for Electric Power Generation, Electric Power Research Institute Final Report AF-642 (January, 1978).

15. H. Grekel, L. V. Kumkel, and R. McGalliard, Chem. Eng. Prog., **61**(9), 70 (1965).

16. T. Nicklin, F. C. Riesenfeld, and R. P. Vaell, Application of the Stretford Process to the Purification of Natural Gas, paper presented at the 12th World Gas Conference, Nice, France (1974).

17. D. K. Beavon, R. H. Hass, and B. Muke, *Oil Gas J.* **77**(10), 76–80 (1979).

18. W. Groendoah, *Chem. Eng. Prog.* **69**(12), 29–34 (1973).

19. Bechtel Incorporated, Final Report: Seattle Solid Waste Ammonia Project Study, City of Seattle, Office of Management and Budget (1977).

20. Anonymous, *C & E News*, **54**(26), 16 (1976).

21. R. Detman, Factored Estimates for Western Coal Commercial Concepts, Interim Report, Energy Research and Development Administration (October, 1976).

22. G. A. White, T. R. Roezkowski, and D. W. Stanbridge, Predict Carbon Formation, in: *Hydrocarbon Process.* **54**(7) pp.130–136. (1975).

23. M. Pettman and G. Humphreys, Improved Designs to Save Energy, in: *Hydrocarbon Process.* **54** (6), 77 (1975).

24. M. B. Sherwin and D. Blum, Liquid Phase Methanol, Electric Power Research Institute, Final Report AF-202, (August, 1976).

25. Catalytica Associates, Inc., Evaluation of Sulfur-Tolerant Catalytic Processes for Producing Peak-Shaving Alcohol Fuels, Electric Power Research Institute Final Report AF 687 (February, 1978).

26. M. D. Bowen E. D. Smyly, J. A. Knight, and K. R. Purdy, in: *Advanced Thermal Processes for Conversion of Solid Wastes and Residues* (J. L. Jones and S. B. Radding, eds.), American Chemical Society, Washington, D.C. (1978), pp. 94–125.

27. J. B. Howard," Pilot Scale Pyrolytic Conversion of Mixed Wastes to Fuel, Vols. I, II," Draft Report submitted to EPA under Contract 68-03-2340 (1978).

28. F. F. Boucher, E. Knell, Pyrolysis of Industrial Wastes for Oil and Activated Carbon Recovery, Environmental Protection Agency Report, NTIS PB-270-961 (1977).

29. P. V. Roberts, J. O. Leckie, and P. H. Brunner, in: *Advanced Thermal Processes for Conversion of Solid Wastes and Residues* (J. L. Jones and S. B. Radding, eds.), American Chemical Society, Washington, D.C. (1978), pp. 392–410.

30. T. E. Lindemuth, in: *Advanced Thermal Processes for Conversion of Solid Wastes and Residues* (J. L. Jones and S. B. Radding, eds.), American Chemical Society, Washington, D.C. (1978), pp. 371–391.
31. E. Del Bel, S. Friedman, and P. M. Yavorsky, in: *Synthetic Fuels Processing, Comparative Economics* (A. H. Pelofsky, ed.), Marcel Dekker, Inc., New York (1977), pp. 443–459.
32. S. M. Kohan and R. L. Dickenson, Production of liquid fuels and chemicals by thermal conversion of biomass feedstocks, paper presented at the 72nd AIChE Annual Meeting, San Francisco, (November, 1979).
33. Bechtel National, Inc., Final Technical Progress Report: Albany, Oregon Liquefaction Project, Department of Energy Report prepared under Contract EG-77-C-03-1338 (1978).

9

The Occidental Flash Pyrolysis Process

PING WU CHANG and GEORGE T. PRESTON

1. Introduction

In 1968, the Occidental Research Corporation, formerly the Garrett Research and Development Company, initiated a program to recover metals and glass from municipal solid waste (MSW) and to convert the organic portion to a fuel oil by Flash Pyrolysis®. The efficacy of the Flash Pyrolysis was established in a 1.4 kg/hr scale laboratory reactor.[1,2] Waste feed, in addition to MSW, includes tree bark, rice hulls, animal manure, rubber,[3] and sewage sludge.[4] The pyrolysis process was then successfully tested in a 3.6 ton/day pilot plant where the major process variables were investigated, material handling problems resolved, and sufficient products produced to permit characterization.

The pyrolytic oil, Pyrofuel®, was intended to be sold as a substitute for No. 6 residual fuel oil. Combustion tests indicated that Pyrofuel or blends of Pyrofuel with other liquid fuel oil is an acceptable liquid fuel for utility boilers.

A 181-ton/day plant was designed and built to demonstrate the process under a contract with the Environmental Protection Agency and San Diego County of California. The pyrolysis section of the plant failed to demonstrate a steady state production of Pyrofuel from MSW during the course of the contract. The plant, however, received and processed refuse at the design rate in the sections for refuse preparation and glass, ferrous metal, and aluminum recovery.[5]

PING WU CHANG • Occidental Research Corporation, P.O. Box 19601, Irvine, California 92713. GEORGE T. PRESTON • Electric Power Research Institute, Palo Alto, California 94303.

2. Process Description

The incoming raw refuse is prepared for pyrolysis in the front end system, and the inorganic materials are recovered. The refuse is shredded and partially dried. Prepared feed is then pyrolyzed in the Flash Pyrolysis reactor. The pyrolytic oil, Pyrofuel, is recovered from the pyrolytic vapor in the oil-collection system. Process heat is supplied by burning char which is the solid product from pyrolysis.

2.1. Front End System

As illustrated in Figure 1 and Table 1, MSW is weighed and delivered to the tipping floor. The incoming refuse is presorted for oversized bulky materials such as refrigerators and dangerous materials such as munitions. The feed is then shredded to a 10-cm top size. The shredded refuse is fed by a conveyor to a magnetic separator which removes more than 95% of ferrous metals. A zig-zag type air classifier of Occidental's design is used to separate the resultant iron-free stream into a light fraction of about 95% wet organics and a heavy fraction comprising mostly inorganics.

The heavy fraction is further treated to recover glass, nonferrous metals and entrapped organic material. In the first section of a two-section trommel screen, material smaller than 1.2 cm is removed. This fraction contains 50% glass and is ground in a rod mill to about 44–840 μm. Glass is recovered from this ground stream in a multistage froth flotation system. The float material after drying is 99.5% glass and corresponds to 70% of the glass in the raw waste.

The second section of the trommel has 10-cm holes. The oversize is recycled to the primary shredders. The undersize from this section contains 10% metals and is fed to the aluminum separation plant. The aluminum separation is achieved by Occidental's RECYC-AL® nonferrous metals recovery system[6,7]. A magnetic field is generated to cause conductive materials deflected off the side of the belt to a collection bin. The product is 90–95% aluminum.

The light fraction from the zig-zag air classifier is dried in a rotary drum dryer to 3% moisture with hot inert gas from the pyrolysis section. The dried material, which has more than 10% inorganics, is screened again at 1200 μm. The inorganic content of the oversize is reduced to 4%. The undersize, which contains organics, is fed to an air table. Three fractions are generated by the air table. A light organic fraction joins the screen oversize. A heavy, glass-rich fraction is fed to the glass plant. A small intermediate fraction is either land-filled or can be fed to the char burner for energy recovery.

The overs from the screen are subjected to a second shredding operation. The pyrolysis feed size is 80% smaller than 1200 μm. Energy consumption is

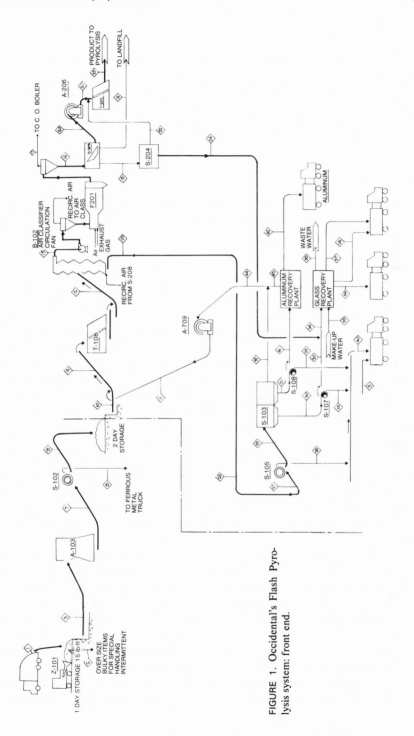

FIGURE 1. Occidental's Flash Pyrolysis system: front end.

TABLE 1. Front-End Equipment List

A-103	Primary shredder	S-105	Magnetic separator
A-206	Secondary shredder	S-107	Magnetic separator
A-709	Aluminum plant shredder	S-108	Magnetic separator
B-102	Blower	S-201	Cyclone
F-201	Rotary dryer	S-203	Vibrating screen
S-101	Zig-zag air classifier	S-204	Air table
S-102	Magnetic separator	S-208	Cyclone
S-103	Trommel	Z-101	Tipping floor

high, about 120–150 kJ/kg. This secondary shredded material which resembles dry vacuum cleaner fluff is referred to as fluff.

2.2   Pyrolysis System

As illustrated in Figure 2 and Table 2, the fluff is pneumatically fed into the Flash Pyrolysis reactor. Process heat is supplied by hot ash particles which mix with the fluff at the reactor inlet.

Pyrolysis takes place at 510°C for a short residence time. No air, oxygen, hydrogen, or catalyst is used in the reactor. The fluff is pyrolyzed into four products: oil, gas, water, and char. The solid residue of pyrolysis, char and ash, is separated from the product vapor in cyclones. Char is combusted in a separate vessel, the char burner, with air. Essentially all of the carbonaceous matter in the char is burned for process heat, leaving a high ash residue. This heats the ash which is recycled back to the pyrolysis reactor at about 760°C. This ash is circulated at a rate five times the fluff feed rate. Excess ash is removed. As the gas–ash–fluff stream is turbulent and the particles are small, excellent heat transfer is achieved, and organics are pyrolyzed very rapidly.

The pyrolysis vapor is quenched rapidly to stop cracking reactions by spraying a light fuel oil into the gas. Pyrofuel, which is immiscible in the oil, settles to the bottom of a decanter and is removed to storage tanks. Water is partially soluble in the Pyrofuel and is used to adjust the Pyrofuel handling properties.

The product gas is used as a transport gas and as fuel for preheating the combustion air and various process heating needs. All process off-gas is filtered in a baghouse before discharge.

Limited waste stream data are available for gas, water, and solid. The afterburner together with the baghouse remove odor and particulates efficiently and SO_2, NO_x, and HCl concentrations meet federal standards. The waste-water stream must be treated prior to disposal via a municipal sewage system due to the high chemical oxygen demand (COD) count of greater than 100,000 ppm. The solid to be landfilled has a high density and represents a small volume.

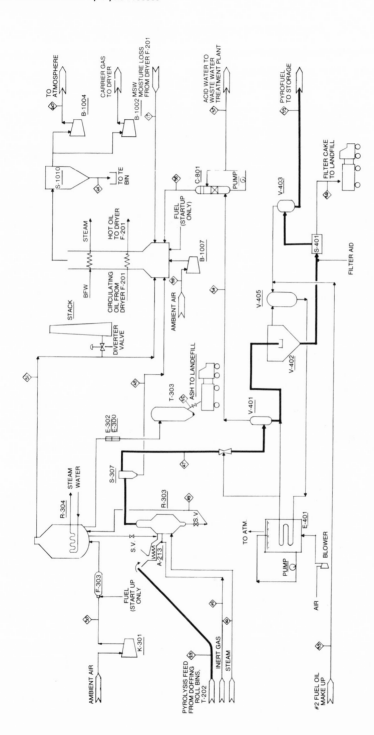

FIGURE 2. Occidental's Flash Pyrolysis system: pyrolysis section.

TABLE 2. Pyrolysis Section Equipment List

B-1002	Afterburner
J-401	Venturi
R-304	Char burner
R-305	Pyrolysis reactor
S-307	Cyclone
S-1010	Baghouse

3. Material and Energy Balances

The pyrolysis yield distribution is shown in Table 3. These yields and compositions are functions of reaction temperature and residence time. Data given in Figure 3 are single-pass pyrolysis reaction yields and not the overall process yields. As noted previously, gas and char are burned for process heating. The net materials leaving the pyrolysis system are Pyrofuel, water, ash, and flue gas.

The product summary of the overall process is shown in Table 4. The yield of Pyrofuel is 25% based on undried raw refuse. The recovered aluminum, glass, and ferrous metals constitute another 11%. The balance of the incoming refuse consists of off-gas, waste liquid, and unrecoverable materials which are landfilled.

The overall energy balance is shown in Figure 3. The process is energy self-sufficient and a net thermal efficiency of 31% can be achieved. The energy benefit ratio is 6.93.

Detailed material and energy balance information of the essential streams

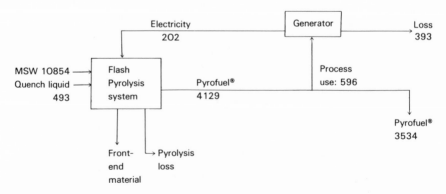

FIGURE 3. Overall energy balance based on higher heating values. Values are given in kJ/kg feed. MSW = municipal solid waste; net thermal efficiency = Pyrofuel product/(MSW + quench liquid) = 31%; energy benefit ratio = Pyrofuel/process use = 6.93.

TABLE 3. Products of Occidental Flash Pyrolysis

Yield at 950°F (500°C), based on dry weight
of feed to pyrolysis reactor

Char, 20%		C	48.8 wt %
		H	3.3
Higher heating value		N	1.1
8,200 Btu/lb		S	0.4
19,100 kJ/kg		Cl	0.3
		Ash	33.0
		O	13.0
			100.0
Oil, 40%		C	57.0 wt %
		H	7.7
Higher heating value		N	1.1
10,600 Btu/lb		S	0.2
24,600 kJ/kg		Cl	0.3
		Ash	0.5
		O	33.2
			100.0
Gas, 30%		H_2	12 mole %
		CO	37
Higher heating value		CO_2	37
380 Btu/scf		CH_4	6
15.0 MJ/Nm³		C_2H_4	3
		C_2H_6	1
		C_3	1
		C_4+	2
		H_2S	0.8
		HCl	0.2
			100.0
Water, 10%			

TABLE 4. Product Summary

Incoming refuse		Products	
Paper	34.13	Pyrofuel	
Garbage	2.13	(20% moisture)	20.93
Garden waste	12.40	Aluminum	0.84
Wood	1.73	Glass	4.56
Plastic	1.73	Ferrous material	5.35
Rubber	0.40		
Leather, textiles	1.07		
Ferrous metals	5.33		
Nonferrous metals	0.93		
Glass	5.73		
Miscellaneous	10.00		
Water	24.40		
TOTAL	100.00		

TABLE 5. Front-end

Stream number and name Component	1 Incoming refuse	2 Oversize reject	3 Primary shred feed	12 Air classifier feed	15 Air classifier overs	16 Screen feed
Organic	53.61	0.87	52.74	63.44	50.95	50.95
Ferrous metal	5.33	0.59	4.74	0.48	—	—
Nonferrous metal	0.93	—	0.93	1.22	0.12	0.12
Glass	5.73	—	5.73	6.15	2.03	2.03
Miscellaneous	10.00	0.30	9.70	13.25	4.37	4.37
Water	24.40	0.24	24.16	30.11	22.58	2.30
Total weight	100.00	2.00	98.00	114.65	80.05	59.79
Energy, kJ/kg dry	14357	10368	14452	15247	17064	17064
Ash, wt% dry	28.72	44.70	28.34	24.62	14.61	14.61
Moisture, wt%	24.40	12.21	24.65	26.26	28.20	3.85

ᵃBasis: 100 wt incoming refuse.

is given in Table 5 for the front-end and in Table 6 for the pyrolysis section. These data are sensitive to the incoming refuse composition and the operating parameters.

4. Product Characterization

4.1. Pyrolytic Oil

The 3.6 ton/day pilot plant generated sufficient pyrolytic oil, Pyrofuel, for product characterization and combustion tests.

The oil produced by Flash Pyrolysis is intended to be sold to electric utility power generating stations as a substitute for No. 6 residual (Bunker C) fuel oil. Table 7 shows some properties of pyrolytic oil compared to No. 6 fuel oil. There are several important differences. First, the sulfur content of pyrolytic oil is significantly lower. Second, since the carbon and hydrogen contents are lower, and the oxygen content is higher than No. 6 fuel oil, the heating value of pyrolytic oil is lower—about 77% of that of a typical No. 6 fuel on a volume basis. Finally, pyrolytic oil is more viscous than residual fuel oil. However, its viscosity is a stronger function of temperature than that of most No. 6 fuel oils. Therefore, although pyrolytic oil must be stored and pumped at 70°C, it can be atomized efficiently at 116°C, only 11°C higher than the atomization temperature for the usual No. 6 fuels.

Mass Balance[a]

21	23	25	8	40	36	37	42
Secondary shred	Pyrolysis feed	Air classifier heavy	Ferrous metal	Ferrous metal	Glass product	Landfill	Aluminum
41.12	45.77	12.47	—	0.01	0.16	0.17	0.10
—	—	0.47	4.47	0.47	—	—	—
—	—	1.10	—	—	—	0.11	0.71
0.20	0.36	4.12	—	—	4.40	0.52	0.03
0.88	1.66	8.88	—	—	—	4.34	—
1.84	2.05	7.53	—	—	—	5.86	—
44.04	49.84	34.57	4.47	0.48	4.56	11.00	0.84
18286	18046	11385	2261	2233	163	7551	4274
8.06	9.24	45.90	87.77	87.91	99.00	71.23	79.33
4.18	4.11	21.77	1.06	1.01	3.41	53.29	3.16

Three special characteristics of pyrolytic oil also result at least in part from the high oxygen content. First, pyrolytic oil is about 80% soluble in water. As a consequence, water is added to it to improve its handling properties by decreasing its viscosity. The second characteristic is that the oil is somewhat acidic, and therefore corrosive to mild steel. The acidity derives partially from carboxylic acids formed in pyrolysis and partially from HCl arising from organic chloride compounds, e.g., PVC in the refuse. The third characteristic, which is believed to stem from high oxygen content, is that the viscosity of pyrolytic oil increases, irreversibly degrading its handling properties, if it is

TABLE 6. Pyrolysis System Mass Balance[a]

Stream number and name / Component	23 Pyrolysis feed	49 Recycle char	53 Makeup liquid	57 Make water	58 Make glass	55 Pyrofuel
Gas	—	—	—	—	6.20	—
Pyrofuel®	—	—	—	—	—	16.75
Water	2.05	—	—	10.52	0.42	4.18
Quench liquid	—	—	0.89	—	—	—
Dry solid	47.79	14.39	—	—	—	—
Total weight	49.84	14.39	0.89	10.52	6.62	20.93

[a]Basis: 100 wt incoming refuse.

TABLE 7. Typical Properties of No. 6 Fuel Oil and Occidental's
Pyrolytic Oil

Property	No. 6	Pyrolytic oil
Analysis (wt %)		
C	85.7	57.0
H	10.5	7.7
S	0.7–3.5	0.2
Cl	—	0.3
Ash	0.05	0.5
N	2.0	1.1
O	2.0	33.2
Specific gravity	0.98	1.30
Energy content		
Btu/lb	18,200	10,600
kJ/kg	42,300	24,600
Btu/gal	148,800	114,900
kJ/l	41,500	32,000
Pour Point °F	65–85	90[a]
°C	18–29	32[a]
Flash point °F	150	133[a]
°C	66	56[a]
Viscosity		
SSU[b] at 190°F	340	1,150[a]
N sec/m² at 88°C	0.064	0.23[a]
Pumping temperature		
°F	115	160[a]
°C	46	71[a]
Atomization temperature		
°F	220	240[a]
°C	105	116[a]

[a]Pyrolytic oil containing 14% water (market quality).
[b]SSU = Saybolt universal viscosity.

maintained at elevated temperature for any appreciable length of time. Thus, a user should keep the oil at or below about 70°C until just before atomization.

Successful tests of the combustion characteristics of pyrolytic oil were performed by Combustion Engineering, Inc., who reported the following:

Pilot scale laboratory tests indicate that pyrolytic oil or blends of pyrolytic oil with No. 6 fuel oil can be successfully burned in a utility boiler with a properly designed fuels handling and atomization system.

Ignition stability with the pyrolytic oil and with the blends was equal to that obtained with No. 6 alone. Also, stack emissions, when burning pyrolytic oil and blends, indicated negligible amounts of unburned carbon at excess oxygen levels over 2%.

The combustion of pyrolytic oil blended with No. 6 fuel oil was successful despite that the two are not miscible, i.e., the blend is a dispersion of very small droplets. Firing such a blend in a utility boiler has two important advantages. First, laboratory studies have shown that blends greatly dimish the corrosive effect of the pyrolytic oil on mild steel. Second, pyrolytic oil can serve to dilute the sulfur content of a No. 6 oil which otherwise would be marginally unacceptable.

4.2. Glass

Over 70% of the glass contained in the received refuse is recovered as a sand-sized, mixed-color cullet of well over 99% purity. This cullet is suitable for use as a raw material for glass container manufacture. Industry tests have shown that cullet requires about 15% less energy to melt than do the raw materials used in making container glass. The cullet can be shipped at a drained-dry moisture content of 5–10%, or it can be dried before shipment, depending on freight rates and the customer's handling requirements.

Furnace tests by the Glass Containers Corporation established that the small particle size necessitated only minor modifications in their cullet handling procedures. Hand-blown cruets and molded beverage containers, made from the cullet by Owens-Illinois, were free of stones or other defects. In addition, manufacturers' tests indicated that the mixed-color cullet can be used to the extent of at least 20% in an amber container batch and at least 30% in a green container batch without significantly affecting the color of the product bottles.

4.3. Ferrous Metal

Separation of magnetic metals in the Occidental system is accomplished by commercially available electromagnet separators. The major contaminants are entrained organics and tin and lead from tin cans. Thus, an attractive market for this material is a detinning processor who chemically recovers the tin and sells the detinned steel to a steel mill, usually as No. 1 Dealer Bundles.

4.4. Aluminum

Because the nonferrous metals recovered by the RECYC-AL separation system include other materials in addition to aluminum beverage can stock, the product is best suited for sale to a producer of secondary aluminum alloys. The major impurities are zinc and copper. Since many alloys contain zinc and copper in small amounts, the secondary producer can blend the refuse-derived aluminum with scrap aluminum from other sources to achieve the desired alloy compositions.

5. Further Applications

5.1. Solid Refuse Derived Fuel

The front end of Occidental's resource recovery system is proven and is ready for commercial implementation. The recovered products are ferrous metal, glass cullet, aluminum, and a dry organic solid boiler fuel. The advantages of this approach over the complete Occidental system including pyrolysis are that it is available now, and that, lacking the additional thermal processing, it is inherently capable of a higher net energy recovery from the refuse. The disadvantage of a dry solid fuel approach is that it is a low-density and high-ash fuel compared to Pyrofuel.

5.2. Flash Pyrolysis of Industrial Wastes

Occidental submitted a final report to the U.S. EPA presenting the results of a laboratory, pilot plant, product evaluation and engineering evaluation contract to study the pyrolytic conversion of Douglas fir bark, rice hulls, grass straw and animal feedlot waste to synthetic fuel oil and char products.[3] Good quality synthetic fuels were obtained from all feedstocks except animal waste, which yielded a high nitrogen oil, and a byproduct char containing high concentrations of sodium and potassium salts.

Combustion tests conducted on the pyrolytic oils from tree bark and rice hulls were conducted at an outside laboratory to determine flame stability and pollutant emissions which would be associated with their use in industrial and utility boilers. The results showed that pyrolytic oils from these waste materials can be successfully used as a substitute or supplemental fuel. Other tests indicated that bark char and rice hull char could probably be processed to a modestly attractive, throwaway activated carbon. The tree bark char was shown to be suitable for the manufacture of good-quality barbecue briquettes.

5.3. Gasification

Much attention has been given recently to the concept of producing methane, methanol, ammonia, or other valuable chemical raw materials by thermal processing of municipal refuse or other organic wastes. Early research at Occidental established that simply by operating the bench-scale pyrolysis reactor at the higher temperature of 790°C, good yields of a high heating value gas are obtained. In a typical experiment, 80% weight conversion to gas was obtained.[8] The rapid heating of the fine organic particles and the short pyrolysis residence time are the critical factors in achieving these favorable results.

References

1. K. Pober and H. Bauer, From garbage—oil, *Chemtech, 7,* 164–169 (1967).
2. G. T. Preston, Resource recovery and Flash Pyrolysis of municipal refuse, in: *Clean Fuels from Biomass, Sewage, Urban Refuse and Agricultural Wastes,* proceedings of the Institute of Gas Technology Symposium, Orlando, Florida (Jan. 27, 1976).
3. F. B. Boucher, E. W. Knell, G. T. Preston, and G. M. Mallan, Pyrolysis of industrial wastes for oil and activated carbon recovery, EPA-600/2-77-091 (May, 1977).
4. E. J. Cone, Disposal of sewage sludge and municipal refuse by the Occidental Flash Pyrolysis Process, in *Advanced Thermal Processes for Conversion of Solid Wastes and Residues* (J. L. Jones and S. B. Radding, eds.) ACS Symposium Series 76, American Chemical Society, Washington, D.C. (1978), pp. 287–306.
5. Occidental Research Corporation, Final Report to EPA on San Diego Project (March 1979).
6. B. Morey, J. P. Cummings, and T. D. Griffin, Recovery of small metal particles from non-metals using an eddy current separator: Experience at Franklin, Ohio, presented at AIME 104th Annual Meeting, New York (1975).
7. B. Morey and S. Rudy, Aluminum recovery from municipal trash by linear induction motors, presented at AIME 103rd Annual Meeting, Dallas, Texas (1974).
8. G. M. Mallan and L. E. Compton, Gasification of Carbonaceous Solids, U. S. Patent, Re. 29,312 of Patent No.3,846,096 (1977).

10
Carboxylolysis of Biomass

T. E. Lindemuth

1. Basic Process Description

Biomass liquefaction or carboxylolysis is essentially the production of a liquid fuel by the reaction of pulverized biomass in a slurry medium with carbon monoxide in the presence of an alkaline catalyst. Reaction conditions require high pressure (150–250 atm), moderately high temperature (300–350°C), and residence time of 10–30 min.

1.1. Feedstock Applicability

The process is potentially applicable to any biomass material that can be dried, pulverized, and slurried. This includes wood, grasses, agricultural residues, municipal solid waste, and peat. Early development work stemmed from studies on coal with carbon monoxide and steam.[1] The relatively low reactivity of this system suggested that other feedstocks might be more attractive.

1.2. Potential Products

Work to date has shown that the expected product is a viscous liquid of rather low volatility. The boiling point range is generally between 200°C and 350°C. Depending on reaction conditions, the viscosity of the oil may be as low as 100 cp, but may also polymerize to a semi-solid at room temperature. These properties suggest that the product oil would be most suitable as a fuel for

T. E. LINDEMUTH • Bechtel National, Inc., 50 Beale Street, San Francisco, California 94119. Mailing address: P. O. Box 3965, San Francisco, California 94119.

industrial or institutional purposes. The low volatility would make its use as a transportation fuel impractical. Since even the solid material has almost no sulfur or ash, it could be used like coal, without the usual air pollution problem. Energy density of the product (solid or liquid) is approximately ten times that of the raw wood.

2. Background

As previously mentioned, early work on carboxylolysis was an outgrowth of experiments on coal liquefaction. In the U.S., the work was done mostly at the U.S. Bureau of Mines, Pittsburgh Energy Research Center (PERC).[2-4] This work indicated that both cellulose conversion and percent oil yield increased with increasing CO partial pressure. Initial experiments were conducted in batch autoclaves and these were later followed by continuous flow testing on a bench scale. The properties of oil–biomass slurries were studied during these tests. It was found that a 30% slurry of ground cellulose in oil was the highest concentration continuously pumpable. Additional experiments sponsored by The Environmental Protection Agency (EPA)[5] indicated that hydrogen might be a more effective reducing gas.

2.1. Albany PDU Design and Construction

Based on the results from the PERC studies, conceptual designs for both a commercial plant and a 3 ton/day process development unit (PDU) were developed.[6,7] This facility was erected in Albany, Oregon during 1975–1977. The simplified flow scheme for the PDU is shown in Figure 1. This flow scheme essentially duplicated bench scale apparatus used earlier at PERC but on a larger scale. In addition to the primary flow scheme, which uses feed in the form of an oil-based slurry, the Albany facility is equipped with a system for wood pretreatment to increase slurry concentration, and lock hoppers for dry solids feeding. Figure 2 is a picture of the feed preparation and reaction section of the Albany PDU. Following commissioning during 1977, a continuing series of development operations has taken place.

2.2. PDU Operational Results

The object of early testing on the PDU has been to verify the batch data from PERC on a continuous basis. The experiments were aimed to elucidate reaction characteristics and monitor major equipment performance. This was done by studying the effects of system temperature, pressure, CO concentration, residence time, and catalyst contacting. In addition, in order to further

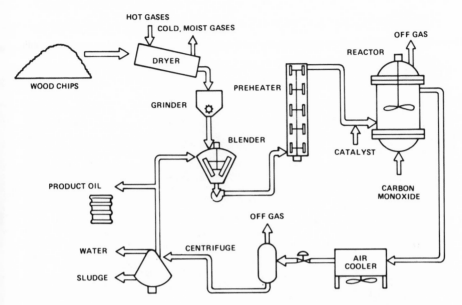

FIGURE 1. Albany PDU schematic (as built).

develop commercial design concepts, the mechanical performance of major process components was recorded.

Under the conditions tested, once-through wood conversion ranged from 50% to 80%. Conversion in this case is defined as the percentage of wood solids that disappear with a residence time of 5–10 min at temperatures above 275°C.

The product oil characteristics are compared to wood in Table 1. As can be seen, the product oil has nearly twice the energy content on a weight basis, and ten times on a volume basis.

The oil produced thus far has been viscous, with an occasional tendency to polymerize to a semisolid. This thickening of the recycle product oil has been the cause for terminating the test runs and is the subject of continued investigations.

2.3. Supporting Research

Currently, supporting research is being done at Battelle Northwest Laboratories in Richland, Washington and at the Lawrence Berkeley Laboratories (LBL) in Berkeley, California. The Battelle work is aimed at reaction mechanism development, PDU product characterization, and catalyst system

FIGURE 2. Albany feed preparation and reaction system.

TABLE 1. Comparative Analysis of Wood
and Carboxylolysis Oil

	Wood	Oil
% Water	50	—
% Carbon	24	77
% Hydrogen	3	6.5
% Nitrogen	0.1	0.4
% Oxygen	22.9	8.4
Heat content kcal/kg	2300	8500
Specific gravity	0.4	1.2

improvements. At LBL, new bench-scale work is being planned for investigating alternate reaction schemes and product upgrading.

3. Conceptual Process Description

Based on experimental work at PERC and Albany, a conceptual design of a commercial-size plant was developed by Bechtel in 1978.[8] While it was recognized that many process design questions remained at that time, the conceptual design served as a basis for preliminary economic analysis, as well as a guide for continued experimental work.

A major ground rule in the preparation of this conceptual design was to describe a process that used wood as its only feedstock and produced liquid fuel as its only product. In addition, an effort was made to minimize potential pollution problems by internal use of process byproducts.

Figure 3 illustrates the flow scheme for a 1000 ton/day (wet) plant. Biomass is the only raw material, and it is used for the liquefaction process, syngas production, and auxiliary heat production. The major processing sections of the plant are reaction, product separation, and syngas generation.

3.1. Reactor

Biomass in the form of wood chips is first dried, ground, and then slurried with recycled oil. Next, the wood–oil slurry is preheated to a moderately high temperature and pressure in the presence of a sodium carbonate catalyst solution, carbon monoxide, and hydrogen. The reaction mixture is then held in the reactor for a period of time to allow liquefaction to occur. Gases are separated from the oil and combusted to provide process heat for the preheater. Some of the catalyst is recovered and recycled from the product oil. The oil is recycled

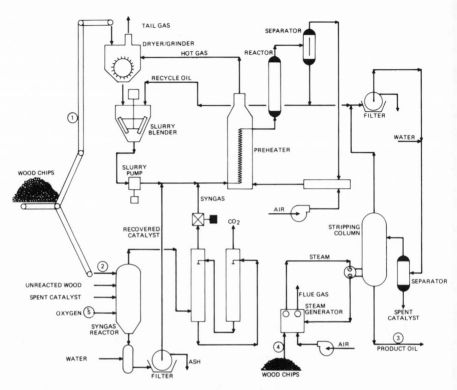

FIGURE 3. Waste-to-oil flow diagram. 1 = wood to liquefaction, 500 T/D; 2 = wood to syngas, 448 T/D; 3 = product oil, 170 T/D; 4 = wood to process heat, 26 T/D; 5 = oxygen to syngas, 151 T/D.

to be slurried with ground wood, and the rest is sent to the product separation section.

3.2. Product Separation

The liquid phase from the reactor section contains unreacted wood, catalyst, and insoluble solids that need to be removed. The stream is first diluted with a solvent and sent to a filter drum for solids removal. The recovered solids are sent to the gasification section for disposal. Water is added to the filtrate to remove the water-soluble catalyst. Oil and water are separated through a phase separation. The catalyst bearing water is sent to the gasification section for catalyst recovery. Solvent is stripped from the oil stream in a distillation column. The solvent is recycled for oil dilution and the product oil is transported to its final commercial use.

3.3. Gasification

Wood is partially oxidized to form carbon monoxide and hydrogen (syngas) used in the liquefaction process. The syngas reactor is also used to regenerate and concentrate the catalyst for recycle. The gas stream produced in the reactor is stripped of all carbon dioxide by an absorption and a desorption column. The purified carbon monoxide and hydrogen gas stream is compressed and sent to the reactor.

3.4. Design Basis

The following liquefaction stoichiometry was assumed [Equation (1)]:

wood + carbon monoxide → synthetic oil + carbon dioxide

$$C_6H_{9.1}O_{4.12} + 2.625\ CO \rightarrow C_6H_{9.1}O_{1.495} + 2.625\ CO_2 \qquad (1)$$

It was assumed that the only reaction which occurs is a reduction of the wood by CO. Other reactions such as dehydration (during heatup and cooldown) and polymerization or condensation reactions were ignored for the sake of this design. It was also assumed that as much H_2 is consumed as is generated in the water–gas shift reaction, causing no net change in hydrogen content. It is realized that this stoichiometry is highly simplified, but the available data do not justify introducing anything more complicated.

Wood is used for conversion to oil, as a CO source, and for steam generation. The design basis for CO production is the Union Carbide Purox® process. The Purox process is based upon the partial oxidation of organic material, such as wood, to produce a medium-Btu fuel gas composed of CO, H_2, CO_2, and H_2O.

Catalyst recovery is a unique feature of the process. The catalyst-bearing water from oil washing is sent to the syngas furnace where the catalyst is regenerated and discharged as a molten slag with other inorganics from the wood. The catalyst is recovered from the discharged slag by redissolving the soluble portion with water.

Oil separation from unreacted wood involves product oil dilution with a low boiling solvent for viscosity reduction. The diluted oil–wood–solvent mixture is filtered to remove unreacted wood and other solids. Previous developments in removing unreacted coal from liquified-coal product indicates that a precoated vacuum drum filter should be successful for removing unreacted wood particles.

3.5. Conceptual Plant and Equipment Description

The following paragraphs further describe the equipment and layout of the conceptual liquification plant.

3.5.1. Wood Preparation

Approximately one-half of the total wood fed to the plant is consumed to form oil; the remainder is used to produce carbon monoxide, hydrogen, and process heat. The total wood consumption is assumed to be 1000 TPD for a commercial-sized plant. A 100-day storage of wood chips (50 mm nominal) is specified so as to compensate for severe winter months where transportation may be difficult.

A belt conveyer system beneath the wood pile is used to transport wood to a distribution tower at the rate of 1000 TPD. From this tower wood is distributed to three sections of the plant. Approximately one-half of the wood is sent to the syngas production section. A small fraction of wood is sent to a wood-fired boiler for steam generation, and the remaining fraction is sent to the wood preparation section for drying, grinding, and oil slurrying.

The wood sent to the wood preparation section is conveyed to a feed bin where it is conveyed into a dual dryer/grinder system. Flue gas from the reactor section is diluted with air to 500°C and is used as the drying gas. The wood is dried to a 10% moisture content. The ground wood is pneumatically conveyed through an air classifier sized for 100 microns. The ground wood is separated from the gas steam by a cyclone followed by a bag filter. The resulting 90°C tail gas is vented to the atmosphere.

Ground wood collected from the cyclone and bag filter is metered by a star valve, to a continuous weighing mechanism, and into a wood–oil blender. Here, the ground wood is slurried with hot recycled oil from the reactor section. A wood concentration in the slurry of 30% has been chosen, since Albany experience indicates that 30% is the maximum wood concentration that will ensure slurry flowability. This slurry is sent to the reactor for oil conversion. Table 2 lists the equipment for wood preparation.

TABLE 2. Wood Preparation Equipment

Conveyor: wood pile to distribution tower 1200 mm × 150 m
Front-end loaders (2): mobile equipment
Distribution tower hopper system
Conveyors (2): tower to syngas bins 900 mm × 90 m
Conveyors (2): tower to grinder/dryer 900 mm × 90 m
Conveyor: tower to wood-fired boiler 600 mm × 60 m
Dryer/grinder package: feed bin, dryer/grinder, classifier, cyclone, bag filter, blowers (2), and surge bin
Wood scale: 300 TPD
Wood–oil blender with agitator: 35,000 liter
Wood–oil blender slurry feed pumps and drives (2): 750 liter/mm, 200 hp
Oil–wood recycle positive displacement pump and drive (2): 600 liter/min, 100 hp

---

3.5.2. Syngas Production

The purpose of this section of the system is to provide a source of purified carbon monoxide and hydrogen reactants for the reactor.

Wood is conveyed by belt to a surge bin which feeds into a hopper and then into the top of the syngas shaft furnace. The syngas furnace was developed and demonstrated by Union Carbide. It is presently being marketed as the Purox process. The Purox process is basically a partial oxidation reaction by which organic material can be gasified to form CO, H_2, CO_2, and H_2O. Pure oxygen supplied by an on-site oxygen plant is introduced into the bottom of the furnace where a partial oxidation reaction occurs at $1650°C$. The small amount of ash in wood will leave with other added inorganics (i.e., Na_2CO_3) as molten slag in a catalyst waste stream from the products separation section. The catalyst-bearing molten slag is water quenched and sent to the catalyst recovery section.

The syngas from the shaft furnace passes through a cyclone to remove entrained solids for recycle and/or disposal. The syngas is sent through a series of three water scrubbers to reduce gas temperature to $40°C$, thereby condensing light hydrocarbons and decreasing the water content of the gases. The scrubbers also remove any particulate matter from the syngas. Syngas initially enters a spray tower where the gases are quenched with water. The cooled gases exit through the top of the spray tower into a venturi scrubber where any remaining particulates are removed. The water droplets are coalesced and separated from the syngas by a nonreversing cyclone. The syngas passes upwards from the cyclone into a flooded mesh demister pad; then it passes through a large surge tank where any entrained water droplets are separated. The surge tank also serves to dampen any pressure fluctuations due to process upsets.

The scrubbed syngas is compressed from atmospheric pressure to 14 atm via two centrifugal compressors (1880 hp). The compressed syngas is sent to a K_2CO_3–CO absorber column. The CO_2-free gas stream is then sent to the reactor section. The resulting potassium bicarbonate solution is reduced in pressure and reheated to release CO_2 in a packed stripping column. CO_2 and H_2O vapor are vented to the atmosphere. The regenerated K_2CO_3 solution is recycled to the absorber column. Table 3 lists the equipment for syngas production.

3.5.3. Reaction Process

The wood–oil slurry, taken at $200°C$ from the wood preparation section, is pumped to 230 atm via a water-cooled diaphragm pump. Catalyst solution is combined with the wood–oil slurry. CO–H_2 gas mixture from the syngas section is compressed to 270 atm and introduced into the wood–oil catalyst slurry. The reactant mixture is charged into the reactor which essentially con-

TABLE 3. Syngas Production Equipment

Wood surge bins (2): 3000 mm diameter × 15 m
Star valves (2)
Feed hoppers (2), with screw conveyers
Shaft furnaces (2): 5 m diameter × 15 m
Slag quench tank (2): 2500 mm diameter × 3.6 m
Slag quench tank centrifugal pumps (2): 75 liter/min
Cyclones (2), with star valve-screw conveyers (2)
Solids hopper (2)
Spray towers (2): 1800 mm diameter × 9 m
Venturi scrubbers (2)
Cyclone/demister pads (2)
Surge tanks (2): 3.6 m diameter × 7.6 m
Compressors (4): 1740 BHP
Absorber column: 1800 mm diameter × 30 m; packing, 1800 mm high, 2-inch saddles
Regenerator column: 2500 mm diameter × 30 m; packing, 1600 mm high, 50 mm saddles
Pumps (3): 2000 liter/min, 200 liter/min, and 40 liter/min
Heat exchangers: (3) 185 m²; (2) 95 m²
Separator tank: 900 mm diameter × 1800 mm
Reboiler heater: 2 million Btu/hr

---

sists of two parts. The first part of the reactor consists of a single vertical helical coil preheater in a cylindrical brick-lined furnace, 4 m in diameter and 15 m high. The second part consists of a large vertically oriented pressure vessel, 1.2 m in diameter and 15.2 m high.

The reaction mixture enters the heater portion of the reactor and is heated to 340°C. The heated mixture is sent to a hold vessel for reaction at 340°C and 230 atm for 20 min. The reacted mixture is reduced in pressure via two flash tanks in series. This means of pressure reduction will minimize the tendency of the oil to foam. Gases from the flash tanks are sent through a surge tank to a series of three water scrubbers similar to those discussed in the syngas production section. The scrubbed gases are combusted with air at 980°C and diluted with recycled gases from the wood preparation section. The 815°C gases are sent to the reactor heater coil furnace and exit at 630°C. The furnace exit flue gas is divided into two streams. One stream is sent to the wood preparation section for wood drying. The other stream is sent to the syngas production section to supply process heat to the K_2CO_3 regenerator column reboiler.

After the reaction mixture is flashed down, the oil stream from the second flash tank is split into a recycle oil stream and a raw product oil stream. The recycle oil stream is sent to the wood preparation section for slurrying with wood. The raw product oil stream is sent to the product separation section.

PERC and Albany data seem to indicate that a plug flow reactor may be advantageous with respect to reaction kinetics. In addition, experience at

Albany has shown that operating equipment with moving parts under the temperatures and pressures required is expensive and unreliable. Table 4 lists major equipment in the reactor section equipment.

3.5.4. Product Separation

From the reaction section, the impure liquid phase is partially cooled with water and routed to a flash tank. The water drops the oil phase temperature to 120°C and is recovered as low-pressure steam. The cooled oil is then diluted with xylene to reduce its viscosity. This mixture, which contains $\frac{1}{2}$–$\frac{1}{3}$ oil is filtered through a precoat vacuum filter. The solids from the filter are stripped of hydrocarbons in a hollow-flyte heated screw conveyor. The dried solids are sent to the syngas furnace.

The filtered oil–solvent stream is next contacted with water to extract the water-soluble catalyst. The two remaining phases are separated in a coalescing separator. The aqueous phase is routed to the syngas furnace for catalyst recovery. The organic phase goes to a stripping column for solvent removal and recycle. The bottoms from the column are the plant product. Table 5 lists the major product separation equipment.

3.5.5. Plant Balance

In addition to the major equipment sections already described, the conceptual design included catalyst recovery, wastewater treatment, and steam generation. The facility was laid out on a plot approximately 350 × 200 m. An artist's rendering of this layout is shown in Figure 4.

TABLE 4. Reactor Section Equipment

Syngas compressor: motor driven, 1800 Hp
Preheater: helical coil, flue gas heated in refractory lined vessel, inconel coil
Reactor: 100 mm ID[a] × 15 m high, steel construction with 316 SS[b] cladding internal
Syngas burner
Flash tanks (2): 1800 mm ID × 5.5 m
Surge tanks (2): 1800 mm ID × 5.5 m
Air blowers (2): 700 and 280 m³/min
Oil–water separator: 4000 liter
Spray tower: 1800 mm ID × 9 m high
Venturi scrubber
Cyclone separator
Water storage tank: 75,000 liter
Water pump: 400 liter/min
High-pressure slurry pumps (2): 700 liter/mm, 600 Hp

[a]ID = internal diameter.
[b]SS = stainless steel.

TABLE 5. Product-Separation Equipment

Oil pumps (2): 200 liter/min
Oil cooler/flash tank: 2m ID × 4m
Waste heat boiler: 2000 kg/hr steam production
Oil surge tank: 75,000 liter
Solvent pumps (2): 600 liter/min
Oil pumps (2): 400 liter/min
Vacuum drum filter: 10 m^2 area
Heat exchanger: 10 m^2, stainless steel
Hollow flyte screw conveyor
Coalescing separator: 2000 liter
Stripping column: 2.5 m ID × 26 m
Solvent storage tank: 75,000 liter
Column reboiler
Miscellaneous pumps: 400 I/M (2), 200 I/M (2), and 40 I/M (2)
Heat exchanger: 100 m^2
Electrostatic coalescer/settler: 4000 liter
Oil storage tanks (6): 75,000 liter

4. Process Efficiency

The net thermal efficiency (NTE) for a conversion process can be defined as the energy content of the product divided by the gross plant energy input (including equivalent thermal energy for electric power generation). Based on the flows shown in Figure 3, the NTE for a commercial plant is expected to be between 50% and 60% when all losses are accounted for. Considerable uncertainty remains with respect to product separation energy needs which will affect the final plant efficiency.

5. Conceptual Economics

The conceptual design just discussed was used by Bechtel[8] to develop a preliminary economic analysis. This analysis was based on plant capacities of 1000 ton/day and 5000 ton/day. In addition to developing capital equipment costs, this analysis covered subcontracts, construction labor, indirect field costs, and engineering services. The total capital investments are approximately $50 million and $150 million for plants of 1000 ton/day and 5000 ton/day, respectively. It should be borne in mind that while there appears to be substantial economy of scale in this capacity range, the cost of biomass delivered to the plant is likely to increase as capacity grows.

Operating costs, as illustrated in Table 6 were also developed. This anal-

FIGURE 4. Artist's rendering of a 1000-ton/day waste-to-oil plant.

ysis, based on 100% capital financing, showed present day, break-even costs of product oil ranging between $25/bbl and $55/bbl, depending on plant capacity and delivered biomass price. As such, it is likely that continued process improvements through the development program at Albany, Oregon, and the rising price of imported oil may make liquid fuel from biomass carboxylolysis economically competititve.

TABLE 6. Biomass Carboxylolysis Conceptual Plant Estimated Operating Costs

	Cost in $1000s	
Item	1000-ton/day plant	5000-ton/day plant
Operating labor	730	1,460
Maintenance, labor, and materials	1,120	2,240
Supervision	420	600
Administration and overhead	550	790
Supplies	670	1,350
Utilities	1,060	3,180
Local taxes and insurance	1,750	5,380
Feedstock ($20/ton)	3,300	16,500
Total Operating Cost	9,600	31,500
Fixed Cost @ 10%/Year	5,180	15,400
Total Annual Cost	14,780	46,900

References

1. Anonymous, Solubilization of low rank coal with carbon monoxide and water, *Chem. Ind. (London)* No. 47 (Nov. 22, 1969).
2. H. R. Appell, Y. C. Fu, E. G. Illig, F. W. Steffgron, and R. D. Miller, Conversion of cellulosic wastes to oil, *U.S. Bur. Mines Rep. Invest.* 8013 (1975).
3. H. R. Appell, Y. C. Fu, S. Friedman, P. M. Yavorsky, and I. Wender, Converting Organic Wastes to Oil, *U.S. Bur. Mines Rep. Invest.* 7560 (1971).
4. S. Friedman, H. H. Ginsberg, I. Wender, and P. Yavorsky, Continuous Conversion of Urban Refuse to Oil Using Carbon Monoxide, U.S. Bureau of Mines, 3rd Mineral Waste Utilization Symposium, IIT Research Institute, Chicago, Illinois (March, 1972).
5. J. A. Kaufman and A. H. Weiss, Solid Waste Conversion; Cellulose Liquefaction, Worcester Polytechnic Institute for U.S. Environmental Protection Agency, EPA Contract 670/2-75-031 (Feb., 1975).
6. Dravo Corporation, Blaw Knox Chemical Plants Division, Economic Feasibility of Converting Wood Wastes to Oil, prepared for U.S. Bureau of Mines, Bruceton, Pennsylvania (June, 1973).
7. The Rust Engineering Co., U.S. Bureau of Mines Wood-to-Oil, Pilot Plant—Final Design Report, prepared for U.S. Bureau of Mines (Feb., 1974).
8. Bechtel National, Inc., Biomass Liquefaction Project, Albany, Oregon: Final Technical Progress Report, for U.S. Department of Energy, Washington, D.C. (April, 1978).

11
The Tech-Air Pyrolysis Process

CARL F. POMEROY

1. Introduction

Research on the Tech-Air pyrolysis system began in 1968 as a method for disposing of peanut hulls. The work at the Engineering Experimental Station of Georgia Tech University involved several steps, from experiments with a bench-scale reactor to the construction and operation of a demonstration plant. Each step involved the testing of several different feed materials such as peanut hulls and sawmill wastes.[1,2]

Results from the demonstration plant indicated it may be possible to develop this pyrolysis process commercially. The American Can Company, owner of Tech-Air, has granted several biomass conversion licenses and is currently (April, 1981) negotiating for additional licenses with a number of parties within the United States and overseas. In addition, Tech-Air is preparing to execute an exclusive U.S. license for the use of an improved version of the system for producing activated carbon.[3,4]

A brief summary of the development of the Tech-Air pyrolysis system, along with available process operations and product yield data are presented.

2. Bench-Scale Reactor

Work on the bench-scale reactor was undertaken to demonstrate the technical feasibility of disposing of peanut hulls by pyrolysis without the pollution

CARL F. POMEROY • Plains Resources Inc., Suite 2000, 2200 Classen Blvd., Oklahoma City, Oklahoma 73106.

problems usually associated with incineration. The system, approximately five-feet tall, was first operated on a batch basis and then on a continuous basis with a manual input feed. The material to be tested was put into a metal pyrolysis tube which was then capped. The closed unit was then placed in a Lindberg tube furnace where it was heated to the desired temperature for a predetermined length of time. The pyrolysis gas was fed to a condenser where the condensable organics and water were removed from the gases.

Experiments were conducted with a variety of feeds under different operating temperatures. The gases were analyzed by gas chromatography. Results for cotton gin waste and for a mixture of pine bark and sawdust have been published previously.[1] A mass balance for the pine bark and sawdust mixture is given in Figure 1.

Hundreds of pounds of peanut hulls and other feed materials were tested during several months, and data were obtained which demonstrated the technical feasibility of an automated converter with a vertical, porous-bed design.

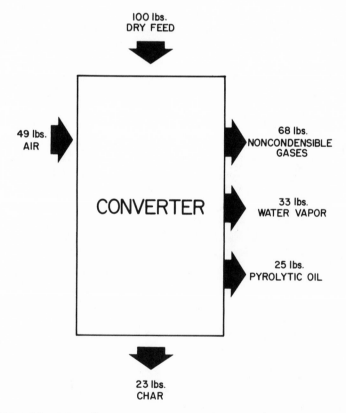

FIGURE 1. Mass balance[1] for the Tech-Air pyrolysis system with pine bark and sawdust.

3. Prototype Plant

The next step in the development of the Tech-Air process was to build and operate a prototype reactor, designated Blue I, approximately 11 ft in height. This unit incorporated the vertical-bed, gravity-fed, counter-current-flow pyrolysis chamber design with a continuous char output system. The pyrolysis reaction chamber was mounted on top of a water-cooled collection chamber. The char feedout was accomplished by use of a horizontal screw at the base.

The process was initiated by heat provided from a natural gas burner and was sustained by admitting a small amount of air low in the bed. The automated unit was constructed in 1971 and operated with a continuous feed rate of 4000 lb/hr. The plant was designed to prove the technical feasibility of a larger automated unit and was not intended for operating a long time. Various materials were tested as feedstocks over the approximately twelve months of operation. Successful continuous operation with simple means of control was demonstrated, allowing the construction of a larger automated unit.

4. Pilot Plant

An improved pilot plant unit, Blue II, was designed and was constructed in 1971 at the Engineering Experimental Station on the Georgia Tech campus. The plant was modified and upgraded several times and was completely rebuilt in 1975, allowing the processing of 500–800 lb of feed per hour (depending on the feed density).

The pyrolysis chamber was built for longer life at higher temperatures, with the use of refractory walls. The unit included a waste receiving bin, a belt conveyor for feed through a rotary air lock into the top of the convertor, the pyrolysis convertor and the char handling system, an off-gas cyclone, an air cooled condenser with a bypass demister, a draft fan, and a vortex after-burner. The off-gas draft fan permitted the operation of the pyrolysis chamber below atmospheric pressure, thus allowing the installation of simple weighted doors for pressure relief on each individual component.

The unit was tested with various types of wastes, including peanut hulls, wood chips, pine bark and sawdust, nonmetallic automobile waste, municipal wastes, macadamia nut hulls, and cotton gin washes. These tests generated data on the processing characteristics and the relative amounts and quality of the char, oil, and gaseous products.

5. Demonstration Plant

The next step in the development of the Georgia Tech system was to build a 50 ton/day demonstration plant. This was done in a woodyard in Cordele,

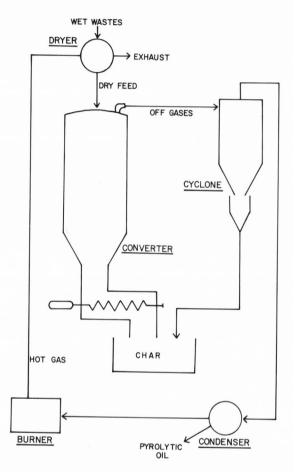

FIGURE 2. Tech-Air pyrolysis system flow diagram.

Georgia to avert the problem of waste transportation to the demonstration plant. The plant operated for several months, using wastes from the sawmill (including a five-month period when it ran continuously for 24 hours a day, five days a week). A flow diagram is given in Figure 2.

Successful operation of the plant demonstrated the feasibility of using the char produced as a raw material for briquettes. In addition, the oil was burned in several commercial applications. Wood waste was a good feedstock for the pyrolysis system. Servicing and cleaning of the off-gas system, due to the accumulation of tars and solids, was the major problem.

A program was initiated to improve the prototype plant and allow further study for continuous operation. This resulted in the construction of a new pilot plant, Blue IV, and the upgrading of the prototype plant. The servicing problem was drastically reduced by installing a scrubbing system using a portion of the condensed pyrolysis oil. Improvements allowed the operation of the plant on a 24-hour-per-day basis for eighteen months until June, 1977. All the char and pyrolysis oil produced during this period was sold on the bulk char and fuel oil markets.

The only available information on the energy balance is presented in Figure 3. No data have been released regarding the energy required to operate the system; the only data concerning process energy required is the heat input to the dryer. The heat requirement represents 9% of the heating value of the feed and is used to reduce the feed moisture from 37% to 4–5%. This required dryer heat would vary with varying feed moisture content. The feed material moisture content ranges from 20% to 55%, depending on weather conditions, season of year, and the proportion of sawdust in the feed.

The amount of heat required to dry the feed material is approximately 1600 Btu/lb water removed. Thus, a feedstock containing 50% moisture would

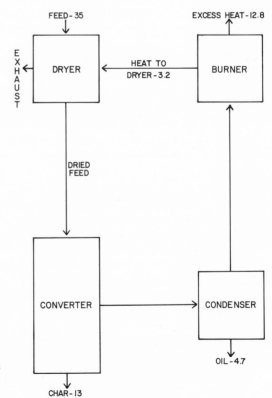

FIGURE 3. Partial energy balance for Tech-Air pyrolysis demonstration plant. Units are 10^6 Btu/hr.

require 800 Btu/lb wet feed for drying. This dryer energy requirement represents a loss of 1600 Btu/lb dry feed.

The energy balance indicates that 37.1% of the heating value of the feed is available in the char, 13.4% in the condensed oil, and 45.5% in the noncondensed gases. The remaining 4% represents process heat loss and the sensible heat of the products.[1]

Mass balance data are available for a mixture of pine bark and sawdust and these are presented in Figure 1. These data are typical for yields from wood wastes. A feed of 100 lb dry pine bark and sawdust is fed into the converter with 49 lb air to yield 23 lb char, 25 lb pyrolytic oil, 68 lb noncondensible gases, and 33 lb water vapor. These yields are an example and depend on the operating conditions and feedstock.[1]

6. Process Description

The specific process operating conditions depend on the nature of the feed material, method of process control, and desired products. A general range of operating conditions, representing various modes of operation tested with this process, is given.

The pyrolysis converter operating temperature ranges from 800 to 1400° F.[5] The process is designed to be flexible enough to operate at the temperature required to produce sufficient gas to dry the particular feedstock being used. If a feedstock with a relatively high moisture content is used, the converter is operated at a relatively high temperature to allow for production of a relatively high volume of gas. The feed material for the demonstration plant varies from 25% to 55% moisture and is dried to reduce the moisture content to 4–7% prior to being fed into the converter. This process allows a maximum feed moisture content of approximately 10% at this point.

In addition to being dried, the feed must be shredded to a size below 1 inch for continuous flow through the porous bed converter. Some materials require agitation for free flow and to prevent bridging in the converter.

The dryer is a compartmented, screw-conveyor type. The inlet temperature from the burner ranges from 400 to 600° F while the exhaust temperature ranges from 130 to 140° F. The dried feed is then conveyed to a storage bin which provides feed to the pyrolysis converter as required. Feeding introduction is accomplished through a rotary airlock to prevent pyrolysis gases from escaping.

The dried feed material is thermally decomposed into a char and hot gases in the pyrolysis unit. The hot pyrolysis gases flow upward through the vertical porous converter chamber while the char is discharged at the bottom of the unit into a sealed conveyor. The hot char is cooled with a water spray and passed through a rotary airlock into a char conveyor. The char is then carried to a storage bin where it is loaded by gravity flow for shipment.

The char output is accomplished by a mechanical screw-type mechanism. This device permits automation of the process by allowing the char output to be set at a predetermined rate with the feed input controlled by a bed level sensor. An alternate method of control is to select a char output rate and an output pyrolysis gas temperature while letting the bed level vary to maintain the selected temperature.

The pyrolysis gases exit the top of the converter at a temperature ranging from 350–500°F. An induced draft fan controls the pressure (near atmospheric) in the pyrolysis chamber and directs the flow of the gases leaving the converter. The hot gas stream contains noncondensible pyrolytic gases, vaporized pyrolytic oil, water vapor, and particulates. This gas stream is sprayed with cooled pyrolysis oil immediately upon leaving the converter. Spraying removes the particulates and condenses the pyrolytic oil by cooling the stream to a temperature between 180 and 200°F. Maintaining this temperature range is important to allow condensation of the oil without condensing an excessive amount of water vapor. The cooled gas stream then passes through a rotary demister to remove any entrained oil droplets. A portion of the gas is used for the dryer, with the remainder available for other possible use. A flare stack is used to dispose of excess gas and to allow flaring during startup. The oil from the condenser and demister is passed through a continuous filter before being pumped to a holding tank. The filter cake is conveyed back to the input feed system and fed back into the converter.

7. Product Yields

The Tech-Air pyrolysis system can result in a range of yields of char, oil, and gas. For a given feedstock, the yields of the various products are primarily controlled by the maximum temperature in the converter. In turn, this bed temperature is controlled by varying the air-to-feed ratio. If the air input rate is fixed, the ratio can be changed by varying the rate of feed input. The feed input is, in turn, controlled by the bed level sensor. Pilot plants have been operated with char yields ranging from 8 to 45 wt % of the dry feed. Table 1 gives yields for various modes of operation, using a feedstock consisting of a mixture of pine bark and sawdust.

This flexibility allows the process to be operated under conditions of maximum char, oil, or gas yield. This enables the production of pyrolytic fuels to meet on-site energy requirements (such as gas needed for drying) and maximize the products with the greatest market value.

If there is a need or desire to produce only oil and gaseous products with-

TABLE 1. Product Yields for a Mixture of Pine Bark and Sawdust[a]

Case	Feed (lb)	Air input (lb)	Char (lb)	Gas (lb)	Oil (lb)
Maximum char	100	31	35	78	18
Maximum oil	100	34	25	86	23
High gas	100	33	20	94	19

[a]Based on 100 lb dry feed.

out the production of char, the system's operation can be changed to provide for turning the char to the pyrolysis converter. This mode of operation allows the system to be used where no utility for the char is readily available, a technique which has been demonstrated in one of the pilot plants employing pine chip feed. The char yield was set at 20 wt % of the dry feed. The char was screened, and the course particles were reinserted into the converter. The finer particles passing through the screen and representing 3.8 wt % of the feed were disposed as ash.

8. Product Characteristics

The particular characteristics of the pyrolysis products are dependent on the choice of feedstock and operating conditions. The higher heating value of char produced from a mixture of pine bark and sawdust generally ranges from 12,300 Btu/lb at high char yields to 13,500 Btu/lb at low char yields. Density of the char is usually in the range of 10 to 13 lb/ft^3.

The pyrolysis gas is normally available for use at a temperature of approximately 200°F and has a moisture content of 30%. Typical gas properties at these conditions include a higher heating value of 225 Btu per actual cubic foot and a density of 0.05 lb/ft^3. The heating value at standard conditions can be calculated from these data, resulting in a value of 285 Btu/SCF on a wet basis and 408 Btu/SCF on a dry basis. Pyrolysis gases typically contain CO, requiring special piping and other handling precautions.

Pyrolytic oil is very viscous and difficult to handle. It is corrosive to mild steel because of its moisture content and high percentage of oxygen. A 26% water content is normally maintained to reduce viscosity. At this moisture content, the visosity ranges from 276 SSU at 68°F to 90 SSU at 150°F. Higher temperatures cause the oil to degrade with an irreversible increase in viscosity. The oil has a higher heating value of 9081 Btu/lb and a density of 9.88 to 10.27 lb/gallon. This is equivalent to a heating value of approximately 90,000 Btu/gallon, 60% of the heating value of No. 6 fuel oil.

9. Product Uses

Char produced from the Tech-Air pyrolysis plants has been sold for use in the manufacture of charcoal briquettes. It can be used as a fuel directly or by mixing it with coal. The char mixture results in lower sulfur and nitrogen oxides emissions than those obtained from coal alone. Tests have also demonstrated the successful use of pulverized char and No. 6 fuel oil. The char can also be used for producing activated carbon.

Pyrolysis oil has been blended with No. 6 fuel oil and burned directly as a boiler fuel. Produced oil has been sold for use in a cement kiln, a power boiler, and a lime kiln. Laboratory work has also indicated its potential utility as a chemical feedstock.[6]

The pyrolysis gas is a fuel and is best utilized to provide process heat and other site-specific applications. Preliminary studies have been conducted with a dry gas mixture that simulated the composition of pyrolysis gas as a fuel in an internal combustion engine. The gas mixture had a higher heating value of 180 Btu/ft^3. The power output of the engine was slightly over 60% of the output with gasoline as fuel.

10. Process Efficiency

There are a number of methods for determining the efficiency of a process. The method chosen here is to define the Net Thermal Efficiency (NTE) as:

$$NTE = \frac{\text{heating value of products} - \text{process energy}}{\text{heating value of feed}} \qquad (1)$$

Several different NTEs can be calculated using various definitions for "heating value of products" and "process energy." A discussion of the method to compare various processes has been given by Radovich et al.[7]; details of the calculations have been given by Pomeroy.[8] Data available on process energy for

TABLE 2. Product Energy Yields for a Mixture of Pine Bark and Sawdust[a]

Case	Feed (Btu)	Char (Btu)	Available gas (Btu)	Oil (Btu)	Dryer gas (Btu)
Maximum char	8700	4300	1628	593	1757
Maximum oil	8700	3300	2050	1193	1757
High gas	8700	2700	1743	2093	1757

[a]Based on 1 lb dry feed.

TABLE 3. Net Thermal Efficiencies and Energy Benefit Ratios for a
Mixture of Pine Bark and Sawdust

Case	Net thermal efficiency (%)	Energy benefit ratio
Maximum char	75.0	3.71
Maximum oil	75.2	3.72
High gas	75.1	3.72

the demonstration plant for drying a 37% moisture content feed to 4–5% moisture are given in Figure 3. A typical 50% moisture content feed would require more energy for drying. There would necessarily be some electrical energy required to operate the system. Most processes are designed to maximize liquid or gaseous products. It is difficult to compare those processes with one that is designed to produce a large amount of char.

Considering these limitations, the maximum limit for the NTE is (96 − 9)/100 = 87%. Since the only products available for sale are the char and pyrolytic oil, and since the gas is just being used for drying, the NTE is only 50.5%. It should be noted that these values are for the case with a 37% moisture content feed.

A further comparison between processes can be made by using a ratio that considers the quality of the fuel produced relative to the quality of fuel consumed. One way to make this comparison is to define an *energy benefit ratio* (EBR) ad the high-quality energy produced divided by the high-quality energy consumed. The only high-quality fuel produced for sale is the pyrolytic oil. Dividing the heating value of the oil produced by the gas consumed in drying gives an EBR of 4.12.

Table 2 presents a summary of the product energy yields for a mixture of pine bark and sawdust with modes of operation giving maximum char, maximum oil, and high gas outputs. The NTEs and EBRs for these cases are shown in Table 3, indicating that the total energy available from the products is essentially independent of the mode of operation.

References

1. J. A. Knight, M. D. Bowen, and K. R. Purdy, Pyrolysis—A method for conversion of forestry wastes to useful fuels, presented at the Conference on Energy and Wood Products Industry, Forest Products Research Society, Atlanta, Georgia (November 15–17, 1976).
2. J. A. Knight, J. W. Tatom, M. D. Bowen, A. R. Colcord, and L. W. Elston, Pyrolytic conversion of agricultural wastes to fuels, American Society of Agricultural Engineers, St. Joseph, Michigan. Presented at ASAE Meeting, Stillwater, Oklahoma (June 23–26, 1974).
3. American Can Company, American Lane, Greenwich, Connecticut, personal communication.

4. Silvicultural Biomass Farms, MITRE Technical Report Number 3747, Vol. 1–5, The MITRE Corporation/METREK Division, McLean, Virginia (May, 1977).

5. J. Alich, E. L. Capener, R. K. Ernest, N. Korens, and K. A. Miller, Program Definition for Fuels from Biomass, Stanford Research Institute, report to California Energy Resources, Conservation, and Development Commission, SRI Project #5527, Menlo Park, California (October, 1976).

6. J. Jones and S. Radding (eds.), *Advanced Thermal Processes for Conversion of Solid Wastes and Residues,* ACS Symposium Series 76, American Chemical Society, Washington, D.C. (1978).

7. J. M. Radovich, P. G. Risser, T. G. Shannon, C. F. Pomeroy, S. S. Sofer, and C. M. Sliepcevich, *Evaluation of the Potential for Producing Liquid Fuels from Biomaterials,* University of Oklahoma, EPRI AP-974, Electric Power Research Institute, Palo Alto, California (January, 1979).

8. C. F. Pomeroy, A Technical and Economic Evaluation of Selected Biomass-to-Fuels Conversion Processes, M. S. Thesis, University of Oklahoma, Norman, Oklahoma (April, 1979).

12

The Purox Process

ANIL K. CHATTERJEE

1. Introduction

The organic fraction of municipal, commercial, and industrial solid waste materials constitutes a potential energy resource. The Union Carbide Corporation (UCC) has developed a patented process—the Purox® system—to convert such wastes to usable medium-Btu fuel gas.

The Purox process is an oxygen-blown gasification process. The advantages of using oxygen instead of air are twofold. First, intense heat can be generated in the oxidation zone. This intense heat is used to melt and consolidate all the inorganics of the solid waste to flowable slag. Second, the absence of nitrogen dilution in the fuel gas, the Purox process produces the fuel gas having a higher heating value (HHV) of 11.8–15.4 MJ/NM^3 (300–390 Btu/scf). This substitute fuel gas can be used directly as fuel for combustion in furnaces, kilns, boilers, and other types of direct combustion equipment. It can also be used to produce methanol, ammonia, and light hydrocarbon fuels.

One important requirement for the operation of the Purox process is that an economical and reliable source of oxygen be available. UCC is noted for their development of a cryogenic air separation process to produce oxygen. For large commercial Purox process plants, such an oxygen plant is necessary. For a medium-capacity plant, UCC's Lindox Pressure Swing Absorption (PSA) system may be used.

UCC has a 184-tonne/day (200-short ton/day) Purox system plant at their facility in South Charleston, West Virginia. This pilot plant facility has been used to gain operating experience, collect performance data, gather

ANIL K. CHATTERJEE • SRI International, Menlo Park, California 94025.

scaleup design parameters, determine the desired maximum size of their modular unit, collect techno–economic data, and demonstrate the range of mixed feedstocks (e.g., refuse and sludge) that can be handled in the process.

2. Description of Process

The Purox reactor is a vertical shaft furnace and hence is a fixed bed updraft type of gasifier. The refuse is ram-fed into the reactor from a side feed opening located near the top of the reactor vessel. The fuel gas produced in the reactor is tapped from the top of the vessel. The reactor was originally designed to receive raw, unprocessed solid waste materials. However, operations with shredded processed refuse produce a fairly uniform bed and better operational capability.

The Purox process has three distinct process streams:

1. Preprocessing plant
2. Gasification and gas cleaning plant
3. Oxygen plant

The other operational subsystems are (1) processed feed material storage and receiving, (2) wastewater treatment, (3) utilities and services, (4) product fuel-gas handling, and (5) residue disposal.

The methanol or ammonia synthesis plant, which is an additional subsystem, is not discussed in this chapter.

2.1. Preprocessing Plant

This section discusses the design of a typical front-end processing facility that may be used for the Purox process. However, the actual sequence of processing operations and the selection of the type and operating features of the equipment do not depict the existing front-end processing facilities of UCC's South Charleston, West Virginia plant.

A typical schematic flow diagram of a preprocessing train[1] is shown in Figure 1. Refuse can be dumped on the tipping floor of the receiving building or the refuse can be received in a pit. For the design of the facility where the refuse is received on the tipping floor of the building, the storage capacity of the building would be one day's refuse pit; the capacity of the pit is generally calculated on the plant's rated capacity to process refuse from 4 P.M. on Friday to 10 A.M. on Monday. This design basis is required for a reactor operating 24 hr/day, 7 days/wk. Most refuse hauling is stopped after 4 P.M., and no refuse is generally delivered to the plant before 10 A.M. on weekdays. Very few cities

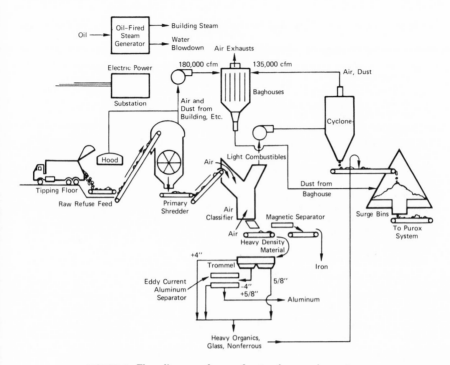

FIGURE 1. Flow diagram of purox front-end processing system.

have Saturday and Sunday pickups. However, most designers prefer the following plant operating schedules:

Receiving the refuse at plant	8 hr/day, 6 days/wk
Front-end processing	16 hr/day, 6 days/wk
Reactor operation	24 hr/day, 7 days/wk

The equipment systems required for a preprocessing plant is given in Table 1.

The preprocessing operation begins when the packer trucks, after weighing their loads, dump the refuse on the tipping floor. The front-end loader operator will coordinate the dumping operation and will sort out large objectionable items like bedsprings, carpet rolls, beddings, and large or heavy machine parts. The front-end loader operator pushes the refuse to the sunken apron conveyor leading to the shredder. The shredder reduces the refuse to a size between 10.2 and 20.3 cm (4–8 in). A high-speed belt conveyor receives the shredded refuse and transports it through a feed conveyor to the air classifier. The air classifier overhead stream is the light fraction of the refuse and

TABLE 1. Preprocessing Plant Equipment

Truck-weigh scale with remote or local printout of the packer truck load.
Printout with the date, identification of the packer truck, time of arrival, tare weight, and the
 load weight.
Front-end loader for the tipping floor.
An overhead crane for the refuse pit design system.
Conveyor system to transport the as-received refuse to the shredder.
Storage bin for controlled feeding of the shredded refuse to the air classifier.
Conveyor system to carry the shredded refuse to the air classifier.
Magnetic separator to remove the ferrous metals from the air classifier heavy fractions.
Trommel screen to separate the glass- and aluminum-rich streams.
An aluminum separation plant.
Glass recovery plant (if economically justified)
Surge bin to hold the light fractions of the air-classified organics.
Baghouse systems to clean all exhausts from the conveyers, shredders, and cyclones.
Power substation and distribution system.
Well-equipped maintenance shop.

---

consists mainly of organics. Some light metallics and glass would also escape with the light fraction stream. The light fraction stream is then blown into a cyclone separator, where the organics are separated and carried to the surge bin. The air carrying the light fractions to the cyclone is discharged to the atmosphere through a bag filter system. The baghouse receives, in addition to this stream, the dusts from the shredder and from the charging conveyor to the shredder. The heavy fraction or the underflow stream of the air classifier contains mostly inorganics such as metals, glass, stone, dirt, and ceramics, and some heavy organics such as wood, shoe soles, leather, and heavy plastics.

The air classifier heavy stream first moves under a drum magnet where the major portion of the ferrous metals are removed; the balance of the stream is conveyed to the trommel screen. In the trommel screen, the first drum may have 1.3-cm (0.5-in) openings and the second drum may have 10.2-cm (4-in) openings. The glasses will be separated in the first screen and the aluminum-rich stream will pass through the second screen. The glass-rich stream is 1.3 cm (0.5 in) in size whereas the aluminum-rich stream ranges from 1.3 cm (0.5 in) to 10.2 cm (4 in) in size. The 10.2-cm (4-in) stream is generally recycled to the tipping floor or taken to the reactor surge bin.

The aluminum-rich stream contains aluminum, ferrous metal, glass, and other nonferrous and organic refuse. A second-stage magnetic separator may be used to remove the ferrous metals. The aluminum contents of the stream may be separated by an eddy-current separator. The reject stream of the aluminum plant is carried to the surge bin, where it mixes with the light organic fractions of the refuse. The surge bin stores and supplies the feed to the Purox reactor at a controlled rate.

2.2. The Basic Process

The terms associated with thermochemical conversion processes include the following:

1. Pyrolysis Process. In a true pyrolysis process, a carbonaceous material is thermally cracked or decomposed to produce fuel gas, organic liquid, and char. This process is sometimes called destructive distillation, and it is completed in an atmosphere devoid of oxygen, steam, or hot CO. Currently, a thermal decomposition process in which the heat energy necessary for thermal cracking is supplied directly from partial combustion of the char or gases, is being called Pyrolysis Process. A carbonaceous solid, organic liquid, or gas can undergo pyrolysis or thermal cracking. A pyrolysis process is an endothermic reaction and requires an external heat source. In the Purox reactor, this heat energy is available from the oxidation of the refuse char which is located at the bottom of the reactor hearth.

2. Slagging Reactor. In a slagging reactor, the hearth temperature of the reactor is above the melting point of the ash. In such reactors, the ashes are tapped as molten slag to a quench tank.

3. Gasification. Gasification is the reaction that occurs between the refuse char with hot CO_2 and water vapor to produce a fuel gas. Char gasification processes, sometimes called reduction processes, are endothermic and require a supply of heat energy to complete the process. As in pyrolysis reactions, this heat is supplied by the combustion of char by oxygen injection to the Purox reactor.

4. Liquefaction. A pyrolysis process produces pyrolytic oil and gas. When the emphasis is on an oil rather than a gas yield, the process may be termed a liquefaction process.

2.3. Purox Gasification Scheme

Figure 2 is a schematic diagram of the Purox gasification scheme.[2] The main components of the process train are:

1. Feed conveyors to carry preprocessed refuse from the surge bin to the ram feeder hopper.
2. A ram feeder to feed the preprocessed refuse into the reactor.
3. Provision for preheating the oxygen by burning fuel gas in the tuyers.
4. Slag quench tank with drag chain conveyor for the removal of slag.
5. Fuel gas burner at the slag tap to keep the molten slag in flowable condition.
6. Water spray gas scrubber and recycling scrubber water cooler.
7. Wet electrostatic precipitator.
8. Solid–liquid separator to remove fly-ash from the oil and water emul-

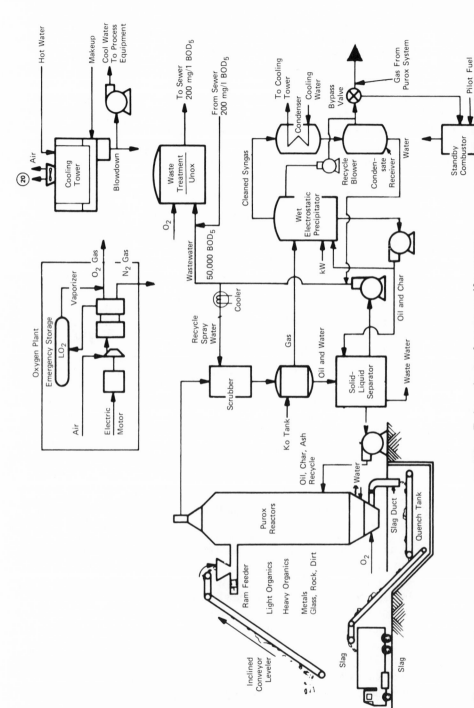

FIGURE 2. Train equipment for purox gasifier system

sifier. The system consists of a slurry tank, a vacuum filter, and a filter tank with associated pumps and piping.

9. Water-cooled condenser to separate the entrained water from the fuel gas.
10. Wastewater treatment facility.

The Purox gasification reactor is a counterflow system in which the refuse feed travels from the top toward the bottom and the hot gas ascends through the zones of oxidation, gasification, pyrolysis, and devolatilization. In the pyrolysis reactions, the carbonaceous materials of the refuse are thermally cracked in an oxygen-free atmosphere to produce the fuel gas. The composition and the quality of the fuel gas will depend on the composition of the refuse and its moisture content. The sequence of typical reactions in the reactor is noted in Chapter 13 of this book.

The fuel gas leaves the reactor at a temperature ranging from 93 to 315°C (200 to 600°F) with a wet bulb temperature range of 77–82°C (170–180°F).

Oxidation reactions occur between the preheated oxygen blown through the water-cooled tuyers of the reactor hearth and the char that is produced in the pyrolysis–gasification reactions. The heat generated in the oxidation process supplies the heat energy necessary to dry the refuse, complete pyrolysis–gasification reactions, keep the hearth at a temperature above 1482°C (2700°F), carry sensible heat in the fuel gas, and offer heat losses from the reactor.

The hearth is refractory-lined, the hearth walls and the slag tap components are water-cooled, and the hearth floor contains the pool of molten slag. The molten slag is generally tapped to a quench tank. The quench tank cools the slag and provides a positive seal to the reactor. The quench water is generally recycled and cooled to maintain the desirable quenching temperature. The molten slag forms into black aggregate material after the quenching operation.

The glass content of the refuse's inorganics help to reduce the melting point and viscosity of the molten slag. It is desirable to keep at least half the amount of glass in the feed to maintain the fluidity of the slag. The molten slag will freeze in the slag runner unless it is kept hot throughout the length of its travel in the runner. For this reason, slag tap burners, using fuel gas, are used to keep the slag in a flowable condition.

2.4. Gas Cleaning

The reactor offgas contains fly-ash, tar vapors, high-, moderate-, and low-molecular-weight organic liquids, many organic acids, and aromatics. Unless this gas is fired directly to a closely coupled combustion chamber, it must be cleaned. In the pilot plant facility at South Charleston, West Virginia, the gas is first cleaned in a spraying water scrubber. In the scrubber, most of the solid particulates are removed and, at the same time, the gas is cooled to near its wet-bulb temperature. The solids are collected in a knockout tank and the relatively clean fuel gas then enters the electrostatic precipitator. Here the last traces of pyrolytic oils and solids are removed. To produce dry fuel gas, the gas is then passed through a water-cooled shell-and-tube type vertical condenser to condense the entrained moisture from the gas. The gas is then ready for use as a medium-Btu-heating-value fuel for various combustion processes.

The collections in the knockout tank are discharged to the solid–liquid separation system. The system consists of slurry and filtrate tanks, a rotary drum vacuum filter with a polypropylene monofilament filter cloth to serve as the filtering media, a vacuum pump, and a vacuum receiver. The solid filter cakes containing oil and water are recycled to the reactor.

2.5. Wastewater Treatment

UCC uses its own Unox wastewater treatment technology. The Unox process uses an oxygen-activated sludge system to reduce the biological oxygen demand (BOD) level of the wastewater. In this system, oxygen is injected simultaneously with the process wastewater to the multistage dispersed-growth biological reactors. The treated water with an acceptable BOD level is then discharged into the city sewer.

2.6. Oxygen Plant

The Purox process required a pure oxygen supply at the reactor and the waste treatment facility. UCC's cryogenic generating plant includes an air liquefaction and fractional distillation process. The on-site oxygen-producing plant will require storage, handling, and transportation facilities for the process oxygen.

2.7. Gas Compression and Drying

After cleanup, the fuel gas is approximately at atmospheric pressure. For pipeline transmission, therefore, the gas must be compressed and dried. If an objectionable level of H_2S remains in the gas, a sulfur removal facility is needed.

2.8. Process Equipment

A vertical shaft furnace has dimensional limitations. Beyond a certain diameter, the flow characteristics through the refuse bed of the reactor become unsteady. Channelling incidents occur, and poor gas quality is achieved. These problems have prompted UCC to market the system in modular form. The maximum processing capability of a single reactor is about 350 ton/day. A modular design has the following advantages:

1. Satellite plants could be built close to the refuse generation point.
2. The breakdown of a single module will not idle the entire plant.
3. A routine maintenance schedule could be adopted for each module.
4. The capacity of a refuse processing plant could be augmented with the growth of the community.
5. The equipment system for a single module may be an off-the-shelf item of normal equipment design specification. Custom designing of the equipment may be eliminated.
6. Plant operation will be more efficient.
7. The truck traffic to the refuse plant can be kept to an acceptable level.

The single most important disadvantage of a modular plant is that the total investment cost becomes higher than that for a single large system.

The simplified dimensional specifications of the Purox Pilot Plant at South Charleston, West Virginia, having a processing capacity of 200 ton/day, are given in Table 2.[2]

3. Performance

A typical mass balance[1] for the preprocessing system is given in Table 3. A simplified mass balance for the Purox reactor[1] system is given in Figure 3. From the mass balance, a typical energy balance for the Purox reactor[1] system can be calculated (Table 4 and Figure 4).

The heat energy produced in the oxidation zone of the reactor supplies the sensible and latent heat to vaporize the free moisture of the feed refuse, the heat of pyrolysis, the sensible and latent heat to melt the inorganics, heat losses from the Purox reactor, and the sensible heat content of the Purox fuel gas.

Neglecting the energy equivalent of the shaft horsepower needed to run the overall Purox system, a conversion efficiency (refuse-to-fuel gas) can be calculated from the energy balance data [Equation (1)]. The overall plant thermal efficiency can also be calculated from these data [Equation (2)].

$$
\text{Conversion efficiency} = \frac{\text{Energy content of Purox fuel gas}}{\text{Total energy content of the refuse}}
$$

$$
= \frac{0.745}{0.917} = 81.24\% = \left(\frac{706.46 \times 10^6 \text{ Btu/hr}}{869.35 \times 10^6} \right) \tag{1}
$$

TABLE 2. Specifications of the Purox Pilot Plant

Plant section	Specifications
Shredder	A Heil (Tollemache) vertical shaft hammermill. The drive motor is 149.200 kW (200 hp), with a capacity of 13.61 tonne/hr (15 ton/hr).
Magnetic separator	Eriez drum type, suspended 10 in above the shredded refuse conveyor. The electromagnet operates at 480 V and is contained in a 102-cm (4-in) drum.
Feed conveyor	Apron type and equipped with refuse leveler.
Ram feeder	Contains hopper, vane, and two hydraulic rams.
Reactor	A vertical shaft furnace 305 m (10 ft) in diameter by 13.4 m (44 ft) high. The upper section is cylindrical and lined with a 10.2-cm (4-in) refractory wall, whereas the lower section is conical and refractory-lined. Fuel gas and oxygen are injected into the hearth through water-cooled tuyers. The hearth outer shell is water-cooled.
Quench tank	Measures 193 cm (76 in) deep and is filled with water. It provides a seal to the slag tap of the reactor. The water is recirculated to keep it from steaming.
Slag conveyor	A Taunton Engineering Company's drag chain conveyor driven by a 2.2-kW (3-hp) electric motor. The speed of the conveyor is 11.17 cm/sec (22 ft/min).
Scrubber	6.3 liter/sec (100 gal/min) of water is sprayed into the scrubber. The fuel gas passes through a gas–liquid separator 132 cm (54 in) in diameter by 18 cm (72 in) high. The gas exits horizontally and enters the electrostatic precipitator.
Electrostatic precipitator (ESP)	A research Cottrell tube-type unit with 60 tubes provides 9.76×10^8 cm^2 of collection area. Nominal gas flow rate is 2.95 m^3/sec. Power to the ESP is supplied with 450 V, A.C. This voltage is stepped up and rectified to supply the required D.C. input to the precipitator electrodes.
Condenser	A vertical shell-and-tube type heat exchanger. The gas is cooled to 26.6°C (80°F). The condensed water is piped to a scrubber slurry tank.
Solid–liquid separator	Consists of slurry and filtrate tanks, vacuum filter, and vacuum pumps. The vacuum filter is a rotary drum type. Polypropylene monofilament cloth is used as filter media.

TABLE 3. Typical Mass Balance

Component	Tonnes/hr	Tons/hr	Component	Tonnes/hr	Tons/hr
Input			Output		
Organics	70.62	77.87	Ferrous stream		
Ferrous metal	8.16	9.00	Fe metal	7.38	8.13
Aluminum	0.54	0.60	organics	0.30	0.33
Glass and ceramics	8.70	9.60	glass	0.06	0.07
Other nonferrous metals	0.30	0.33	Total	7.74	8.53
Dirt and stones	2.36	2.60	Aluminum	0.36	0.40
Total input	90.71	100.00	Feed to Purox Reactor		
			organics	70.34	77.54
			Fe metal	0.79	0.87
			aluminum	0.17	0.19
			glass	8.64	9.53
			other nonferrous	0.30	0.33
			miscellaneous	2.37	2.61
			Total output	82.61	91.07

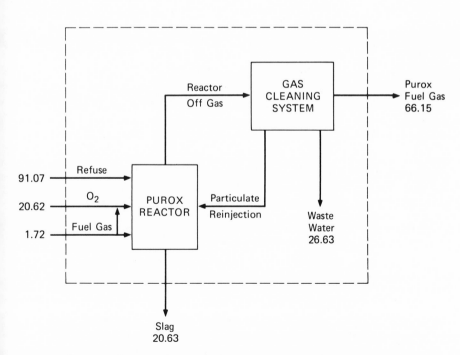

FIGURE 3. Typical mass balance purox system. Note: all data are in tons per unit time.

TABLE 4. Typical Energy Balance[a]

Component	Tonnes/hr	Tons/hr	MJ/kg	Btu/lb	MJ/hr	Btu/hr in millions
Input						
Refuse	82.62	91.07	11.1	4773	0.917	869.35
Fuel gas	1.56	1.72	69.77	30,000	0.108	103.20
Oxygen	19.61	21.62	0	0	0	0
Total					1.025	972.55
Output						
Heat loss to air from the reactor					0.022	20.77
Heat loss from the molten slag					0.034	32.38
Heat loss from hearth cooling					0.029	27.42
Heat loss in gas cleaning					0.029	27.58
Heat loss in waste treatment					0.167	157.94
Heat equivalent to product gas					0.745	706.46
Total					1.026	972.58

[a]Heat equivalent of shaft horsepower used in the process: 0.186 MJ/hr (176.46 × 10⁶ Btu/hr).

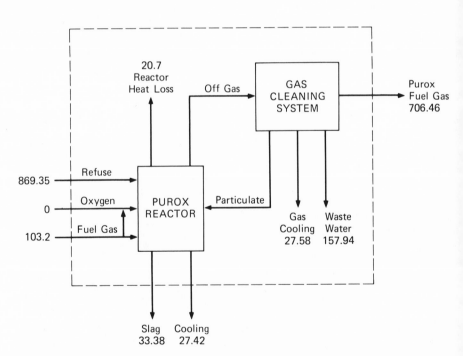

FIGURE 4. Typical purox system energy balance. Note: all data are in millions of Btu per unit time.

$$\text{Overall plant thermal efficiency} = \frac{\text{Energy content of product fuel gas} - \text{Energy spent to produce gas}}{\text{Energy content of the refuse}} \quad (2)$$

where

$$\text{Energy content of product fuel gas (per unit time)} = \text{scf of product gas produced} \times \text{HHV of the gas,} \quad (3)$$

$$\text{Energy spent to produce fuel gas (per unit time)} = \text{Energy equivalent of electric power used plus the fuel gas used in slag tap and tuyers, and} \quad (4)$$

$$\text{Energy content of refuse (per unit time)} = \text{HHV of refuse} \times \text{pounds of refuse processed in reactor to produce fuel gas.} \quad (5)$$

$$\text{Overall plant thermal efficiency} = \frac{(0.745 - 0.186 - 0.108)}{0.917} = 0.491$$

4. Case History

Some of the test data[1] from a typical run of the Purox Pilot Plant are tabulated in Table 5.

The substitute fuel gas given in Table 5 was produced from the thermal decomposition of the feedstock given in Table 6.

A typical slag analysis is given in Table 7.[1]

5. Economics

5.1. Construction and Operating Costs

The Purox system has not been built in a public facility and hence background information on construction and operating costs cannot be estimated. The EPA's final report, "Engineering and Economic Analysis of Waste to Energy Systems,"[2] presented an estimate of the capital and operational costs for a 1361-tonne/day (1500-ton/day) Purox Plant to produce synthesis fuel gas based on mid-1975 prices. There are various methods by which these prices

TABLE 5. Test Data from a Typical Run

	Volume percent, dry basis	
Composition of fuel gas	Typical	Range
H_2	23	21–32
CO	38	29–42
CH_4	5	4–7
C_2H_2	0.7	0.2–1.5
C_2H_4	2.1	1.3
C_2H_6	0.3	0.1–0.5
C_3	0.6	0.2–0.3
C_4	0.5	0.1–0.6
C_5	0.4	0.1–0.6
Higher hydrocarbons	0.5	0.25–1.4
CO_2	27	20–34
N_2 + Ar (with oxygen)	1.5	0–2.0
H_2S	0.05	0.02–0.06
CH_3OH	0.1	0.05–0.15
Organic vapor	0.2	0.1–0.4

Water content at $100°F$ = 6% by volume
Heating values
  HHV: 11.815–15.36 MJ/Nm³   (300–390 Btu/scf)
  LHV: 11.03–14.18 MJ/Nm³   (280–360 Btu/scf)

can be updated to the mid-1979 or any other year's cost-index level. Frequently, engineering cost estimates are made based on the chemical engineering plant cost index, published by the journal *Chemical Engineering*. For example, the *Chemical Engineering* price index for 1975 equipment and construction cost is 182.4 and for mid-1979 it is about 238. A ratio of 238/182.4 is then the multiplier for updating the construction cost items. The breakdown of the capital cost estimate is given in Table 8.

TABLE 6. Feedstock Composition

Ultimate analysis of refuse	Wt %	Ultimate analysis of refuse	Wt %
H_2O	26.00	S	0.10
C	25.90	Cl	0.13
H	3.60	Metals	8.00
O	19.90	Glass	11.00
N	0.47	Ash	4.00

HHV: 10.44 MJ/kg (4492 Btu/lb)

TABLE 7. Typical Slag Composition

Component	Wt% as oxides of the component
Silicon	59.2
Aluminum	10.5
Calcium	10.3
Sodium	8.0
Iron	6.2
Magnesium	2.2
Potassium	1.0
Miscellaneous	2.6

A breakdown of the construction cost is presented in Table 9.

A breakdown of the site improvement costs estimated in the EPA report[2] is given in Table 10.

Union Carbide estimated the cost of the 1361 tonne/day (1500 ton/day) Purox system as $17,703,000 based on mid-1975 dollars. UCC planned to supply four 317.45 tonne/day (350 ton/day) reactors for the 1361 tonne/day (1500 ton/day) plant, with the reservation that a fifth unit may be added to the system. The Purox system equipment and the numbers of each piece of equipment are noted in Table 11. These data are also taken from the EPA report,[2] and the list shows the concept of modular equipment system design.

5.2 Product Cost Analysis

This section describes a method of calculating the Purox System product cost. For this estimate, the capital and operating cost values are based on mid-1975 dollars as presented in Reference 2.

TABLE 8. Estimate of Capital Costs[a]

Cost item	Millions of dollars (1975 dollars)
Construction	53.75
Interest during construction	4.30
Startup costs	2.56
Working capital	1.79
Total	62.40

[a] Source: Reference 2.

TABLE 9. Estimate of Construction Costs[a]

Cost item	Thousands of dollars (1975 dollars)
Taxes	none (assumed)
Land value ($0.50/ft² for 21.5 acres)	468
Site improvement	880
Front-end preprocessing train	10,066
Processed feed materials storage and handling facilities	3,476
Purox equipment including oxygen and wastewater treatment plant	17,703
Installation cost of Purox system	10,675
Fuel gas pumping and drying equipment system	2,410
Subtotal	45,678
Contingency at 10%	4,567
Engineering and construction management at 10%, less 6% engineering on Purox equipment	3,505
Total	53,750

[a] Source: Reference 2.

The assumed data base for calculating the product cost is as follows:

- Capacity of the Purox system plant 1361 tonnes/day (1500 ton/day)
- Yearly plant utilization factor 0.92
- Drop charges of raw refuse $11.0/tonne ($10/ton)
- Amortization @ 8.5% for 20 years (capital recovery factor) 0.10567
- Sale price of scrap ferrous metals $44.1/tonne ($40/ton)
- Sale price of scrap aluminum metals $330.76/tonne ($300/ton)
- Plant capital cost $62,400,000
- Annual operative costs $7,016,000
- Ferrous metal recovery 110.65 tonnes/day (122 ton/day)
- Aluminum metal recovery 5.44 tonnes/day (6 ton/day)
- Synthesis fuel gas production (yearly) $0.25 \times 10^9 \ Nm^3$ (9.31×10^9 scf)
- Heating value of the fuel gas $12.81 \ MJ/Nm^3$ (370 Btu/scf)
- Yearly refuse receipt rate 456,839.4 tonne (503,700 tons)

TABLE 10. Estimate of Site Improvement Costs

Cost items	Thousands of dollars (1975 dollars)
Clearing and grubbing	40
Utilities	240
Excavation, backfill, and disposal	270
Fine grading	40
Paving	200
Landscaping	40
Fencing	50
Total	880

TABLE 11. Equipment of the Purox System

Equipment	Number[a]	Equipment	Number[a]
Purox System		Air Separation Plant 363.1	
Purox reactors (350-ton/day		tonne/day (400 ton/day)	
units)	5	Interchanger	
Feed conveyor	5	Liquid oxygen storage	
Refuse feeder	5	Air compressor and drive	
Refuse converter	5	After coolers	
Slag quench tank and		Expander turbines	
conveyor	5	Oxygen vaporizer	
Electrostatic precipitator	5	Air surge tank	
Condensate pumps	10	Blowdown silencer	
Condensers	5	Drain vaporizer	
Solid-liquid separation system	5	Thaw system	
Combustors	3	Booster compressor	
Wastewater treatment system		Circuit breaker panel	
8.33 Mliters/day (2.2 Mgal/		Instrumentation	
day)		Local and remote panels	
Surface aerator, electric		Controllers, control valves,	
motors, gear boxes skids, and		transmitters, and analyzers	
special valves			
Purge blower			
Instrumentation and controls			

[a]Where there is no number given, this means that there is none available for the particular equipment described.

The sequence of calculations for determining the product cost includes the following:

1. The amortization cost per million Btu of product gas per year, which is given by

$$\text{Amortization cost (per million Btu)} = A/B \qquad (6)$$

where A is the heat value of the product gas per year and is equal to 12.81 $\times$ 10^9 $\times$ 0.25 MJ = 9.31 $\times$ 10^9 $\times$ 370 Btu, and B is the capital cost, equal to $62.4 \times 10^6 \times 0.10567$.

2. The unit (MJ) operating cost of product gas per year, which is given by

$$\text{Operating cost} = C/A \qquad (7)$$

where C is the yearly operating cost and is 7.016×10^6.

3. The unit (MJ) revenue from the sale of ferrous scrap of product gas per year, which is given by

$$\text{Steel revenue} = E/A \quad \text{(per MJ)} \qquad (8)$$

Here E is the annual revenue from the sale of recovered ferrous metal and is equal to $D \times$ $40/ton, where D is the ferrous metal recovered per year: D = 37190 tonne = (122 ton/day) $\times$ (365 days/yr) $\times$ (0.92 utilization factor).

4. The revenue from the sale of aluminum scrap, which is given by

$$\text{Aluminum revenue} = G/A \quad \text{(per MJ)} \qquad (9)$$

Here G is the revenue from the sale of aluminum scrap and is given by $F \times$ $300/ton, where F is the aluminum scrap recovered per year and is equal to 5.44 $\times$ 365 $\times$ 0.92 tonne = (6 ton/day) $\times$ (365 days/yr) $\times$ (0.92 utilization factor).

5. Drop charge per MJ per year, which is given by

$$\text{Drop charge} = I/A \qquad (10)$$

Here I is the yearly drop charge and equals ($10/ton) $\times$ H, where H is the refuse received per year and is equal to 503,700 tons.

6. Net cost per MJ of product fuel gas, which is given by

$$J = \text{Items} (1 + 2 - 3 - 4 - 5) = \text{dollars}/10^6 \text{ Btu} \qquad (11)$$

From Item 6, calculations can be made to evaluate the processing cost [Equation (12)].

$$\frac{\text{Processing cost}}{\text{Tonnes of raw refuse}} = \frac{J \times A}{H} \tag{12}$$

As the drop charge decreases, the net cost per MJ and, consequently, the processing cost per tonne of raw refuse will rise. For example, if the drop charge is decreased to \$5.51/tonne (\$%/ton), the processing cost will increase (in the above case) by approximately 40%. For a municipally operated Purox plant, where the drop charge is zero, the processing cost per tonne of raw refuse will be substantially higher than the value calculated on the basis of \$11.03 tonne (\$10/ton) of drop charge. If the costs of ferrous and aluminum materials rise and if some revenue can be realized from the sale of glass products, then some adjustments must be made in the processing cost per ton of raw refuse. However, the capital cost for the glass recovery plant will increase, and hence a cost–benefit analysis will be needed to determine the necessity of a glass plant.

The information given in Items 1–6 can be used to calculate the fair drop charge fee to arrive at the break-even processing cost value. Cost estimates for small- or large-size plants can be made from Equation (13):

$$\frac{P_1}{P_2} = \left(\frac{C_1}{C_2}\right)^n \tag{13}$$

where P is equal to the price in millions of dollars, C is equal to the capacity in tonnes/day, and n is the scaling factor.

For example, if the total cost estimate for a 1361-tonne/day (1500-ton/day) plant is \$53.75 million, then the cost for a 1904-tonne/day (2100-ton/day) plant with a scaling factor, say, of 0.95, will be as calculated in Equation (13a):

$$\frac{P_1}{53.75} = \left(\frac{1904}{1361}\right)^{.95} \tag{13a}$$

Therefore, $P_1 = \$74 \times 10^6$.

It should be understood that the scaling factor for one set of plant equipment will not be the same for a different type of equipment. For example, the scaling factor for the reactor system will be different from that of the oxygen plant or gas cleaning system.

Purox units are to be sold in multiples of 317.45 tonne/day (350 ton/day). For a 1361-tonne/day (1500-ton/day) capacity, five modules are recommended, and for a 1904-tonne/day (2100-ton/day) capacity, seven modules are recommended. Because of the modular design, the capital cost of a Purox reactor system increases almost linearly with the size of the plant. A summary of net unit cost to produce the Purox fuel gas as a function of plant size is given in Table 12.

6. Applications of Purox Fuel Gas

The major components of Purox fuel gas are carbon monoxide and hydrogen, having a negligible nitrogen content. The simplest application of the Purox

TABLE 12. Summary of Net Unit Cost to Produce Purox Syngas as a Function of Plant Size[a,b]

| | Plant size | | | | |
| Item | 700 tons/ day | 1500 tons/day | | 2100 tons/day | |
	2 modules	4 modules	5 modules, one spare	6 modules	7 modules, one spare
Capital ($000)	30,960	54,790	62,400	77,800	86,220
Amortization ($000/yr) 8½%, 20 yrs	3,272	5,790	6,594	8,221	9,111
Overhead and maintenance ($000/yr)	4,514	6,952	7,016	9,372	9,410
Total cost of operation ($000/yr)	7,786	12,742	13,610	17,593	18,521
Utilization factor	0.80	0.85	0.92	0.85	0.92
Refuse feed (tons/yr)	204,400	465,400	503,700	651,500	707,500
Product gas (10^6 Standard cubic feet/year)	3,776	8,597	9,310	12,036	13,060
Product gas (10^6 Btu/yr)	1,397,000	3,181,000	3,444,000	4,453,000	4,836,000
Aluminum and steel credits (10^3/hr)	946	2,147	2,243	3,003	3.261

Drop charge, $/Mg	Net cost, $/ton				
0	33.46	22.77	22.56	22.39	21.57
5	28.46	17.77	17.56	17.39	16.57
10	23.46	12.77	12.56	12.39	11.57

Drop charge, $/ton	Net cost, $/$10^6$ Btu				
0	4.90	3.33	3.30	3.28	3.16
5	4.17	2.60	2.53	2.55	2.42
10	3.44	1.86	1.84	1.82	1.69

[a] English units.
[b] Source: Reference 2.

fuel gas will be in a furnace as an auxiliary or supplementary fuel. For a closed coupled furnace, the gas cleaning step may be omitted. However, for pipeline transmission, the gas must be compressed and dried.

Alternative applications of the Purox fuel gas may include the following: (1) pipeline synthesis fuel gas (compression and drying); (2) methanation (catalytic conversion); (3) electric power generation (gas turbine-steam, combined cycle; (4) fuel-grade methanol synthesis (catalytic conversion), and (5) anhydrous ammonia (converting Purox fuel gas to pure hydrogen, adding purified nitrogen from an oxygen plant, and then converting the mixture to ammonia).

The Purox system can be used by municipalities to convert MSW to medium-Btu fuel gas and also for codisposal of MSW and sewage sludge. In EPA-sponsored test runs at UCC's pilot plant facilities, filtered sewage sludge was mixed with shredded and magnetically separated MSW, and the mixture was used as feed material for the Purox reactor. The pilot plant runs showed that over the range of reactor feed compositions, having dry sludge solids to MSW ratios of 0.025–0.074, the Purox converter was able to produce useful fuel gas and reduce the solids of the sludge to inert slag.

The economics of the codisposal of sludge and refuse should be determined in large-scale plants. In the pilot plant study, at the sludge to refuse ratio of 0.05, the net sludge disposal cost of \$110/tonne of sludge was projected based on a refuse drop charge of \$14.32/tonne.[1]

References

1. "The Codisposal of Sewage Sludge and Refuse in the Purox System," EPA Report No. 600/ 2-78-193, published under Contract S803769-01-3, prepared for Municipal Environmental Research Laboratory, Office of Research and Development, U.S. Environmental Protection Agency, Cincinnati, Ohio, by Union Carbide Corporation, Linde Division, Tonawanda, New York (1978).
2. "Engineering and Economic Analysis of Waste Energy Systems," EPA Final Report under Contract 68-02-2101, prepared for Industrial Environmental Research Laboratory, Office of Research and Development, U.S. Environmental Protection Agency, Cincinnati, Ohio, by the Ralph M. Parsons Company, Pasadena, California (1977).

Suggested Reading

J. Jones and S. Radding (eds.), *Advanced Thermal Processes for Conversion of Solid Wastes and Residues,* American Chemical Society Symposium Series 76, American Chemical Society, Washington, D.C. (March 13, 1978).
Anderson, J. E., The oxygen refuse converter, Proceedings of the 1974 National ASME Conference, Incinerator Division Miami, Florida (April 1974), p. 337.

Fisher, T. F., Kaskohns, M. L., and Rivero, J. R., American Institute of Chemical Engineers, 80th National Meeting, Boston (September 7–10, 1975).
Moses, C. T. and Rivero, J. R., Design and operation of the Purox system demonstration plant, Fifth National Congress on Waste Management Technology and Resource Recovery, Dallas, Texas (December, 1976).

13
Gasification

ANIL K. CHATTERJEE

1. Introduction

The concept of converting lignocellulosic resources to fuel gas and energy is not new. The gasification of wood and other lignocellulosic feedstocks was being extensively practiced during the early part of this century to produce low-heating-value fuel gas.

Gasifiers have been called converters, reactors, gas generators, and gas producers. In principle, any unit that converts a carbonaceous feedstock to fuel gas may be called a gasifier. The fuel gas produced from a gasifier is hot and often contains substantial quantities of solid particulates. The gas may therefore be used directly to fire stationary combustors such as a boiler, kiln, or process furnace, or it can be cooled and cleaned before it is used as substitute fuel gas for internal combustion engines.

Gasifiers may be designed to process a wide variety of feedstocks such as forestry products, agricultural residues, or aquatic crops and municipal solid waste (MSW). MSW is a heterogeneous waste that contains large quantities of biomass-derived products such as paper, cardboard, wood, textiles, and leather.

2. Types of Gasifiers

Gasifiers are generally classified on the physical condition of the carbonaceous material in the reactor.[1] The major categories of flow reactors include

---

ANIL K. CHATTERJEE • SRI International, Menlo Park, California 94025.

235

those with:

1. Fixed beds (actually moving packed beds).
2. Stirred beds.
3. Fluidized beds.
4. Entrained (moving) beds.
5. Tumbling beds.

Figures 1–5 show these categories of gasifiers. The gasifiers can be of air- or oxygen-blown type. An air-blown gasifier typically produces low heating value gas 2.98–5.59 MJ/m³ (80–150 Btu/ft³) whereas an oxygen-blown gasifier generates medium heating value gas 7.45–13.04 MJ/m³ (200–350 Btu/ft³). An exception to this rule is illustrated in Figure 3. With a dual reactor system, char may be burned and heated (along with sand) in an air-blown reactor. The heated sand and char are then moved into another reactor where an incoming carbonaceous material is pyrolyzed to produce a medium-Btu gas. Various designs of gasifiers have been built, ranging in capacity from 0.03 to 84,400 MJ/hr (100,000 to 80 million Btu/hr) of fuel-gas output.

Gasifiers can be designed for atmospheric or higher pressure operations. The reactor throughput rate increases with pressure and therefore a reactor

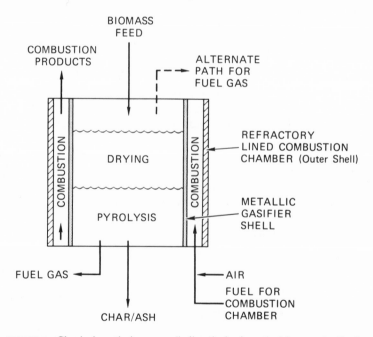

FIGURE 1. Classical pyrolysis process (indirectly fired, vertical flow, packed bed).

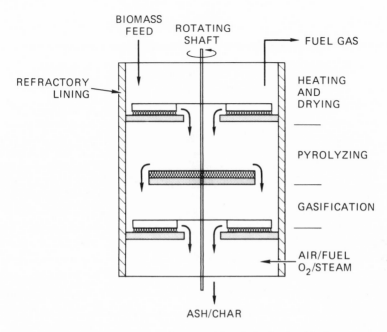

FIGURE 2. Multiple hearth reactor (directly fired, stirred bed).

designed to operate at higher than atmospheric pressure requires a smaller reactor vessel. The fuel gas produced from a pressurized reactor does not require further gas compression for transmission to the user, who may be located at some distance from the reactor.

However, a high pressure reactor must incorporate specific design features. The reactor must be a pressure vessel, and specialized feeding devices are needed. The pipings and the related components must be designed to withstand the pressure level of the high pressure gas.

At present, most reactors that are designed to gasify solid wastes or biomass, operate at atmospheric pressure conditions. In the following discussion, only three types of reactors are covered—the fixed-bed, the fluidized-bed, and the entrained-bed gasifiers.

2.1. Fixed Bed

A fixed bed gasifier is generally a vertical shaft furnace, similar in design to a cupola or blast furnace. It receives the feed either from the top or the side at a fixed height. Inside the gasifier, the fuel feedstock is supported either on a fixed grating or on a sand bottom. A fixed bed gasifier may further be cate-

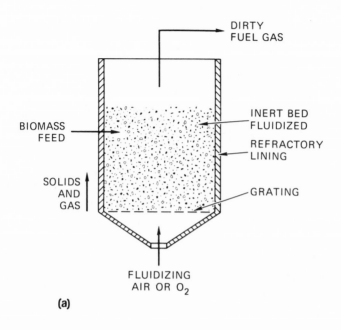

(a)

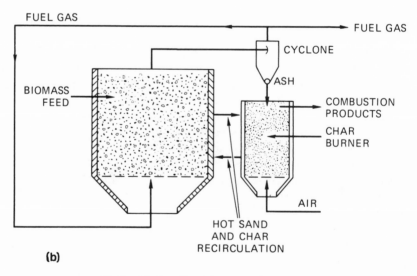

(b)

FIGURE 3. Fluidized-bed gasifier. (a) Directly fired; (b) indirectly fired.

SOLIDS
AND
FUEL GAS

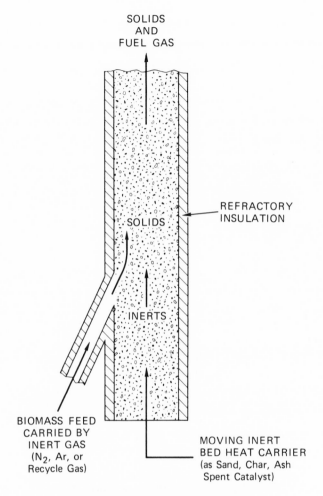

FIGURE 4. Entrained-bed reactor.

gorized as: countercurrent flow (updraft), cocurrent flow (downdraft), or a cross flow (crossdraft) unit. Figures 6–8 show the schematics of these three varieties of fixed-bed gasifiers.

In an updraft gasifier, the feedstock descends from the top to the bottom and the air ascends from the bottom to the top. The air for the updraft gasifier is generally injected upwards through the gratings. The oxidation zone lies at the bottom of this type of gasifier and gasification occurs through zones of decreasing temperatures as the gas rises through the reactor fuel bed. For the downdraft gasifier, the air is injected through a single or a series of tuyère nozzles, equally spaced around the furnace. The tuyère air is injected downwards toward the bottom of the furnace. In crossdraft gasifiers, the air is fed into the gasifier through a horizontal nozzle. The resulting fuel gas is discharged through a vertical grate on the opposite side of the air-injection location. The feed to the gasifier is admitted from either the top or the side.

In an updraft gasifier, because the reaction gases flow counter to the path of the incoming cool feedstock and exit at a relatively low temperature, the fuel gas produced by an updraft gasifier typically has high tar content.

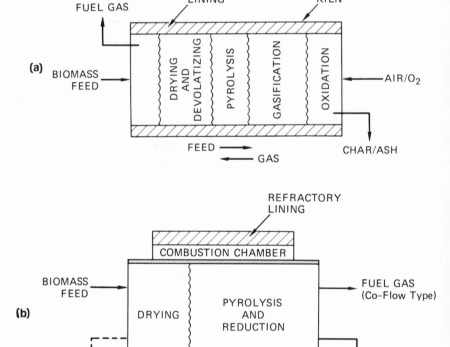

FIGURE 5. Rotary-bed gasifier. (a) Directly fired; (b) indirectly fired.

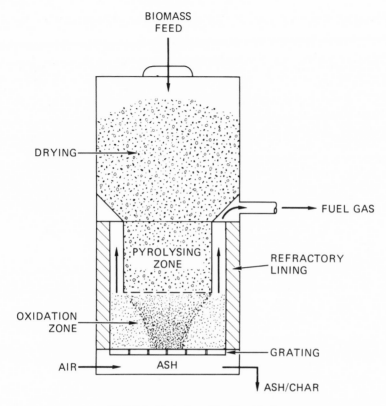

FIGURE 6. Updraft gasifier (fixed bed).

Downdraft gasifiers produce cleaner gas with relatively reduced tars, compared with updraft or crossdraft units. The tars and other vapors pass through
a high temperature zone last, resulting in thermal cracking of the tar vapors.
As the gases pass through the solid char bed, the entrained fly-ash and dirt are
cleaned. The reactions in a crossdraft gasifier are similar to the downdraft unit.

A fixed bed reactor has fairly discrete reaction zones. The depth of the
thermochemical conversion zones vary with the chemical composition of the
feed, its particle size, moisture content, temperature, and mass-flow rate of
the air blown into the hearth. These reactors are highly efficient in converting
lignocellulosic materials to fuel gas. Because of the packed bed feature of this
reactor, the offgas often has a relatively lower particulate carryover relative to
a tumbling bed or a fluidized reactor. An updraft reactor of a given size can
be operated at various loads by varying the air rate to the gasifier. Because of
the nitrogen dilution effect and the cooling of the fuel gas that occurs in the
drying zone, the offgas has a low heating value.

2.2. Fluidized Bed

A typical fluidized bed reactor is a large refractory lined vessel as shown in Figure 3. The fluidized medium may be sand, char, or ashlike inert material. At the bottom of the fluidized bed reactor, sand or other inert material is supported on a perforated plate. The grid network of the perforated plate ensures uniform distribution of the fluidizing medium. The fluidizing of the sand is accomplished by supplying the controlled flow rate of air or oxygen through the perforated network of the bed support plate. The velocity of the air or oxygen through the openings of the grid plate should be designed to keep the sand in suspension.

When the sand or the inert bed attains a fluidized state, each inert bed particle is separated from the others, and the whole inert bed will look like a

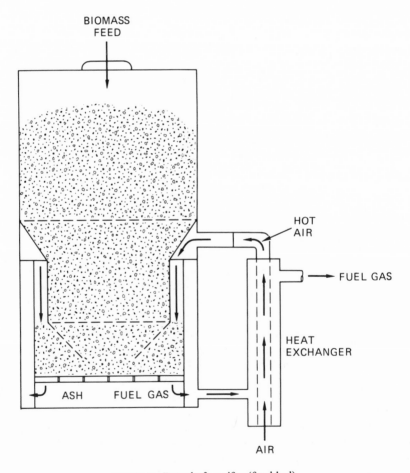

FIGURE 7. Downdraft gasifier (fixed bed).

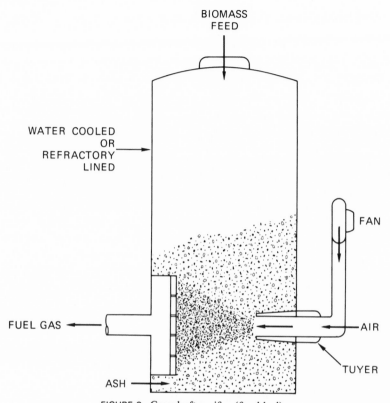

FIGURE 8. Crossdraft gasifier (fixed bed).

turbulently mixed fluid bath. In steady-state condition, each particle of the fluizided inert bed has the same temperature level. A temperature rise at any location in the bed will be transmitted quickly throughout the bed, similar to the phenomenon in a fluid field. In a fluidized bed reactor, the temperature profile typically is uniform from the bottom to the top of the bed. Simultaneous oxidation and gasification occur at essentially constant fluidized bed medium.

When the feed material is introduced into the bed, it appears to sink or float in the bed according to its size or density. At this point the feed materials are in contact with many hot inert materials, resulting in rapid heating and pyrolysis of the feed materials. The heat of pyrolysis is supplied by the high rate of char oxidation. The gases from the oxidation reaction process are effectively mixed by the fluidized bed material. This improves the flow of gases to and from the hot solid surface of the feed materials.

Because of the highly turbulent mixing characteristics of the fluidized bed reactors, the capacity is generally a function of the volume of the fluid bed. Traditionally, the height-to-diameter ratio of a fluidized-bed reactor is 10:1. For other types of gasifiers, this ratio is usually in the range of 3:1.[2]

244 Anil K. Chatterjee

The essential requirements of the fluidized bed are to allow dispersion of the feed material into the bed medium, to promote sufficient residence time for the rapid heating of the solid materials to their oxidation temperature, and to provide the necessary heat of pyrolysis.

The ideal bed temperature should be low enough to prevent slagging of the residues below $1150°C$ $(2100°F)$ and high enough to promote partial oxidation of the feed material above $850°C$ $(1600°F)$. Normally, the temperature of the fluidized bed reactor offgas is approximately the same as its bed temperature.

One of the desirable features of the fluidized bed reactor is that feedstocks with wide range of physical properties can be successfully processed. High ash and moisture contents of the feedstocks pose no serious problems and may not require any additional auxiliary equipment. However, problem feedstocks such as those with high moisture levels, slow burning rates, large particle size, and high residue levels, if preprocessed in the form of fine shredding and limited drying, would certainly offer improved operating efficiency of the reactor. The reactor has high turndown capability and the fuel-gas quality is fairly uniform.

For unprocessed feedstocks, high carbon loss may occur in the residue removal operations. In some instances, a second stage combustor may be needed to oxidize the unreacted carbon in the residue. For this reason, although the fluidized bed reactors offer excellent carbon–oxygen contact, preprocessing and/or post gasification operations may be required for best operations.

The widespread use of the fluidized bed reactors has been limited by the low operating efficiency and high capital costs.

The fluidized-bed reactor will require high fan horsepower to overcome the large pressure drop from 0.076 to 0.2 kg/cm^2 (30–80 in water column) across the fluidized bed and the air distribution systems. A fluidized-bed reactor handling dry or high-energy-level feedstocks will require high excess air to control the temperature of the oxidation process to a level that will prevent slagging of the residue or softening of the bed media. The fluidized bed reactor is designed with a large oxidation chamber to provide a relatively modest velocity profile through the bed, in the range of 0.003–0.006 m/second (6–12 ft/sec). In comparison, the air velocity through a spreader stoker combustion chamber of a boiler is designed to operate at 0.024 m/sec (50 ft/sec).

2.3. Entrained Bed

An entrained bed gasifier is sometimes called a moving bed reactor (Figure 4). Occidental Reserach Corporation's flash pyrolysis reactor is an entrained bed reactor that uses spent refinery cat cracker catalyst and char as the heat transfer media. In the design of the flash pyrolysis process, Occidental planned to use the char from the pyrolysis operation as the replenishing transport bed media.

The entrained bed reactors have high heat transfer characteristics. The transport media serves as an efficient heat transfer resource. Special preprocessing work to the feedstock is generally required before they are fed to the reactor. The transport medium is generally maintained between 482 and 593 °C (900–1100 °F) for maximum fuel-oil yield.

Because these reactors have excellent solid–gas contact, they can handle wide varieties of biomass. The reactors have a high unit-volume capacity, and their offgases contain very little uncracked hydrocarbons.

3. Fuels for Gasifiers

Under thermodynamic equilibrium conditions, a properly designed gasifier can use any organic solid wastes or biomass feedstock to produce the fuel gas.

The following waste or biomass feedstocks have been gasified to produce low-Btu fuel gas (Table 1).

Some of the desirable qualities of a proper biomass feedstock for gasification are given in Table 2.

All gasifiers, other than the Torrax unit, generally require some preprocessing of the feed. In particular, some agricultural and forestry feedstocks must be preprocessed before they are acceptable to the gasifier. Some of the preprocessing steps are as follows:

1. For Municipal Solid Wastes
   - Size reduction: One- or two-stage shredding
   - Metals removal: Ferrous, by magnetic separation; nonferrous, by eddy current or equivalent
   - Glass removal: Air classification, trommeling, screening
2. For Forestry Products
   - Hogging, chipping, drying and/or densifying to produce pellets or cubes.
3. For Agricultural Products and Residues
   - Chipping, drying, and densifying

Some of the advantages and disadvantages of preprocessing operations are given in Table 3.

4. Thermochemistry

In a typical gasifier, numerous thermochemical reactions occur. The degree of equilibrium that is attained in the reactions determines the quality of the fuel gas. A visualization of the events that occur inside the gasifier may clarify these reactions.

TABLE 1. Low-Btu Fuel Gas from Biomass

Organic elements of municipal solid waste (MSW) including paper, linen, rubber, grass, tree trimmings, wood, and food wastes.

Forestry residues such as wood scraps, bark, tree trimmings, sawdust, and charcoal.

Agricultural residues such as corn cobs, flax waste, rice husks, rice stalks, cane sugar bagasse, coffee husks, and nut shells (e.g., walnut, coconut).

TABLE 2. Desirable Properties of Biomass for Gasification

Average moisture content of less than 50%

Average heating value (HHV) of not less than 9.8 MJ/kg (4200 Btu/lb).

Average feedstock size range greater than 1.27 and less than 7.62 cm ($>$ ½ in and $<$ 3 in).

Ash fusion temperature of not less than 1149°C (2100°F).

Low ash content (6–10%).

Easy ignition characteristics.

Uniform chemical composition.

Can form structurally strong char.

Relatively simple to collect, store, and handle.

Economically justifiable transportation cost.

Available in adequate quantities to consistently meet the gasifier load demand.

TABLE 3. Preprocessing Operations

Advantages
    Uniform distribution of the feed to bed in the gasifier is expected.
    Channeling action in the flow paths does not occur.
    Bridging incidents inside the gasifier are reduced.
    A mechanical feeder can be used to load the gasifier.
    The gasifier top sealing device can be operated without incidents.
    Uniform gas quality is ensured.
    Steady rate of slag, ash, or char production may be possible.
    Storage and handling are made easier.
Disadvantages
    Additional cost for the preprocessing equipment may be needed.
    Heavy maintenance costs (material and labor) are associated with preprocessing equipment operation.
    Long-term breakdown of preprocessing equipment may shut down gasification operation.
    Overall operating cost of the preprocessing equipment is high.

Assume that the gasifier is a vertical shaft furnace filled with a biomass feedstock such as wood chips. The top of the gasifier is closed so that no air can enter the gasifier. The wood chip pile is supported by a grating located at the bottom of the gasifier. There is provision for air–steam injections to the grating. The wood is ignited by a match or blow torch; a controlled quantity of combustion air is supplied underneath the grating. In the gasification process the quantity of air that is supplied to the hearth is always substoichiometric.

Shortly after ignition, the gasifier vessel will have four distinct reaction zones as seen in Figure 9. Close to the bottom of the gasifier is the oxidation zone, which is followed by the reduction, pyrolysis, and drying zones.

The thermochemical reactions in the gasification process may be written as shown in Table 4.

A typical reaction is shown in Figure 9. The zonal equilibrium conditions are dependent on the temperature and pressure. The degree to which equilibrium is approached depends on solid–gas interactions, and residence time. For example, Equation (6) shows the exothermic reaction of hydrogen with char. The amount of CH_4 formed depends on the reaction temperature and pressure; high pressure and low temperature favor the formation of CH_4.

In the reactions in the oxidation zone, the carbon of the char reacts with the oxygen of the combustion air to produce hot CO_2. This exothermic reaction is essential to supply the heat energy required to complete the reactions in the reduction, pyrolysis, and drying zones. Therefore, a certain amount of the carbon elements of the fuel stock will always be needed for combustion instead of gasification. For this reason, straight gasification efficiency, when calculated by neglecting the sensible heat of the fuel gas, is seldom found to be more than 70%. However, a gasification process that produces pyrolytic oil and char can achieve an overall thermal efficiency in excess of 70%.

The oxidation reaction $C + O_2 \rightarrow CO_2$ is very fast and is strictly limited to mass transfer characteristics. For a well-designed gasifier, this zone becomes fairly narrow in depth, 7.6–12.7 cm (3–5 in). For some gasification processes, steam is added to the combustion air so that the water gas reaction of $C + H_2O \rightarrow CO + H_2$ could occur. This reaction is slightly exothermic if the water vapor is condensed to water, but it becomes endothermic if the water vapor is not condensed. The heat of reactions of these two cases are shown in Equations (4) and (5). The reaction $H_2O \rightarrow H_2 + \frac{1}{2}O_2$ depends upon an appropriate temperature.

Because the combustion air that is injected into the oxidation zone is less than stoichiometric in quantity, the product of combustion that rises through the gasifier feedstock contains little if any free oxygen. This enables the pyrolysis reactions to occur. In the pyrolysis process, the feedstock containing carbonaceous material is thermally decomposed in an oxygen-free atmosphere to produce a fuel gas of low- to medium-heating value. In the classical definition of pyrolysis, heat is supplied to the fuel indirectly as shown in Figure 1.

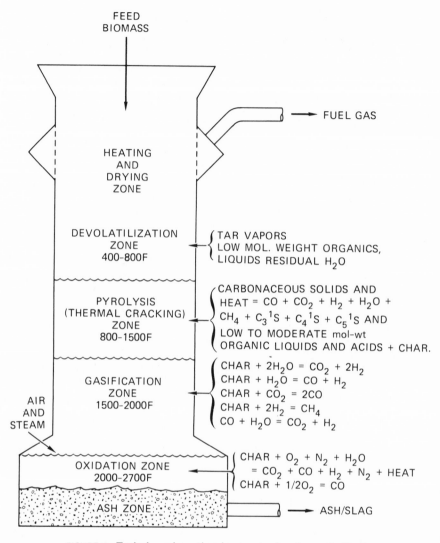

FIGURE 9. Typical zonal reactions in a reactor (moving packed bed).

However, in recent years, some solid waste processing systems have been branded as pyrolysis processes in which starved air combustion occurs. The Andco–Torrax Process is a typical starved air combustion process. On the other hand, Occidental Research Corporation's solid waste conversion process is a pyrolysis process. In the combustion process, the products of combustion are generally CO_2, H_2O, N_2, and excess O_2. In the gasification process, the product gas stream contains CO_2, CO, H_2, CH_4, C_2H_4, C_3H_6, NH_3, H_2S, N_2,

TABLE 4. Thermochemical Reactions

Zones	Reactions	ΔH^a (Btu/lb mole)	ΔH^a (kJ/kg mole)	Equation number
Oxidation	$C + O_2 \rightarrow CO_2$	+169,686	394	(1)
	$H_2O \rightarrow \frac{1}{2}O_2 + H_2$	+104,076	241	(2)
	(for steam injection case)			
Gasification	$C + 2H_2O \rightarrow CO_2 + 2H_2$	+38,466	89	(3)
	$C + H_2O \rightarrow CO + H_2$	+56,718	132	(4)
	$C + CO_2 \rightarrow 2CO$	+74,970	174	(5)
	$C + 2H_2 \rightarrow CH_4$	−32,327	75	(6)
Pyrolysis	$CO_2 + H_2 \rightarrow CO + H_2O$ (liquid)	−815	1.89	(7)
	$CO_2 + H_2 \rightarrow CO_2 + H_2O$ (gas)	+17,723	41	(8)
	$CH_4 + \frac{1}{2}O_2 \rightarrow CO + 2H_2$	+14,380	33	(9)
Drying	Surface moisture + heat → High-to-moderate molecular weight organic liquids + char + CH_4 + H_2 + H_2O + CO + CO_2			

$^a\Delta H$ indicates the heat released (minus) or absorbed (plus) when a compound is formed from its elements. $+\Delta H$ = endothermic; $-\Delta H$ = exothermic reactions.

H_2O, tar vapors, and low-molecular-weight organic liquids.[1]

The composition of the fuel gas produced by the gasification process will depend on the degree to which the various reactions have attained equilibrium. Normally all combustion reactions are reversible, and this reversibility increases with a rise in temperature. The equilibrium point in any combustion reaction can be moved by varying the temperature and pressure conditions. Examples of three common reversible combustion reactions are:

$$CO + \frac{1}{2}O_2 \rightleftarrows CO_2 \qquad (10)$$

$$H_2 + \frac{1}{2}O_2 \rightleftarrows H_2O \qquad (11)$$

$$CO_2 + H_2 \rightleftarrows CO + H_2O \qquad (12)$$

The concentrations of the reacting gases can be expressed in partial pressures for each gas. Therefore, the equilibrium constants for the three reactions are expressed as:

$$K_{H_2O} = \frac{(H_2O)}{(H_2)(O_2)^{1/2}} \qquad (13)$$

$$K_{CO_2} = \frac{(CO_2)}{(CO)(O_2)^{1/2}} \qquad (14)$$

$$K_{water\ gas} = \frac{(H_2O)(CO)}{(CO_2)(H_2)} \qquad (15)$$

A calculation of the equilibrium constant is useful for determining the effect of reaction zone temperature, pressure, feed moisture content, ratio of fuel to oxidant, for predicting the degree of dissociation of a reaction gas, and for estimating the degree of equilibrium reached in the reaction. The equilibrium constants of some of the combustion reactions are plotted in Figure 10. As an illustration, suppose that a gas sample is collected from a certain zone of a gasifier where the reaction is practically complete. If, during gas sampling, the temperature and the moisture content of the gas are measured, then from the dry gas analysis and its known moisture content, the by-volume percentage of the moist gas can be calculated. From this value, the equilibrium constant of the water–gas reaction ($CO_2 + H_2 \rightarrow CO + H_2O$) can be calculated as:

$$K_{(H_2O)} = \frac{\% \text{ volume } (CO \times H_2O)}{\% \text{ volume } (CO_2 \times H_2)} \tag{15a}$$

With the calculated value of the equilibrium constant, the reaction temperature is checked from the K versus T curve (Figure 10). If this temperature is below the measured temperature, the reaction has not reached its equilibrium. An example of this procedure is illustrated below.

The gas analysis (dry basis) is given in Table 5.

Assuming the sample gas contains 10% H_2O, one can calculate the gas composition on a wet basis by dividing each dry basis value by 1.1. The calculated gas analysis (wet basis) is given in Table 6.

The equilibrium constant is calculated in Equation (15b):

$$K = \frac{[CO] \, [H_2O]}{[CO_2] \, [H_2]} = \frac{18 \times 1.1}{9.9 \times 16.54} = 1.0 \tag{15b}$$

Corresponding to this value of K, from Figure 10, it is seen that the temperature is $825°F$. If, for example, the temperature was measured at $950°F$, i.e., the measured temperature is higher than the calculated temperature, then the reaction is at a stage that has not yet reached the equilibrium state.

5. Design Considerations

A gasifier system design will depend on the combustor to be used for the fuel gas and the type of biomass to be used in the gasifier. For example, if the fuel gas is to be used to generate steam in a boiler or to dry biomass resources such as the organic fraction of municipal solid waste or wood chips, then the gasifier can be an updraft closed coupled unit. A fluidized bed gasifier can also

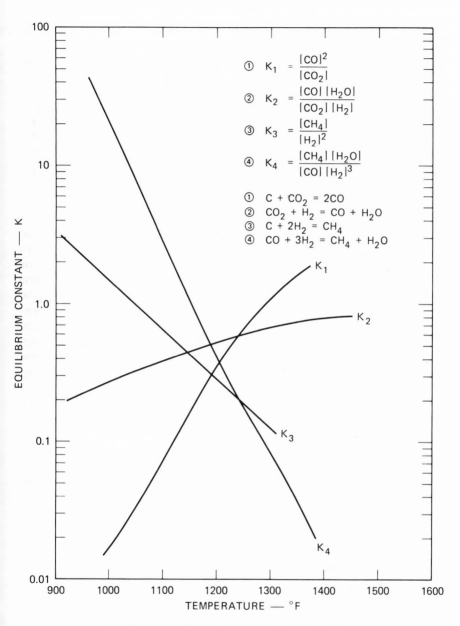

FIGURE 10. Equilibrium constants for various reactions.

TABLE 5. Typical Fluegas Analysis (Dry Basis)	
	Dry basis (volume %)
CO_2	10.9
CO	19.8
H_2	18.2
CH_4	0.8
N_2	49.8
O_2	0.2
Other hydrocarbons	0.3

TABLE 6. Typical Fluegas Analysis (Wet Basis)	
	Wet Basis (volume %)
CO_2	9.9
CO	18.0
H_2	16.54
CH_4	0.73
N_2	45.27
O_2	0.18
Other hydrocarbons	0.28
H_2O	9.1

be used very successfully for the boilers. The high sensible heat of the fluidized bed reactor offgas will aid in achieving high overall thermal energy recovery for the gasifier. However, if this gas is to be used in a dryer, high dilution air will be required to bring the mixture temperature of the combustion products below the ignition point of the biomass resources that are being dried. On the other hand, if the fuel gas is to be used in compression ignition engines, a downdraft gasifier must be used so that the gas is free from the tar vapor and ashes. Gasifiers operating with charcoal will differ from units using wood, solid waste, or densified agricultural biomass. A wood gasification system to produce pyrolytic oil and char or for methanation synthesis will require gas scrubbing, quenching, and demisting units.

Some of the important design considerations for the fixed bed stationary gasifiers relate to:

- Fuel preparations such as shredding, drying, and feeding devices.
- Ash and char cooling and handling systems.
- Refractory wall versus water-cooled wall. A water-cooled wall loses heat but a refractory wall gasifier is a severe maintenance problem. For an unprocessed solid waste, a water-cooled wall gasifier is preferable because severe bridging could occur from the worn out and damaged refractory surfaces of a gasifier.
- Slagging and temperature incursion problems.
- Offgas duct work design.
- Proper oxygen, steam, or air injection design.
- Appropriate diameter-to-height ratio; too large a diameter will cause flow distribution problems, whereas too tall a gasifier may affect the quality of the fuel gas.
- Gas cleaning and cooling system design.
- Overall environmental, safety, and health considerations.
- Adequate supply of biomass feed resources.

The design considerations for mobile biomass gasifiers are quite different. The important and necessary design features of a mobile biomass gasifier include the following:

- Compact and lightweight.
- Easy maintenance and replacement of parts.
- Easy mounting and dismounting features.
- Safety features against fire, toxicity, and burn from hot surface.
- Reliable and easy gas cooling and cleaning devices.
- Sample feeding and ash removal devices.
- Suitable materials of construction to withstand the mechanical vibrations and strains that are generated from moving vehicles.

There are no defined design guidelines for engineers designing biomass gasifiers. However, a designer of biomass gasifiers should be aware of the effects of process variables on its performance. Design variables include oxidant choice (air or oxygen), biomass type and composition, and the pressure and temperature at which the gasifier will operate. In addition, residence time of the feedstock becomes important for certain types of biomass gasifiers. The problem of introducing feedstock to a pressurized gasifier has not been fully solved. The use of biomass densified pellets to produce fuel gas has been investigated, but the benefits in operating costs have not been fully realized.

Many designs have been successful in proving the mechanics and principles of biomass gasification processes. Pilot plant, demonstration, and commercial-size units have been designed and operated for some periods, to prove the feasibility of the biomass gasification process. Continued efforts and refinement in equipment system design are needed to make the biomass gasification process design a commercial entity.

6. Performance Characteristics

Various biomass gasification processes are discussed in other chapters of this book. This section describes a conceptual design performance of a wood gasification process that produces char and pyrolytic oil.[3] The design data base includes the following:

1. The system capacity is 2000 wet short tons/day.
2. Wet wood chips are used as the biomass feed material.
3. The composition of the wood (dry basis percent by weight) is 52.3% carbon, 6.0% hydrogen, 41.2% oxygen, negligible amounts of sulfur and nitrogen, and 0.5% ash.
4. The moisture content of the wet wood chips is assumed to be 50%.
5. The heating value of the dry wood chips is assumed to be 20.7 MJ/kg (8900 Btu/lb).

6. The wet wood chips are dried in a horizontal rotary dryer to a 7% moisture level.

7. The continuous flow reactor has a packed bed (fixed) of wood chips supported on a grate and flowing countercurrent to the upward moving gas.

8. The gasifier feed will consist of wood chips at 7% moisture level and recycled oily filter cakes.

9. The pyrolytic gas will be scrubbed, quenched, and demisted to recover pyrolytic oil.

10. The gas quenching and scrubbing will be achieved by recirculating pyrolytic oil.

11. No water treatment facility will be required. The water used in the gasification process enters into the system with the feed and is used in the pyrolysis reaction.

12. Four gasifiers will be required, each 12 ft 6 in. in diameter and 10 ft high, to accommodate the 2000 wet short tons capacity of the plant.

Figure 11 is a flow diagram for a 2000-TPD wood gasification process system that produces char and pyrolytic oil. The plant receives chipped wood having a moisture content of 50% which is dried in a rotary dryer to a 7% moisture level. The heat energy for the dryer is supplied by the pyrolytic gas produced in the system. At the start of the process, natural gas is used in the dryer. An air-to-fuel ratio of 15:1 is used to cool the combustion-product temperature to a level below the autoignition temperature of the wood chip. A temperature level of 177–204°C (350–400°F) is assumed adequate for the entry point to the dryer. The exhaust from the dryer is vented through a bag filter system.

The dry wood chip feedstock is then fed to the gasifier. The air to the gasifier is supplied beneath the grating, and the gasifier is provided with a steam injection facility. As the plant operates, it will produce char. The char is withdrawn by a water-cooled screw conveyor at 426.7–537.8°C (800–1000°F) and cooled in a rotary char cooler to a temperature of 65.6°C (150°F). The char production from the plant is calculated to be 265.8 tonnes/day (293 tons/day) or 11.07×10^3 kg/hr (24.4×10^3 lb/hr). The offgas from the gasifier, which is the pyrolytic fuel gas, is dirty and contains tars and unsaturated hydrocarbon vapors.

The fuel gas then enters a venturi scrubber system at a temperature of 149–204°C (300–400°F). The scrubbing and quenching of the fuel gas is designed to be performed by oil and not by water. At the initial operation stage, No. 2 diesel oil may be used as a scrubbing liquid. Because the fuel gas contains mostly fine solids, the venturi scrubber may be designed to have a 0.03–0.04 kg/cm^2 (12- to 16-inch H_2O) pressure drop across its throat.

Quenching of the fuel gas generally requires a large amount of oil. The

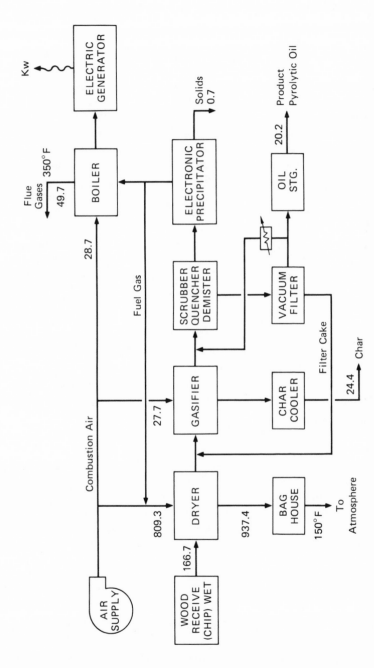

FIGURE 11. Mass balance of oil and char production by pyrolysis of wood; 2000 wet ton/day.[3] Note: all data are in thousands of pounds per hour.

cooling load of the quench oil will equal the latent heat of condensation of the product oil assumed to be 0.465–0.7 MJ/kg (200–300 Btu/lb) as well as the sensible heat load from 149–204°C to 66–93°C.

The condensed organic liquid (pyrolytic oil) is collected in the quench pot, and the noncondensible fuel gas is demisted and further cleaned in an electrostatic precipitator unit. The clean fuel gas can then be used for burners for the boiler of the rotary dryer. The quenched oil contains flyash and scum and is therefore filtered through a rotary vacuum filter. The filtrates (pyrolytic oil) are pumped out. A large portion of this oil is used in the quenching operation to the venturi scrubber. The excess oil, which may contain solids of 0.4–0.5% by weight, is stored and becomes a product for the wood gasification process.

The concentrate of the vacuum filter unit contains 50 wt% of solids and is in the slurry state. The slurry is maintained at a temperature level of 66–93°C (150–200°F) and is pumped to the gasifier. The pyrolytic oil production from this plant is calculated to be 1210 bbl/day. At a density of 181.44 kg (400 lb/bbl), this is equivalent to 9163 kg/hr (20.2 $\times$ 10^3 lb/hr) of oil production.

The system component mass balance data are incorporated in Figure 12. From these data, an overall system mass balance can be prepared, as shown in Table 7 (see also Figure 11).

The overall system energy balance can now be prepared by estimating the composition and heating values of the oil and char from Table 8.

The heating value of the char was calculated from Dulong and Petit's formula of HHV = 14500 C + 62000 (H − $O_2/8$) + 4000 S and the heating value of oil was calculated with the formula HHV = 22820 − 3780 d^2, where d = specific gravity of the oil at 60/60 F. (The oil is assumed to have a specific gravity at 60/60 F of 1.8.)

On the basis of the above heating values, an overall energy balance can be calculated as shown in Table 9 and in Figure 13.

The energy distribution in the gasification process is thus about 39% in the oil and 61% in the char. The overall thermal efficiency = 73.6%.

The conceptual process system is assumed to be self-sufficient in energy consumption. The fuel gas produced from the system supplies the heat energy requirements of the dryer and the boiler. The steam from the boiler is used to generate power to drive blower fans, pumps, and other auxiliary power drives.

7. Combustion Characteristics

The fuel gas produced from the gasification process is rich in carbon monoxide, hydrogen, and hydrocarbon gases. In the combustion of low- and medium-heat value gases, some important considerations are the air-to-fuel ratio, the flame temperature, the ease of ignition of the gas, the flammability limits, and the flame length and its stability.

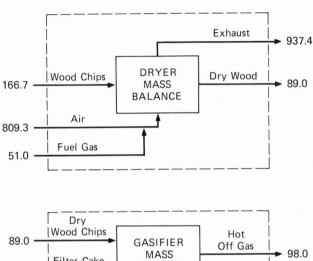

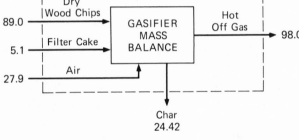

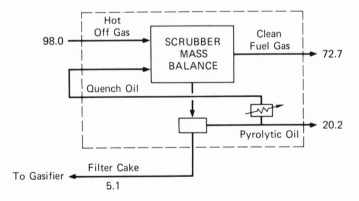

FIGURE 12. Component mass balance (2000 ton/day wood gasification system). Note: all data are in thousands of pounds per hour.

TABLE 7. Overall System Mass Balance

	lb/hr	kg/hr		lb/hr	kg/hr
Input			Output		
Wet wood	166,700	75,615	Pyrolytic oil	20,200	9,162.7
Air to gasifier	27,700	12,565	Char	24,400	11,067.8
Air to dryer	809,300	367,098	Dryer stack gas	937,400	425,204.6
Air to boiler	28,700	13,018	Boiler stack gas	49,700	22,543.9
			Solids from electronic		
			precipitator	700	371.0
Total	1,032,400	468,296	Total	1,032,400	468,296

On a volumetric basis, the low-Btu gas contains one-sixth the energy content of the natural gas. However, for the combustion of the low-Btu gas, the burner size must be increased by only 31%, and its flue-gas volume will increase by only 19%, as compared with the combustion of natural gas. For a well-designed burner, low-Btu gas can be burned with stable flame, although the flame temperature for the low-Btu gas is lower (1760°C versus 1960°C) for natural gas because of high nitrogen dilution. Flame length generally increases with the higher heating value of a gas. However, there seems to be some controversy over this claim by burner manufacturers. Low-Btu gas has wider flammability limits and the ignitability of such gas poses very little problem. To prevent condensation of pyrolytic oils and tars, the fuel piping system must be insulated.

For medium-heat value gas, the flame temperature is approximately 38°C lower and requires only 5% more combustion air than for the natural gas. Consequently, the flue gas volume for the medium-heat value gas is approximately the same as that for the natural gas. Because of the higher hydrogen content of low-heat value gas, some losses will occur in the combustion process due to

TABLE 8. Composition and Heating Values of Oil and Char

Composition	Oil (weight %)	Char (weight %)
Carbon	52.7	77.1
Hydrogen	5.9	5.2
Oxygen	25.2	10.2
Nitrogen	Negligible	Negligible
H_2O	16.2	5.8
Ash	0.04	1.7
Totals	100.0	100.0
Heating value (HHV)	24.58 MJ/kg	31.65 MJ/kg
	(10,572 Btu/lb)	(13,613 Btu/lb)

TABLE 9. Overall Energy Balance

	Millions of Btu per hour	Percent	GJ^a/hr	Percent
Input				
Wet wood at 75,615 kg/hr (166,700 lb/ hr) at 10.35 MJ/kg (4,450 Btu/lb)	742.0	100	783	49.5
Output				
Oil @ 20,200 lb/hr @ 10,572 Btu/lb	213.6	28.8	225	14.2
Char @ 24,400 lb/hr @ 13.613 Btu/lb	332.5	44.8	351	22.2
Dryer stack gas exhaust	87.4	11.8	92.	5.8
Boiler stack gas exhaust	11.2	1.5	12.	0.8
Heat reject to cooling	44.9	5.8	47	3.0
Heat losses through insulation	32.6	3.8	34	2.2
Unaccountable losses	33.7	3.5	36	2.3
Totals	742.0	100	797	100

^{a}GJ = gigajoules.

formation of steam from the hydrogen. The losses from the flue gas will be higher because of increased flue gas volume, and this high volume will also increase the pressure drop across the boiler. Low flame temperature with low-heat value gas combustion will increase convective heat transfer and decrease heat absorption in the furnace volume.

Converting pulverized coal-burning boilers for use with low-heat value gas as fuel will require the least modification, whereas conversion of boilers designed to burn natural gas as fuel for use with low-heat value gas will require extensive modification. The packaged boilers equipped with oil/gas burners have combustion chambers too small to accommodate the combustion products volume of the low-heat value gas. Derating of the boiler by as much as 50% may be needed and economic justification for such retrofitting may be difficult. In retrofitting a boiler that normally burns coal, oil, or natural gas, for use with low-heat value gas as fuel, some of the following modifications may be required:

- Larger sizes of induced and forced draft fans.
- Modification of the convective section of the boiler to accommodate increased flue-gas volume.
- Modification of the burner and windbox to accept larger size burners.
- Installation of enlarged fuel flow pipings, combustion controls, flame safety guards, and ignition devices.
- Strengthening of the ducts and associated boiler accessories to handle a higher volume of cold combustion air and hot flue gas.

Combustion of low-heat value gas is not a novelty. Blast furnace gas has a heating value ranging from 2.6–3.35 MJ/m^3 (70–90 Btu/scf), and this gas

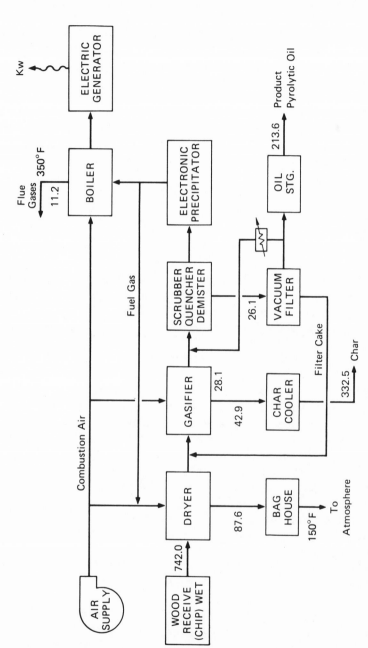

FIGURE 13.   Energy balance of oil and char production by pyrolysis of wood 2000 wet ton/day.[3] Note: all data are in millions of Btu per hour.

has been burned successfully in the iron industry. For burning low-heat value gas, scroll-type burners are generally used. It is very important that a tailor-made burner be used for all low-heat value gases. Some of the important design parameters for the burners are (1) the composition of the low-heat value gas, (2) the temperature and pressure of the gas entering the burner, and (3) contaminants in the form of moisture, tar, and acid vapors.

8. Ancillary Equipment

The gasifier itself is a fairly simple piece of equipment. However, the total gasification process equipment systems are quite varied and require careful design engineering.

Most of the pilot, prototype, and demonstration biomass gasification systems have faced severe equipment failure problems. In many cases, these failures have been attributed to the inadequacy of design engineering data and the heterogeneity of the feedstocks. Biomass gasification industries have attempted to use off-the-shelf blowers, pumps, compressors, heat exchangers, dryers, conveyers, and process instrumentations, but experience has shown that more custom-designed equipment may be necessary.

Some of the ancillary equipment systems required for the gasifier proper are the following:

- Feed receiving, handling, processing, and storage facilities.
- Feed charging equipment.
- Blower for combustion air.
- Steam injection provision.
- Raking, dumping, cooling, storage, and disposal facilities for the ash/char.
- Process control instrumentations.

Where the gasifier will be used as a substitute fuel supply source, the following additional equipment may be necessary:

- Gas cooling and gas cleaning systems.
- Water supply, distribution, and treatment facilities.
- Supplementary oil and natural gas supplies, distribution and application facilities.
- Environmental pollution control systems.
- Oil, gas, and char handling equipment.
- Oil, gas, and char storage or distribution facilities.

9. Utilization of Biomass Fuel

Gasification of biomass to produce low-heat value substitute fuel gas offers a promising potential. SRI International, in Volume 1 of a report entitled

TABLE 10. Biomass Availability in the U.S.

Year	Low moisture	High moisture	Woody biomass	Manure	Base case total
1985	69	32	97	57	245
2000	91	49	126	72	291
2020	102	60	151	77	326

Mission Analysis for the Federal Fuels from Biomass Program,[3] estimated the biomass availability in the contiguous United States (expressed in millions of short dry tons) as given in Table 10, which does not include the resources from the municipal solid wastes.

If it is estimated that 67 million short dry tons of biomass has the energy equivalence of 1 quad (quadrillion Btu = 10^{15} Btu), then even excluding municipal solid waste resources nearly 4 quads of energy could be harnassed from the available biomass in the United States.

References

1. J. Jones, Converting Solid Wastes and Residues to Fuel, in: *Chemical Engineering*, SRI International (June 2, 1978), pp. 87–94.
2. R. Overend, *Gasification—An Overview*, Proceedings of a workshop on air gasification, Solar Energy Research Institute, Seattle, Washington, February 2, 1979.
3. S. M. Kohan and P. M. Barkhordar, *Mission Analysis for the Federal Fuels from Biomass Program*, Vol. IV, SRI International, Menlo Park, California (January, 1979).

Suggested Reading

E. Wang and M. Cheng, A Comparison of Thermochemical Gasification Technologies for Biomass, in *Energy from Biomass and Wastes*, Conference proceedings of the Institute of Gas Technology (IGT) meeting in Washington, D.C. (August, 1978), Institute of Gas Technology, Chicago, Illinois.

Preliminary Technical Information, Low Btu gas production by the power gas producer unit, Davy Powergas Inc., Houston, Texas, (July, 1978).

H. A. Simons, Engineering feasibility study of the British Columbia Research hog fuel gasification system, British Columbia Research, Vancouver, British Columbia (May, 1978).

N. E. Rambush, *Modern Gas Producers*, Benn Bros. Ltd., London (1923).

Y. Y. Hsu, Clean fuels from biomass, NASA/TM/X/1538, National Aeronautics and Space Administration, Houston, Texas (1974).

A. F. Roberts, A review of kinetics data for the pyrolysis of wood and related substances, *Combust. Flame* **14**, 261–272 (1970).

D. Hathaway, Gasification gets another tryout, *B.C. Lumberman* (October 1978).

G. V. Voss, Industrial wood energy conversion, Proceedings of the Fuel and Energy from Renewable Resources Symposium, American Chemical Society, Chicago (August, 1977).

H. F. Feldman, Conversion of forest residues to a methane rich gas, *Energy from Biomass and Wastes,* Conference Proceedings of the Institute of Gas Technology meeting in Washington, D.C. (1978).

H. G. Hopkins, D. J. Basvino, "Syngas from Manure: A conceptual plant design, Bechtel National, Inc., San Francisco (July, 1978). E. M. Wilson, J. M. Leavens, N. Y. Snyder, J. J. Brehany, and R. F. Whitman, Engineering and economic analysis of waste to energy systems, Report No. 5495-1, The Ralph M. Parsons Company (June, 1977).

R. R. Burton and R. C. Bailie, Fluid-bed pyrolysis of solid waste materials, *Combustion* (February, 1974).

J. A. Coffman, Steam gasification of biomass Proceedings of the 3rd Annual Biomass Energy Systems Conference, Golden, Colorado (1979), p. 349.

Richard C. Bailie, *Energy Conversion Engineering,* Addision-Wesley Publishing Co., Reading, Massachusetts (1978).

Advanced Thermal Processes for Conversion of Solid Wastes and Residues, ACS Symposium Series 76 (J. Jones and Shirley Radding, eds.), American Chemical Society, Washington, D.C. (1978).

Retrofit '79, proceedings of a workshop on air gasification, SERI, Seattle, Washington (February 2, 1979), Solar Energy Research Institute Publication No. SERI/TP/ 49–183.

14

The Syngas Recycle Process

HERMAN F. FELDMANN

1. The Concept

The Syngas process is being developed to convert shredded municipal solid waste (MSW) into a methane-rich gas that can be used interchangeably with natural gas for industrial and utility purposes. The basic elements of the Syngas process are illustrated in Figure 1.* Shredded but unseparated solid waste is fed into the first stage of the Syngas system which is the so-called methane production reactor (MPR). In the MPR, the volatile portions of the MSW are gasified. Heavier liquid products occurring in pyrolysis systems are cracked to methane because of the hydrogen in the feed gas to the MPR.

After leaving the MPR, solids fall through a stripping zone where steam entrains the very light (compared to the metal and glass) organic char which is blown into an oxygen or air-blown gasifier. The hot synthesis gas produced in the gasifier is then fed directly to the MPR.

Metal and glass, after falling through the stripping zone, enter a quench pot and are discharged from the reactor system in a water slurry from which they are separated for recovery and recycle. Some of the more significant features in this system are the following:

- Separate control of methane production and gasification reaction zones to exploit the fact that methane production requires considerably milder conditions than gasification.

*Patent applications have been filed on other gasification alternatives which would be applicable to other feeds. However, the basic principle remains the same as for the illustrated system.

HERMAN F. FELDMANN • Battelle, Columbus Laboratories, 505 King Avenue, Columbus, Ohio 43201.

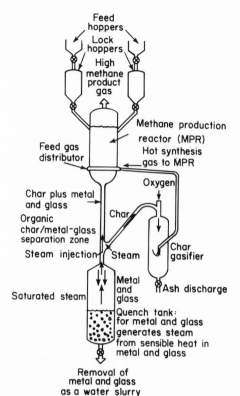

FIGURE 1. Conceptual solid waste gasification system.

- The contacting schemes for handling the metal- and glass-containing shredded MSW will be different than that for gasification of the resulting char.
- Metal and glass can be separated from organic materials after the methane production zone before exposure to oxidation or slagging conditions.
- Physical separation of methane production and gasification zones insures that no methane is burned by oxygen or reformed by steam fed to the gasifier.

These features of the system allow for considerable improvements in system thermal efficiency by increasing the methane yield and allowing the reactive portion of the solid waste to be converted at a lower temperature than is required for the gasification of the less reactive residual char.

Resource recovery and recycle are also greatly facilitated by the Syngas configuration. For example, the temperature in the MPR is low enough to prevent metal and glass from slagging while at the same time allowing the removal of organic contamination from the metal and glass by pyrolysis and cracking. Freeing the metal and glass of organic contaminants increases their recycle value. Separation of metal and glass from the organic char is easily effected by virtue of the great density difference between the metal and glass and the organic char.

2. Experimental Basis for the Process

Because of the availability of data on the gasification of carbonaceous char, the key data needed for the design of the reactor system shown in Figure 1 was for the MPR. Therefore, the experimental program focused on the devolatilization and hydrogasification of the solid waste. Full details of this experimental study have been reported[1] and the experimental data from this program plus literature data on the oxygen–steam gasification of chars produced from coal were used to calculate the overall material and energy balances for an integrated Syngas system.

One of the important findings in the experimental program was that in the MPR the MSW conversion rate is limited by the rate of heat transfer at a temperature level of 1000–1200°F. Because the organic fraction of solid waste is comprised mainly of paper which is relatively thin, heat transfer rates in general are fairly rapid. This means that the MPR reactor is capable of high volumetric throughputs of MSW.

Another finding of the study that is important to the selection of an optimum plant design is that the conversion rates are not greatly influenced by operating pressure. Thus, having a high hydrogen partial pressure is not essential to the process chemistry, and the operating pressure can be selected based on the availability of solid feed systems and ultimate gas utilization.

3. Reactor Design Considerations

Experimentally, the MPR was operated as both a free-fall and moving-bed reactor. However, the tests established that the conversion rate was limited only by heat transfer to the particle and that the conversion level was determined by the H_2 "seen" by the waste which is expressed by the H_2-to-MSW feed ratio. Thus, there is considerable flexibility in reactor design.

For the purposes of estimating reactor throughput, a free-fall system is assumed. One major advantage of a free-fall MPR is that there should be no problems with bridging of solid waste. (For example, even our 2.8-in-I.D.* MPR was operable in the free-fall mode without bridging problems.) The pres-

*I.D. = inside diameter.

ence of metal and glass should assist in maintaining flow through the free-fall zone. Another advantage of the free-fall system is the countercurrent flow of solids and gas which allows increased utilization of sensible heat in the product gases, thereby allowing direct utilization of higher moisture content waste.

Crude measurements of the free-fall velocity of shredded paper, which will have the lowest terminal velocity of any significant fraction of solid waste, indicate that its free-fall velocity is about 4 ft/sec. Operating this system in the free-fall mode at a gas velocity limited to 3 ft/sec allows a gas production rate in the MPR equivalent to about 7 million Btu/hr-ft² of reactor cross section (1100 lb dry organic feed per hr per ft² of reactor cross section). This specific Btu production rate is quite high compared to most coal gasification processes especially fixed- or moving-bed systems. For example, the specific Btu output of a Wellman-Galusha coal gasifier has been reported to be about 1.5 million Btu/ft²-hr.[2]

After the char–metal–glass mixture falls through the free-fall zone, the char is stripped from the metal–glass mixture by a stream of steam and the char is blown into a gasifier into which oxygen is injected to complete gasification of the char.

Before a detailed design of the system is possible, additional data will have to be generated. For example, it will be necessary to make more accurate measurements of entrainment velocities of various solid waste constituents after primary shredding. In addition, the drying rates of solid waste will have to be determined in order to evaluate whether predrying is necessary or if flash drying can be carried out in the top of the MPR itself.

Integration of the MPR and gasifier should provide no technical problem but will require more detailed engineering analysis.

4. Material Balance

Overall material balances for the Syngas process utilize experimental data generated in our small process development unit (PDU) to determine the key factors influencing carbon conversion and product distribution in the first stage MPR. As discussed, the char produced in the MPR is then fed to a second stage entrained gasifier where it is completely gasified with oxygen and steam. Because the char fed to the second stage is mostly devolatilized, the performance of the entrained gasifiers was simulated using published data for entrained systems feeding coal or coal char.[3] A different type of gasifier than the entrained gasifier will probably be selected for a commercial plant. However, for the purposes of estimating material balances, an entrained gasifier presents a conservative estimate because of the higher oxygen consumption compared, for example, to fluid-bed gasifiers.

For the steam–oxygen gasifier, experimental data presented by von Fredersdorff and Elliott[3] for an IGT entrained gasifier was used. A comparison of the IGT entrained gasifier data with that from other entrained gasifiers indicated excellent consistency and it is therefore felt that the results are typical. Because the char from the MPR enters the gasifier at 1000–1200°F, oxygen consumption should be less than that calculated based on the above gasifier tests because the coal used in these tests was not preheated. It should be stressed that the experimental data were generated in a gasifier that was small compared to commercial size. Thus, these results include the effects of heat losses and conservatively represent the performance of a larger system.

After cooling and quenching, the product gas from the Syngas reactor will have the approximate composition and the yield shown in Table 1.

The major uncertainty in the gas composition is the distribution between H_2, CO, and CO_2, which is determined by the water–gas shift reaction ($H_2O + CO \rightarrow H_2 + CO_2$). The above composition is based on assuming an H_2-to-CO ratio of approximately 1. Since the water–gas shift reaction is somewhat exothermic and produces CO_2, which is a noncondensible gas, the greater the extent of this reaction the lower the heating value of the quenched product gas.

Because oxygen consumption is important in determining gas production costs and gas composition, the oxygen consumption projected for the Syngas process is compared with those reported for two fixed-bed gasification processes in Table 2. The Purox process[4,5] utilizes a slagging bottom and is being developed for solid waste feeds, while the Lurgi process[6] is a commercial gasification process utilizing coal as a feedstock.

The lower oxygen consumption of the Syngas process compared to the Purox process is due to the separation of gasification zones, lower gasification temperature, no slagging of metal and glass, and increased methane production. The quoted Lurgi process data are for a western coal and the primary

TABLE 1. Fuel-Gas Yield and Composition
from Syngas Reactor System

Fuel-gas composition	Volume percent
H_2	24.7
CH_4	18.8
CO	24.7
CO_2	30.9
C_2H_6	0.9
Total	100.0

Heating Value = 366 Btu/scf
Yield, scf/lb dry "organics" = 18.8

TABLE 2. Oxygen Requirements per Million Btu of Raw Product Gas

Process	Oxygen requirements (tons oxygen/million Btu, raw product gas)
Syngas	0.014
Purox[4,5]	0.029
Lurgi[6]	0.019[a]

[a]For a high-oxygen western coal.

reason for the reduced oxygen consumption of the Syngas process compared to the Lurgi is the higher oxygen content of the MSW compared to the western coal.

Lurgi data with coals of varying oxygen content demonstrate that increases in the oxygen content of the coal reduce the external oxygen that must be added. Thus, based on the relatively high oxygen content of MSW compared to coal, one would expect to be able to gasify MSW with considerably less external oxygen than is required to convert coal.

5. Integrated Process Flowsheet

An integrated process flowsheet is shown in Figure 2. A typical capacity for such a plant would be on the order of 400 tons/day of MSW, a size that minimizes supply and logistics problems while still allowing the economy of scale. It is assumed that the waste would be delivered from transfer stations where primary shredding would be done.

The plant consists of the Syngas reactor system into which is fed the shredded MSW, steam, and oxygen. Raw products include the wet fuel gas containing entrained ash plus a low yield of organic liquids. (In previously referred to pilot plant experiments[1] less than 1% of the solid waste was recovered as a separate organic phase.)

Dust and particulate material are removed with a cyclone and the dust-free gas is then quenched to remove water and tar. The tar is returned to the gasifier for conversion to synthesis gas. Entrained mist is removed and the final product gas is available for industrial fuel applications where, for the majority of applications, it will be interchangeable with natural gas.

A portion of the product gas is burned to generate the plant steam requirements. It is assumed that an on-site package oxygen plant could be used to provide the oxygen required for gasification and that the oxygen plant would operate on purchased electricity.

Metal and glass are discharged from the collection pot as a water slurry

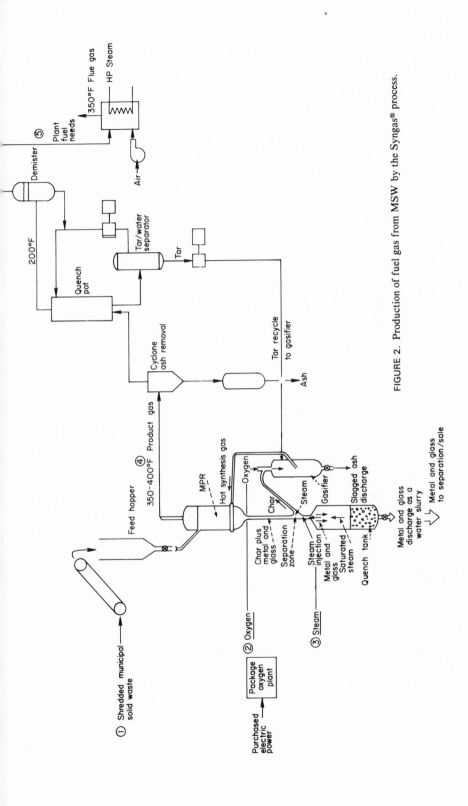

FIGURE 2. Production of fuel gas from MSW by the Syngas® process.

and are separated for sale. The separation portion of the plant is not included in the flowsheet.

As previously discussed, the operating pressure of the plant would be selected on the basis of the requirements of the ultimate gas user and the availability and cost of solids feed systems. Typically operating pressures could range from atmospheric to 10–15 atm, depending on the distance from the gas user and the ultimate utilization of the gas.

6. Process Energy and Material Balances

The development of a detailed process energy balance requires a more extensive process design than is currently available for the Syngas process. However, evaluations of similar types of plants indicate that the major energy consumption will be for the production of oxygen and the generation of process steam. For example, in a recent Solar Energy Research Institute (SERI) publication[7] evaluating a slagging-bottom fixed-bed gasifier, over 90% of the electricity purchased was for the operation of an oxygen plant.

In the process configuration shown, the product is an intermediate-Btu gas and no acid gas removal is included because the gas will conform to EPA standards without sulfur removal. Therefore, plant steam requirements will be primarily for gasification. The steam is assumed to be generated in an on-site boiler fired by a portion of the product gas. Efficiency of the on-site boiler is assumed to be 60%.

The other major process energy requirement is for the production of oxygen. The most likely source of oxygen for a plant this size would be a package oxygen plant operated with purchased electricity. To correct the net product gas production for the energy required by the oxygen plant, an electric power consumption of 300-kw-hr/ton of oxygen[8] is assumed. The amount of product gas required to generate the necessary power, assuming a 30% conversion efficiency to electricity, is then subtracted to estimate the overall thermal efficiency for the plant. As mentioned, the electricity is purchased and not generated by the combustion of the product gas, but making this comparison allows a direct comparison with "grass roots"-type gasification plants.

To make the process flow streams calculations as handy as possible, yield data are reported on a per-ton-of-organic-waste basis. Thus, the yields can easily be applied to various sized systems and to solid wastes containing differing amounts of inorganic constituents.

Table 3 summarizes the material balances based on one ton of "organic" solid waste containing 30 wt % moisture. Because of the low oil yields and anticipated low oil quality, no attempt would be made to market the oil by-product, and it would simply be separated from the aqueous phase and fed into the gasifier for complete conversion to synthesis gas.

TABLE 3. Production of an Intermediate-Btu Gas by the Syngas Process[a]

	Stream number					
	1	2	3	4	5	6
Major elements						
Carbon	53.4					
Hydrogen (as H_2)	78.2(33.4)[b]		26.4			
Oxygen (as O_2)	37.8(16.6)[b]	8.28	12.3			
Gaseous constituents						
H_2				17.1	1.4	15.7
CH_4				13.1	1.1	12.0
CO				17.1	1.4	15.7
CO_2				21.3	1.8	19.5
C_2H_6				0.60	0.04	0.56
C_6H_6				0.20	0.02	0.18
H_2O			26.4	58.7	—	—
Total				128.1	5.76	63.64

[a]Basis: 1 ton of wet organic material containing 30 wt % moisture; flow rates given in lb-moles.
[b]Value in parenthesis is the contribution of the moisture in the solid waste.

After quenching and dust removal, a portion of the product gas, Stream 5, is burned to generate the steam required for gasification. Stream 6 is the clean intermediate product gas that is available for sale.

The simplest way to summarize the energy input and output of the overall process is in the form of a pie chart, as shown in Figure 3.

While the value of 85% of the heating value recovered as saleable gas is high compared to the efficiency of SNG from coal plants, which have efficiencies on the order of 70%, it must be remembered that the MSW gasification

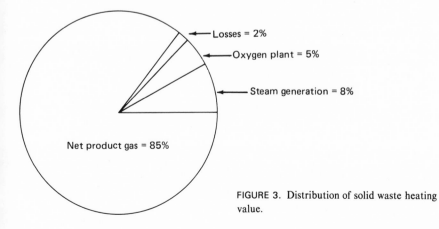

Losses = 2%

Oxygen plant = 5%

Steam generation = 8%

Net product gas = 85%

FIGURE 3. Distribution of solid waste heating value.

plant does not require: (1) water–gas shift to adjust the H_2-to-CO ratio, (2) acid gas removal, and (3) methanation to produce a methane product. In addition, the plant is considerably simpler. Thus, while a detailed process design will probably result in a somewhat different thermal efficiency than reported here, it is expected that the thermal efficiency will nevertheless be considerably higher than the more complex SNG plant.

References

1. H. F. Feldmann, G. W. Felton, H. Nack, and J. Adlerstein, Syngas process converts waste to SNG, *Hydrocarbon Process.*, **55**, 202–204 (1976).
2. R. W. Culbertson and S. Kasper, Economic advantages and areas of application of small gasifiers, presented at the Fourth International Conference on Coal Gasification, Liquefaction, and Conversion to Electricity, University of Pittsburgh, Pittsburgh, Pennsylvania (August 2–4, 1977).
3. C. G. von Fredersdorff and M. A. Elliott, Coal gasification, in: *Chemistry of Coal Utilization,* (H. H. Lowry ed.), John Wiley and Sons, Inc., New York and London (1963), pp. 892–1022.
4. Solid Waste Disposal Resource Recovery, Brochure Published by Union Carbide's Environmental Systems Department, Linde Division, 270 Park Avenue, New York (undated).
5. J. E. Anderson, The oxygen refuse converter—A system for producing fuel gas, oil, molten metal and slag from refuse, presented at the American Society of Mechanical Engineers 1974 National Incinerator Conference, Miami, Florida (May, 1974).
6. D. C. Elgin and H. R. Perks, Results of American coals in Lurgi pressure gasification plant at Westfield, Scotland, *Proceedings of the Sixth Synthetic Pipeline Gas Symposium,* American Gas Association, Chicago, Illinois (October 1974), pp. 247–265.
7. R. E. Desrosiers, Process designs and cost estimates for a medium Btu gasification plant using a wood feedstock, Solar Energy Research Institute-Technical Report 33·151 (1979).
8. S. Katell and P. Wellman, An Evaluation of Tonnage Oxygen Plants, *ACS Fuel Division Preprints,* **14** (3), (1970), pp. 99–104.

Section C
Biochemical Conversion Processes

15

Basic Principles of Bioconversions in Anaerobic Digestion and Methanogenesis

MICHAEL J. MCINERNEY AND MARVIN P. BRYANT

1. Introduction

In methane fermentation, organic matter is anaerobically degraded to CO_2 and CH_4 with only a relatively small yield of microbial cells.[1,2] Thus, a large amount of organic matter is destroyed, but about 90% of the energy available in the substrate is retained in the easily purified gaseous product, CH_4. Methane fermentation is very important in the recycling of carbon and other elements in nature and has long been used for the stabilization of sewage wastes. Its potential for converting biomass resources such as animal manures and other agricultural residues and municipal refuse into the energy-rich fuel, CH_4, has only recently been fully appreciated.

The process involves many different kinds of interacting microbial species, most of which do not directly produce CH_4.[3-6] Bacteria are the main biological agents involved in organic matter destruction and CH_4 production, but fermentative ciliate and flagellate protozoa and some anaerobic fungi may also contribute in some ecosystems. Methane fermentation occurs in many anaerobic ecosystems such as in sewage and organic waste digestors, aquatic sediments of lakes, estuaries, and marine systems, flooded soils, tundra, peat bogs, and marshes, where the main electron acceptor, CO_2, is produced from the degraded organic substrates. It does not occur in environments where other electron acceptors such as oxygen, sulfur, sulfate, or nitrate are readily avail-

MICHAEL J. MCINERNEY AND MARVIN P. BRYANT • Departments of Dairy Science and Microbiology, University of Illinois, Urbana, Illinois 61801.

able. Temperature-wise, two groups of microbes are involved; mainly thermo-philic species are active from about 45° C to 75° C and mesophilic species are active at lower temperatures, but *Methanosarcina* may be active in both ranges.[7] The pH range of the fermentation is from about 5 to 8 but the rate decreases rapidly below pH 7. Fossil fuels are probably remnants of the indi-gestible organic residue such as lignified plant materials that have undergone further geophysical modifications.

2. Stages of the Fermentation

In order to simplify the discussion of the microbiology, chemistry, and kinetics of the fermentation, several schemes have been described that separate it into various stages involving different metabolic groups of bacteria. While it is possible to do this in a scheme, it is emphasized that these bacterial groups cannot be separated in the discussion of their metabolism since the efficient metabolism of each group is dependent on the others.

2.1. Two-Stage Scheme

Since the 1930s, the fermentation has been described as a two-stage pro-cess involving two major metabolic groups of bacteria.[1,8] The acid-forming stage involves a complex of species that hydrolyze the primary substrate poly-mers such as polysaccharides, proteins, and lipids and ferment the product mainly to fatty and other organic acids, alcohol, ammonia, sulfide, CO_2, and H_2. The methane-forming stage involves a complex of species that degrade the acids, alcohols, H_2, and CO_2—formed in the first stage—to CO_2 and CH_4.

In 1967, Bryant et al.[9] showed that the fermentation of ethanol as carried out by *Methanobacillus omelianskii* was in fact carried out by a syntrophic association of two bacterial species; one species produced acetate and H_2 from ethanol only when the second species, a methanogen, was present to utilize the H_2. This, plus the fact that pure cultures of methanogens that degraded pro-pionate and longer-chained fatty acids had not been obtained, suggested that the two-stage scheme was unsatisfactory, i.e., alcohols other than methanol, and fatty acids other than formate and acetate were probably not catabolized by methanogens *per se* but by another metabolic group of bacteria.

2.2. Three-Stage Scheme

Figure 1 illustrates the scheme which best fits current information on methane fermentation. The first stage, as before, involves the fermentative bac-teria. Propionate and longer-chain fatty acids, some organic acids, and alcohols are probably degraded by a second intermediate group of bacteria called the

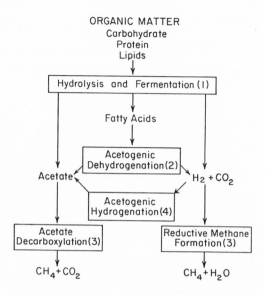

FIGURE 1. Three-stage scheme for the complete anaerobic degradation of organic matter showing the general pathways and the three major metabolic groups of bacteria: (1) fermentative bacteria; (2) obligate H_2-producing, i.e., proton-reducing, acetogenic bacteria; and (3) methanogenic bacteria. Acetate and sometimes other acids may be produced from H_2 and CO_2 by a fourth group of bacteria.

obligate H_2-producing (i.e., proton-reducing) acetogenic bacteria,[3,4] but only a few species have been documented. Some other bacteria produce acetate and sometimes other acids from H_2 and CO_2.[10–12] Finally, methanogens rapidly utilize the H_2 produced by the other bacteria to reduce CO_2 to CH_4, and some cleave acetate to CO_2 and CH_4 which is a quantitatively important reaction since about 70% of the methane produced is derived from the methyl group of acetate.

2.3. Metabolic Groups Involved in Partial Methane Fermentation

Partial methane fermentation occurs in certain parts of the gastrointestinal tract such as the rumen, cecum, and large bowel of many animals.[13] Due to the short retention time of the material in the system, usually only the fermentative bacteria and in some systems the H_2-utilizing methanogens grow fast enough to be maintained in the system. Thus, acetate and longer-chained fatty acids are not significantly degraded, and they accumulate and are absorbed from the tract and used by the animal as energy sources.

3. The Methanogens

The methanogenic bacteria are essential to anaerobic degradation because they are the only organisms that are able to catabolize acetate and hydrogen to gaseous products in the absence of light energy or exogenous electron acceptors such as oxygen, sulfate, or nitrate. Without the methanogens, effective degradation would cease because nongaseous, reduced fatty acid and alcohol products of the fermentative bacteria, that have almost as much energy content as the original substrate, would accumulate.

3.1. Physiology

Methanogens are a unique group of bacteria composed of many different species having quite different cell shapes and structures but all obtaining energy for growth through, as yet, incompletely known mechanisms that lead to the formation of CH_4.[6,14-16] They require very strictly anaerobic conditions for growth and can grow only when oxygen is excluded and the redox potential is maintained below about -300 mV. Nutritional requirements of these organisms are simple, and growth of most species occurs in mineral salts media with CO_2, ammonia, and sulfide as the main carbon, nitrogen, and sulfur sources, respectively.[6,14-17] Ammonia is essential for growth, and none of the species are known to use amino acids or peptides as the main nitrogen source.[17] Sulfide may be essential, but a few species can use cysteine as a sulfur source.[17] Some species require the addition of one or more compounds such as acetate, 2-methylbutyrate, coenzyme M, B vitamins, amino acids, and an unknown factor present in rumen fluid.[6,14-17] Lysis of the cells and possibly excretion of compounds by methanogens may supply essential nutrients such as amino acids, B vitamins or possibly other growth factors to more fastidious nonmethanogenic organisms.[18-20] These organisms contain several coenzymes: coenzyme 420, involved in electron transfer in place of ferredoxin; coenzyme M, involved in methyl-transfer reactions; factor B, a heat-stable cofactor with a molecular weight of about 1000, required for the enzymatic formation of CH_4 from methyl–coenzyme M; and probably others that have not been found in any other organism.[14] Recent evidence suggests that electron-transport-linked phosphorylation may be the source of adenosine 5′-triphosphate (ATP) synthesis.[21,22]

3.2. Phylogeny and Taxonomy

The comparative analysis of the similarity of oligonucleotide sequences of the 16 S ribosomal RNA molecule has been used by Carl Woese and his associates[23] as an approach to the study of the phylogenetic relationships among microbial species. The analysis of these relationships clearly shows that the methanogens constitute a well-delineated but diverse group of organisms that are very distantly related to other life forms. The divergence of the methanogens is the most ancient event so far detected in microbial evolution.[24] Table 1 shows the revised taxonomy of the methanogens based on these analyses as proposed by Balch et al.[15] These observations are reinforced by the unique biochemistry and coenzymes involved in methane formation described before and by the fact that methanogens differ from almost all other bacteria in not containing muramic acid in the cell wall[25] and in having C_{20} phytanyl and C_{40} diphytanyl glycerol ethers in their lipids in place of the usual ester-linked glycolipids and phospholipids.[26]

3.3. Substrates

As a group, methanogens utilize a narrow range of substrates (Table 1) as energy sources and individual species may use only one or two of these compounds. All species except the acetate-utilizing rod[27] and a thermophilic strain of *Methanosarcina* [7] utilize H_2 and CO_2 for growth as shown by Equation (1):

$$4H_2 + HCO_3^- + H^+ \rightleftharpoons CH_4 + 3H_2O$$
$$\Delta G^{\circ\prime} = -135.6 \text{ kJ/reaction}^* \quad (1)$$

The large negative standard change in free energy ($\Delta G^{\circ\prime}$) indicates that the equilibrium of the reaction is far in favor of H_2 use and CH_4 formation. The affinity of rumen methanogens for H_2 use is very high and the K_m for H_2 utilization is about 1 μM.[29] Anaerobic digesters[30,31] have also been shown to have a great capacity to utilize H_2 for CH_4 formation. This rapid use of H_2 by methanogens maintains a low concentration of H_2 in the ecosystems, about 1 μM in the rumen,[32] even though a large amount of H_2 is produced. Formate serves as an approximately equivalent energy source for many species of methanogens.

Acetate is an important substrate for methanogens and about 65–70% of the methane produced from organic matter in sewage digestors[33,34] and lake sediments[35] is produced via the methyl group of acetate. Only a few species of methanogens are believed to degrade acetate and only *Methanosarcina* species have been isolated in pure culture. Acetate is degraded as shown in Equation (2):

$$CH_3COO^- + H_2O \rightleftharpoons CH_4 + HCO_3^-$$
$$\Delta G^{\circ\prime} = -31.0 \text{ kJ/reaction} \quad (2)$$

Although the $\Delta G^{\circ\prime}$ for this reaction under standard conditions is barely sufficient to form one mole of ATP ($\Delta G^{\circ\prime} = -30.6$ kJ),[31] *Methanosarcina barkeri* obtains sufficient energy for growth from the catabolism of acetate in the absence of any other metabolizable substrate.[36–38] The small negative $\Delta G^{\circ\prime}$ may explain the relatively slow growth rate of methanogens on this substrate.

---

*All thermodynamic data ($\Delta G^{\circ\prime}$ or $\Delta E_0'$) given in this chapter were as reported by Thauer *et al.*[28] or calculated from data therein.

TABLE 1. Proposed Taxonomic Scheme of Balch et al.[15] Based on Comparative Cataloging of the 16 S Ribosomal RNA and Substrates Used for Growth and Methanogenesis

Taxa	Type strain	Former designation	Substrates for growth and CH$_4$ production
Order I. *Methanobacteriales* (type order)			
Family I. *Methanobacteriaceae*			
Genus I. *Methanobacterium* (type genus)			
1. *Methanobacterium formicicum* (neotype species)	MF	*Methanobacterium formicicum*	H$_2$, formate
2. *Methanobacterium bryantii*	M.o.H.	*Methanobacterium* sp. strain M.o.H.	H$_2$
Methanobacterium bryantii strain M.o.H.G.		*Methanobacterium* sp. strain M.o.H.G.	H$_2$
3. *Methanobacterium thermoautotrophicum*	ΔH	*Methanobacterium thermoautotrophicum*	H$_2$
Genus II. *Methanobrevibacter*			
1. *Methanobrevibacter ruminantium* (type species)	MI	*Methanobrevibacter ruminantium* strain MI	H$_2$, formate
2. *Methanobrevibacter arboriphilus*	DH1	*Methanobacterium arbophilicum*	H$_2$
Methanobrevibacter arboriphilus strain AZ		*Methanobacterium* sp. strain AZ	H$_2$
Methanobrevibacter arboriphilus strain DC		*Methanobacterium* strain DC	H$_2$, formate
3. *Methanobrevibacter smithii*	PS	*Methanobacterium ruminantium* strain PS	

Taxon	Strain	Species/isolate	Substrates
Family I. *Methanococcaceae*			
Genus I. *Methanococcus*			
1. *Methanococcus vannielii* (neotype species)	SB	*Methanococcus vannielii*	H$_2$, formate
2. *Methanococcus voltae*	PS	*Methanococcus* sp. strain PS	H$_2$, formate
Order III. *Methanomicrobiales*			
Family I. *Methanomicrobiaceae* (type family)			
Genus I. *Methanomicrobium* (type genus)			
1. *Methanomicrobium mobile* (type species)	BP	*Methanobacterium mobile*	H$_2$, formate
Genus II. *Methanogenium*			
1. *Methanogenium cariaci* (type species)	JR1	Cariaco isolate JR1	H$_2$, formate
2. *Methanogenium marisnigri*	JR1	Black Sea isolate JR1	H$_2$, formate
Genus III. *Methanospirillum*			
1. *Methanospirillum hungatei*	JF1	*Methanospirillum hungatii*	H$_2$, formate
Family II. *Methanosarcinaceae*			
Genus II. *Methanosarcina* (type genus)			
1. *Methanosarcina barkeri* (type species)	MS	*Methanosarcina barkeri*	H$_2$, CH$_3$OH, CH$_3$NH$_2$, acetate
Methanosarcina barkeri strain 227		*Methanosarcina barkeri* strain 227	H$_2$, CH$_3$OH, CH$_3$NH$_2$, acetate
Methanosarcina barkeri strain W		*Methanosarcina barkeri* strain W	H$_2$, CH$_3$OH, CH$_3$NH$_2$, acetate

Preferred energy sources such as H_2 or CH_3OH may regulate the degradation of acetate by *M. barkeri* in a manner that resembles catabolite repression where one or more enzymes needed for acetate use is repressed.[38] The role that H_2 plays in regulating acetate degradation in natural ecosystems remains to be determined. Short-term exposure to H_2 does not affect acetate degradation in sludge[29] or in cultures[39] that contain the acetate-degrading rod mentioned earlier but the incubation period was probably too short to significantly affect the level of pre-existing enzyme or enzymes needed for acetate degradation.

Other methanogenic substrates used by *M. barkeri* include methanol and *N*-methyl compounds[40] such as methyl, dimethyl, or trimethylamine. Methyl mercaptan has been shown to be a methanogenic substrate in lake sediments,[41] but the organism responsible for its degradation is unknown.

4. The Fermentative Bacteria

The fermentative bacteria are a very complex mixture of many bacterial species, most of which are obligate anaerobes,[3,5,42] but some facultative anaerobes, such as streptococci and enterics, may be numerous. Anaerobic mesophilic species from genera such as *Bacteroides, Clostridium, Butyrivibrio, Eubacterium, Bifidobacterium, Lactobacillus,* and many others may be found among the predominant organisms. The thermophilic species isolated are often sporeforming anaerobes belonging to the genus *Clostridium,*[43] but, in work on thermophilic cattle waste digestors,[44] mainly gram-negative, nonsporing anaerobes were found.

Much information on the nutritional features of major rumen and intestinal fermentative bacteria[45,46] exists and it is likely that the nutrition of these bacteria from various habitats is similar. Most rumen bacteria have simple nutrient requirements that reflect the chemical composition of the environment.[45,46] Ammonia usually can serve as the main nitrogen source and is essential for some species; only a few species require amino acids, but some may use peptides as their main nitrogen source; sulfide often serves as the main sulfur source, although methionine and cysteine are sometimes required. Heme, a few B vitamins, and saturated fatty acids such as *n*-valerate, isobutyrate, 2-methylbutyrate or long-chain fatty acids may be required by some species.

4.1. Fermentation of Polysaccharides

The first stage of the fermentation is very similar to that occurring in the rumen,[9] and the biochemical activities of rumen and gut microorganisms have recently been extensively discussed.[47,48] Polysaccharides such as cellulose, hemicelluloses, pectin, and starch are hydrolyzed to sugars and oligosaccharides which are then taken up by the bacteria and fermented to a variety of products as shown in Figure 2. The sugars are fermented primarily by the Embden–Meyerhof–Parnas pathway to pyruvate generating electrons, designated as 2H, that are actually reduced nicotinamide adenine dinucleotide (NADH). The kinds of products formed will depend on how pyruvate is further metabolized. It can be catabolized to acetate, CO_2, and H_2 or to propionate (via lactate or succinate), butyrate, or ethanol. Lactate has been found in large amounts in cattle waste[49] and may be important both as an intermediate and as a substrate under a few conditions. Succinate is an important extracellular intermediate because it is formed by some important fermentative species in the rumen and in sludge and is then decarboxylated to propionate by others.[50]

4.2. First Site of H_2 Regulation

Studies[51] on the metabolic interactions among fermentative bacteria and between fermentative bacteria and H_2-utilizing bacteria indicate that the concentration of H_2 in the ecosystem plays an important role in regulating the proportions of the various products produced by the fermentative bacteria. Central to this is the fact that H_2 is produced from electrons generated in the oxidation of reduced pyridine nucleotides [Equation (3)].

$$NADH + H^+ \rightleftarrows H_2 + NAD^+ \qquad \Delta G^{\circ\prime} = + 18.0 \text{ kJ/reaction} \qquad (3)$$

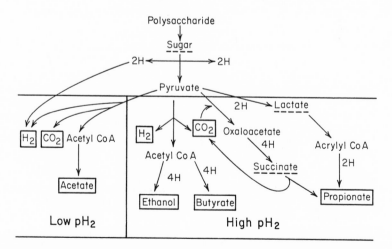

FIGURE 2. Pathways involved in the catabolism of carbohydrates by fermentative bacteria and major end-products formed at low and high partial pressures of H_2 (pH_2). $\square$ = final product; broken underline (- - - -) indicates extracellular intermediate.

The equilibrium of this reaction favors H_2 formation only when the partial pressure of H_2 is very low[51] as it is when H_2 is effectively metabolized by methanogens. At low partial pressures of H_2, the flow of electrons (NADH) generated during glycolysis is towards the reduction of protons resulting in H_2 formation which allows pyruvate to be degraded to acetate, CO_2, and H_2. H_2 production from pyruvate or acetaldehyde is energetically favorable even at a high partial pressure of H_2.[51] As the partial pressure of H_2 is increased as, for example, when the methanogenic system is stressed by shortening the retention time or overloading the system with degradable organic matter, the flow of electrons from NADH shifts from H_2 production to formation of reduced electron sink fermentation products such as propionate and longer-chained fatty acids, lactate or ethanol from pyruvate. Thus, the fermentative bacteria usually produced little or no ethanol or lactate and considerably less propionate and butyrate, and more acetate, CO_2, and H_2 in ecosystems where methanogens are effectively utilizing H_2.

4.3. Fermentation of Other Complex Substrates

Organic wastes usually contain larger amounts of protein and fats and less carbohydrate than do diets fed ruminants. Proteins are hydrolyzed to peptides and amino acids which are then fermented to products as shown in Figure 2 and to products such as isobutyrate, isovalerate, and D-2-methylbutyrate, n-valerate, various aromatic acids such as phenylacetic, phenylpropionic, benzoic, and indolylacetic, and ammonia and sulfide.[3,5,52] Glycerides, phospholipids and other fats are hydrolyzed with the release of long-chain fatty acids, and other products such as glycerol and galactose are fermented to products shown in Figure 2. The long-chain fatty acids are not further degraded by fermentative bacteria, but unsaturated fatty acids such as linolenic, linoleic, and oleic are hydrogenated to their corresponding saturated fatty acids.[3,5]

5. The H_2-Producing Acetogenic Bacteria

Only a few species of H_2-producing, i.e., proton-reducing, acetogenic bacteria have been isolated and studied, but this group is probably composed of many bacterial species, each having various energy-source specificities. As a group they degrade propionate and long-chain fatty acids, alcohols, and probably aromatic and other organic acids from the first-stage of fermentation with production of acetate, H_2, and, in the case of odd-numbered-carbon energy sources, CO_2.

5.1. Ethanol and Lactate Fermentations

The isolation of the S organism from $M.$ $omelianskii$ was the first documentation of a species of this group.[9] The methanogen was thought to oxidize ethanol to acetate and to reduce CO_2 to CH_4 [Equation (4)]:

$$2\ CH_3CH_2OH + HCO_3^- \rightleftarrows 2\ CH_3COO^- + CH_4 + H^+ + H_2O$$
$$\Delta G^{\circ\prime} = -116.4\ kJ/reaction \quad (4)$$

but this fermentation was shown to be carried out by a syntrophic association of two bacterial species. The S organism catabolizes ethanol to acetate and H_2 as shown in Equation (5):

$$CH_3CH_2OH + H_2O \rightleftarrows CH_3COO^- + H^+ + 2H_2$$
$$\Delta G^{\circ\prime} = +9.6\ kJ/reaction \quad (5)$$

The methanogen, on the other hand, uses the H_2 to reduce CO_2 to CH_4 [Equation (1)]. The formation of H_2 and acetate from ethanol is energetically unfavorable unless the H_2 is used to reduce CO_2 to CH_4. H_2 inhibits the growth of the S organism on ethanol and good growth occurs only when the H_2 utilizer is present.

Strains of *Desulfovibrio desulfuricans* and *D. vulgaris*[53] produce H_2 from lactate or ethanol when grown without sulfate in the presence of H_2-utilizing methanogens. Lactate is degraded to acetate, CO_2, and H_2 [Equation (6)] and ethanol is degraded to acetate and H_2 [Equation (5)] if the H_2 produced is rapidly utilized by a methanogen for CH_4 formation.

$$CH_3CHOHCOO^- + 2H_2O \rightleftarrows CH_3COO^- + HCO_3^- + H^+ + 2H_2$$
$$\Delta G^{\circ\prime} = -4.2\ kJ/reaction \quad (6)$$

This rapid use of H_2 by methanogens causes a shift in the equilibria of these reactions which makes the production of H_2 favorable and allows growth of acetogenic bacteria on these substrates. When *D. desulfuricans* is grown in the presence of *M. barkeri* which uses both acetate and H_2 for methanogenesis, lactate is completely degraded to CO_2 and CH_4.[54]

5.2. Fatty Acid-Oxidizing, H₂-Producing Bacteria

The propionate and longer-carbon-chained fatty acids are ecologically much more important as intermediates in anaerobic degradation than are lactate or ethanol,[3,13,39,55] but the species that catabolize the compounds have only recently been documented. Work with highly purified cultures suggests that methanogens are able to degrade these compounds.[8] *Methanobacterium suboxydans* was thought to β-oxidize butyrate and caproate to acetate and CH_4 and valerate to acetate, propionate, and CH_4. Propionate could then be decarboxylated to acetate, CO_2, and CH_4 by *Methanobacterium propionicum*. However, as indicated above, it is now believed that these compounds are degraded by H_2-producing acetogenic bacteria because of the following: (1) pure cultures of fatty-acid-degrading methanogens were never obtained; (2) the ethanol-fermenting culture, *M. omelianskii,* was shown to be a syntrophic association of two bacterial species[9]; (3) propionate and butyrate enrichments utilize H_2 without a lag and, when vigorously sparged with CO_2, H_2 replaces CH_4 as a product in these enrichments but not in acetate enrichments[56]; and (4) short-term exposure to H_2 inhibits propionate and butyrate but not acetate degradation by enrichments[56] and by sewage sludge.[29]

An anaerobic fatty-acid-catabolizing bacterium has been recently isolated in coculture with H_2-utilizing bacteria such as methanogens or desulfovibrio.[57] It β-oxidizes even-numbered-carbon fatty acids such as butyrate [Equation (7)], caproate or caprylate to acetate and H_2

$$CH_3CH_2CH_2COO^- + 2H_2O \rightleftarrows 2CH_3COO^- + H^+ + 2H_2 \qquad (7)$$
$$\Delta G^{\circ\prime} = +48.1 \text{ kJ/reaction}$$

or odd-numbered-carbon fatty acids such as valerate [Equation (8)] or

$$CH_3CH_2CH_2CH_2COO^- + 2H_2O \rightleftarrows CH_3COO^-$$
$$+ CH_3CH_2COO^- + H^+ + 2H_2 \qquad (8)$$
$$\Delta G^{\circ\prime} = +48.1 \text{ kJ/reaction}$$

heptanoate to acetate, propionate and H_2. This bacteria is unable to utilize any common bacterial energy source or combination of electron donor and electron acceptor that would enable it to grow without the H_2-utilizing bacterium; it may be the first documented obligate syntrophic bacterium. A species of bacterium that catabolizes propionate to acetate, CO_2 and H_2 [Equation (9)]

$$CH_3CH_2COO^- + 3H_2O \rightleftarrows CH_3COO^- + HCO_3^- + H^+ + 3H_2 \qquad (9)$$
$$\Delta G^{\circ\prime} = +76.1 \text{ kJ/reactions}$$

has recently been isolated.[58]

5.3. Second Site of H_2 Regulation

The equilibrium of reactions for the degradation of butyrate and propionate under standard conditions [Equations (7) and (9)] does not favor degradation as indicated by positive values for $\Delta G°$'s. But, under the conditions that occur during efficient methane fermentation in natural systems, these compounds are readily degraded. This is because the partial pressure of H_2 is maintained at a very low level in these ecosystems, making H_2 production from these compounds thermodynamically favorable, and, consequently, this is the second site where H_2 plays a key regulatory role.

Figure 3 shows the effect of the partial pressure of H_2 on the $\Delta G°'$ for the reactions involving ethanol, propionate, and butyrate degradation and methane formation. When the partial pressure of H_2 is decreased below about 0.15 atm, the degradation of ethanol becomes energetically favorable (negative $\Delta G°'$ value). But the degradation of butyrate or propionate is not energetically favorable until the partial pressure is lowered to about 2×10^{-3} atm or 9×10^{-5} atm, respectively. Thus only a slight increase in the partial pressure of H_2 will

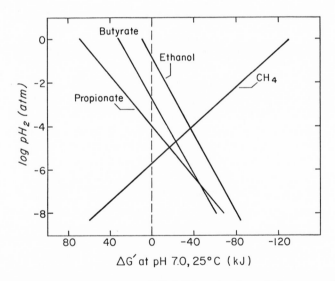

FIGURE 3. Effect of the partial pressure of H_2 (pH_2) on the change-in-free-energy ($\Delta G°'$) for the reactions involving ethanol, propionate, and butyrate degradation [Equations (4), (7) and (9), respectively] with H_2 and acetate production and methane formation [Equation (1)] from H_2 and CO_2. The calculations were based on the assumption that the concentrations of ethanol, acetate, propionate, and butyrate were 1 mM each, that the concentration of bicarbonate was 50 mM, and that the partial pressure of CH_4 was 0.5 atm using the equation, $\Delta G' = \Delta G°' + 1.36$ log [(products)/(reactants)].[28]

stop the degradation of these compounds with propionate degradation being the first reaction to be inhibited; many studies have shown that during digester failure propionate is the first acid that accumulates.[59] The formation of CH_4 from H_2 and CO_2, however, is still energetically favorable even at a partial pressure of H_2 below 10^{-5} atm. The partial pressure of H_2 not only regulates the proportion of endproducts of the fermentative bacteria but also the degradation of these products by H_2-producing acetogenic bacteria and again points out the essential role of methanogens in the fermentation.

6. Stoichiometry, Kinetics, Environmental and Nutrient Parameters of Fermentation

6.1. Stoichiometry

Knowledge of the chemical composition of the substrate is important for predicting digestibility and methane production. The extent of digestion of cellulose and other plant cell-wall components is inversely related to the amount of the indigestible components such as lignin and silica that is associated with these polysaccharides. Equations, similar to those developed for estimating ruminal digestion of forages by Goering and Van Soest,[60,61] can probably be developed for the estimation of the extent of organic matter digestibility and methane production from animal wastes and other organic substrates based on the effects of these indigestible components. Thermochemical treatment increases the biodegradability of the organic substrate by removing lignin from the polysaccharide, thereby making it more accessible to microbial attack, and by degrading some of the lignin into soluble anaerobically biodegradable substances.[62]

Organic matter destruction is directly proportional to methane production and *vice versa*. Buswell and Mueller[63] developed Equation (10) to predict the amount of methane produced based on a knowledge of the chemical composition of the waste:

$$C_nH_aO_b + (n - \tfrac{1}{4}a - \tfrac{1}{2}b)H_2O \rightleftarrows (\tfrac{1}{2}n - \tfrac{1}{8}a + \tfrac{1}{4}b)CO_2$$
$$+ (\tfrac{1}{2}n + \tfrac{1}{8}a - \tfrac{1}{4}b)CH_4 \quad (10)$$

6.2. Kinetic Factors Influencing Efficiency

The efficiency of methane fermentation is often defined as either the extent of organic matter destruction measured as the percent volatile solids (VS) destroyed or by the methane production rate, and which of these two parameters is used depends on whether the process is used for waste treatment or energy production. In a semicontinuously fed, completely mixed system, the efficiency is related to two important operational factors: (1) the hydraulic retention time, and (2) the volumetric organic matter loading rate.

6.2.1. Effect of Retention Time (RT)

The retention time (RT) of the system expresses the volume of fluid in the reactor per volume of fluids passing into and out of the reactor per day. Thus, a 10-liter reactor that is fed 2 liters/day so that the volume of the reactor fluid is constant, would have an RT of 5 days. In a completely-mixed, conventional process, RT is equal to the reciprocal of the microbial growth rate when it is long enough to allow the maintenance of the microbial population. There is a minimum RT below which efficient fermentation ceases due to the washout of the microbial population.[64] As the RT is increased in a system fed a constant concentration of organic matter in the feed, a higher proportion of the organic matter is destroyed, but less total organic matter is available, and the amount of methane produced per liter of reactor per day may decrease.

In studying the effect of retention time on the fermentation of sludge at 35° C,[65] it was found that protein and carbohydrate-fermenting bacteria grew rapidly, and the substrates were rapidly degraded to fatty acids even at RTs of less than one day. However, the fermentation of fatty acids did not occur until the RT was extended to about 5 days or more due to the slow growth of the fatty-acid-fermenting bacteria.

6.2.2. Rate-Limiting Step

As indicated above, and from studies on growth rates of isolated fatty-acid-degrading bacteria,[57,58] the rate-limiting step in methane fermentation often involves the degradation of fatty acids which is related to the efficiency of H_2 utilization by methanogenic bacteria as discussed in Section 5.3. McCarty[65] proposed a theory to explain these results based on the amount of energy released per electron equivalent of energy source fermented. Carbohydrate, protein, and H_2 fermentations release more energy per electron equivalent which results in greater cell synthesis per unit of substrate fermented and allows the organisms to reproduce faster so that they are maintained in the reactor at shorter RTs. The fermentation of acetate and long-chain fatty acids, on the other hand, releases less energy per electron equivalent of substrate fermented which results in lower cell synthesis and slower growth rates. The rate-limiting step depends on the composition of the waste since, in the fermentation of urban organic refuse which contains mainly cellulosic materials such as paper products and little lipid or protein, the hydrolysis of cellulosic materials may be the overall rate-limiting reaction.[66]

6.2.3.   Effect of Volumetric Organic Loading Rate

The volumetric organic loading rate is the rate at which organic waste is supplied to the reactor and can be expressed as the % wt of organic matter added each day to the reactor volume. It is related to the RT and the percentage of organic matter present in the feed as shown in Equation (11):

$$\text{Reactor loading rate (\%)} = (\% \text{ organic matter in feed})/\text{RT} \qquad (11)$$

Thus, the reactor loading rate can be increased at a given RT by feeding more concentrated slurry of organic matter or at a given percentage of organic matter by shortening the RT. At high loading rates, more CH_4 is produced per volume of reactor but less per mass of feed because of less VS destruction. Lowering the volumetric organic loading rate will result in an increase in the % VS destruction, but less methane will be produced per volume of reactor. The thermophilic digestion of cattle waste at reactor loading rates of 2.7% produced methane at a very high rate of 4.5 liters per day per liter of reactor when operated at a short RT of 3 days.[67]

For more extensive discussions on the bacterial growth and kinetics, the reader is referred to papers by McCarty[65] and Lawrence.[64]

6.3.   Nutrient and Environmental Requirements

The fermentation and microbial growth are dependent on the optimal supply of nutrients. Inorganic minerals are the only nutrients required other than the organic substrates utilized as energy sources since, as discussed before, the bacteria have simple nutrient requirements and the various organic materials required by some species, such as B vitamins, a small number of amino acids, or fatty acids, are supplied by other bacterial species. Most of the crude substrates utilized for methane fermentations contain sufficient minerals to satisfy the nutrient requirements for growth of the bacteria, but, in some cases, such as the fermentation of urban refuse,[66] the addition of major minerals such as ammonia phosphate, sulfide, and iron may be necessary.

Near-optimal environmental conditions must be maintained for efficient operation since the rate of the fermentation is quite slow, even under optimal conditions, when compared to other less complete fermentations.[3,59] The pH optimum is about 6.7–7.4, and pH values below 6 or above 8 are very restrictive. Ammonia toxicity may be a problem when feeding wastes such as animal wastes high in nitrogen. For the fermentation of some substrates such as cattle waste[67,68] or poorly degradeable high cellulosic urban waste,[66] thermophilic digestion is much more rapid, but more studies are needed to compare mesophilic and thermophilic temperatures on methanogenesis from other wastes. Excessive quantities of many materials such as ammonia, fatty acids, heavy metals, or more soluble light metals may be toxic to the fermentation.[69,70] The concentrations and conditions under which these materials become toxic are difficult to define as they may be modified by complex interactions or microbial adaptation. Drugs such as monensin, antibiotics, or methane inhibitors fed to animals may be inhibitory to the fermentation if they are present in animal waste and fed to unadapted digestors.

7. Summary

The degradation of organic matter to CO_2 and CH_4 in methane fermentation occurs by the concerted action of three major metabolic groups of bacteria. Fermentative bacteria hydrolyze the primary substrates such as proteins, lipids, and polysaccharides and ferment the products with the production of acetate and other saturated fatty acids, CO_2, and H_2 as major endproducts. The second group called the obligate H_2-producing (proton-reducing) acetogenic bacteria, produce H_2 and acetate (and sometimes CO_2) from endproducts of the first group. The methanogenic bacteria catabolize mainly acetate, CO_2, and H_2 to the terminal products. The maintenance of a very low concentration of H_2 in the ecosystem by methanogens is essential to efficient fermentation because it maintains lowered production of propionate and other reduced products while increasing formation of acetate and H_2 by fermentative bacteria; low H_2 concentration, furthermore, is essential for the catabolism of other fatty acids (and probably aromatic acids) to acetate and H_2 to be thermodynamically favorable.

Organic matter destruction is directly related to methane production as described by the Buswell–Mueller equation. The amount of lignin and other indigestible plant cell-wall components affects the digestibility of cellulose and other polysaccharides. The efficiency of the fermentation is related to the retention time of the material in the reactor and to the volumetric organic loading rate. The rate-limiting step in most fermentations is the degradation of fatty acids. Optimal environmental and nutrient parameters are important for efficient fermentation.

ACKNOWLEDGMENTS

We thank William Balch and Dr. Ralph Wolfe for allowing us to see pertinent materials in press.

References

1. P. L. McCarty, in: *Principles and Applications in Aquatic Microbiology* (H. Henkelekian and N. C. Dondero, eds.), John Wiley and Sons, Inc., New York (1964), pp. 314–343.
2. P. L. McCarty, *Public Works* 95, 107–112 (1964).
3. M. P. Bryant, *J. Anim. Sci.* 48, 193–201 (1979).
4. M. P. Bryant, in: *Microbial Energy Conversion* (H. G. Schlegel and J. Barnea, eds.), Verlag Enrich Gotze KG, Göttingen (1976), pp. 107–118.
5. P. W. Hobson, S. Bousfield, and R. Summers, in: *CRC Critical Reviews in Environmental Control*, Chemical Rubber Company, Cleveland (1974), pp. 131–191.
6. R. A. Mah, D. M. Ward, L. Baresi and T. L. Glass, *Ann. Rev. Microbiol.* 31, 309–341 (1977).
7. S. H. Zinder and R. A. Mah, in: *Abstracts of the Annual Meeting of the American Society for Microbiology*, I5, American Society for Microbiology, Washington D.C. (1979), p. 95.
8. H. A. Barker, *Bacterial Fermentations*, John Wiley and Sons, Inc., New York (1956).
9. M. P. Bryant, E. A. Wolin, M. J. Wolin, and R. S. Wolfe, *Arch. Mikrobiol.* 59, 20–31 (1967).
10. W. E. Balch, S. Schoberth, R. S. Tanner, and R. S. Wolfe, *Int. J. Syst. Bacteriol.* 27, 355–361 (1977).
11. K. Ohwaki and R. E. Hungate, *Appl. Environ. Microbiol.* 33, 1270–1274 (1977).
12. S. Schoberth, *Arch. Microbiol.* 114, 143–148 (1977).
13. R. E. Hungate, *The Rumen and Its Microbes*, Academic Press, New York (1966).
14. R. S. Wolfe, in: *Microbial Biochemistry*, Vol. 21 (J. R. Quayle, ed.), University Park Press, Baltimore, Maryland (1979), pp. 267–300.
15. W. E. Balch, G. E. Fox, L. J. Magrum, C. R. Woese, and R. S. Wolfe, *Microbiol. Rev.* 43, 260–296 (1979).
16. J. G. Zeikus, *Bacteriol. Rev.* 41, 514–541 (1977).
17. M. P. Bryant, S. F. Tzeng, I. M. Robinson, and A. E. Joyner, in: *Anaerobic Biological Treatment Processes, Advances in Chemistry Series 105*, American Chemical Society, Washington D.C. (1971), pp. 23–40.
18. A. J. B. Zehnder and K. Wuhrmann, *Arch. Microbiol.* 111, 199–205 (1977).
19. A. Wellinger and K. Wuhrmann, *Arch. Microbiol.* 115, 13–17 (1977).
20. L. Baresi, R. A. Mah, D. M. Ward, and I. R. Kaplan, *Appl. Environ. Microbiol.* 36, 187–197 (1978).
21. P. O. Mountfort, *Biochem. Biophys. Res. Commun.* 85, 1346–1351 (1979).
22. H. J. Doddema, T. J. Hutten, C. van der Drift, and G. O. Vogels, *J. Bacteriol.* 136, 19–23 (1978).
23. C. R. Woese and G. E. Fox, *Proc. Natl. Acad. Sci. USA* 74, 5088–5090 (1977).
24. G. E. Fox, L. J. Magnum, W. E. Balch, R. S. Wolfe, and C. R. Woese *Proc. Natl. Acad. Sci. USA* 74, 4537–4541 (1977).
25. O. Kandler and H. König, *Arch. Microbiol.* 118, 141–152 (1978).
26. T. G. Tornabene and T. A. Langworthy, *Science* 203, 51–53 (1979).
27. A. J. B. Zehnder and B. Huser, personal communication.
28. R. K. Thauer, K. Jugerman, and K. Decker, *Bacteriol. Rev.* 41, 100–180 (1977).
29. R. E. Hungate, W. Smith, T. Bauchop, I. Yu, and J. C. Rabinowitz, *J. Bacteriol.* 102, 389–397 (1970).
30. H. F. Kasper and K. Wuhrmann, *Appl. Environ. Microbiol.* 36, 1–7 (1978).
31. R. F. Strayer and J. M. Tiedje, *Appl. Environ. Microbiol.* 36, 330–340 (1978).
32. R. E. Hungate, *Arch. Mikrobiol.* 59, 158–164 (1967).
33. P. H. Smith and R. A. Mah, *Appl. Microbiol.* 14, 368–371 (1965).

34. J. S. Jerris and R. L. McCarty, *J. Water Pollut. Control Fed.* **37**, 178–192 (1965).
35. T. E. Cappenberg and R. A. Prins, *Antonie van Leeuwenhoek; J. Microbiol. Serol.* **40**, 457–469 (1974).
36. A. W. Lawrence and P. L. McCarty, *J. Water Pollut. Control Fed.* **41** (pt. 2), R1–R17 (1969).
37. R. A. Mah, M. R. Smith, and L. Baresi, *Appl. Environ. Microbiol.* **35**, 1174–1184 (1978).
38. M. R. Smith and R. H. Mah, *Appl. Environ. Microbiol.* **36**, 870–879 (1978).
39. L. van der Berg, G. B. Patel, D. S. Clark and C. P. Lentz, *Can. J. Microbiol.* **22**, 1312–1319 (1976).
40. H. Hippe, D. Caspari, K. Fiebig, and G. Gottschalk, *Proc. Natl. Acad. Sci. USA* **76**, 494–498 (1979).
41. S. H. Zinder and T. D. Brock, *Nature* **273** (5659), 226–228 (1978).
42. D. F. Toerien and W. H. J. Hattingh, *Water Res.* **3**, 385 (1969).
43. R. H. McBee, *Bacterial Rev. 14,* 51–63 (1951).
44. R. A. Leedle, M.S. Thesis, University of Illinois, Urbana, Illinois (1977).
45. M. P. Bryant, *Fed. Proc.* **32**, 1809–1813 (1973).
46. M. P. Bryant, *Am. J. Clin. Nutr.* **27**, 1313–1319 (1974).
47. R. A. Prins, in: *Microbial Ecology of the Gut* (R. T. J. Clark and T. Bauchop, eds.), Academic Press, New York (1977), pp. 73–183.
48. M. P. Bryant, in: *Duke's Physiology of Domestic Animals* (M. J. Sevenson, ed.), Cornell University Press, Ithaca, New York (1977), pp. 287–304.
49. J. E. Wolt and M. P. Bryant, unpublished results.
50. C. C. Scheifinger and M. J. Wolin, *Appl. Microbiol.* **26**, 789–795 (1973).
51. M. J. Wolin, *Am. J. Clin. Nutr.* **27**, 1320–1328 (1974).
52. S. R. Edsden, M. G. Hilton, and J. M. Waller, *Arch. Microbiol.* **107**, 283–288 (1976).
53. M. P. Bryant, L. L. Campbell, C. A. Reddy, and M. R. Crabill, *Appl. Environ. Microbiol.* **33**, 1162–1169 (1977).
54. M. J. McInerney and M. P. Bryant, in: *Abstracts of the Annual Meeting of the American Society for Microbiology,* Section I43, American Society for Microbiology, Washington, D.C. (1978), p. 88.
55. C. P. Chynoweth and R. A. Mah, in: *Anaerobic Biological Treatment Processes, Advances in Chemistry Series 105* (R. F. Gould, ed.), American Chemical Society, Washington, D.C. (1971), pp. 41–54.
56. D. R. Boone and P. H. Smith, in: *Abstracts of the Annual Meeting of the American Society for Microbiology,* Q82, American Society for Microbiology, Washington, D.C. (1978), p. 208.
57. M. J. McInerney, M. P. Bryant, and N. Pfennig, *Arch. Microbiol.,* **122**, 129–135 (1979).
58. D. R. Boone and M. P. Bryant, *Appl. Environ. Microbiol.* **40**, 626–632 (1980).
59. P. L. McCarty, *Public Works* **95**(10), 123–126 (1964).
60. H. K. Goering and P. J. Van Soest, Forage Fiber Analysis, Agricultural Handbook. No. 379, U.S. Department of Agriculture, Washington, D.C. (1970).
61. P. J. Van Soest, in: *Digestion and Metabolism in the Ruminant* (I. W. McDonald and A. C. I. Warner, eds.), The University of New England Publishing Unit, Armidale, Australia (1975), pp. 351–365.
62. P. L. McCarty, L. Y. Young, J. M. Gossett, D. C. Stuckey and J. B. Healy, in: *Microbiol Energy Conversion* (H. G. Schlegel and J. Barnea, eds.), E. Goltz KG, Göttingen (1976), pp. 179–199.
63. A. M. Buswell and H. F. Mueller, *Ind. Eng. Chem.* **44**, 550–552 (1952).
64. A. W. Lawrence, in: *Anaerobic Biological Treatment Processes, Advances in Chemistry Series 105* (R. F. Gould, ed.), American Chemical Society, Washington, D.C. (1971), pp. 163–189.

65. P. L. McCarty, in: *Anaerobic Biological Treatment Processes, Advances in Chemistry Series 105* (R. F. Gould, ed.), American Chemical Society, Washington, D.C. (1971), pp. 91–107.
66. J. T. Pfeffer and J. C. Liebman, *Resour. Recov. Conserv.* 1, 295–313 (1976).
67. V. H. Varel, H. R. Isaacson, and M. P. Bryant, *Appl. Environ. Microbiol.* 33, 298–307 (1977).
68. R. I. Mackie and M. P. Bryant, *Appl. Environ. Microbiol.* 41, 1363–1373 (1981).
69. P. L. McCarty, *Public Works* 95 (11), 91–94 (1964).
70. I. J. Kugelman and K. K. Chin, in: *Anaerobic Biological Treatment Processes, Advances in Chemistry Series 105* (R. F. Gould, ed.), American Chemical Society, Washington, D.C. (1971), pp. 55–90.

16
Design of Small-Scale Biogas Plants

MICHAEL R. BRULÉ

1. Introduction

Energy shortages and pollution problems have continued to accelerate interest in biogas plants for industry and agriculture. Recently, several commercial large-scale anaerobic-digestion facilities have come on stream in the wake of dwindling natural gas supplies. Small-scale units are also being implemented to supplement energy requirements for dairies and farms. The current status of these methane-from-cow-manure or "moothane" plants is reviewed herein. The intent is to acquaint those interested in, but not intimately familiar with this field, with current work in biogasification systems. Particular emphasis is on small-scale units; design criteria and economic considerations are discussed for an actual working unit.

2. Biomass Conversion to Methane

2.1. The Biogasification Process

"Biogasification," i.e., the gasification of biomass, can be accomplished by any of three processes: anaerobic digestion, hydrogasification, or pyrolysis.[1-3] The latter two processes have undergone more engineering development than anaerobic digestion.[4-7]

MICHAEL R. BRULÉ • School of Chemical Engineering and Materials Science, University of Oklahoma, Norman, Oklahoma 73019. Present address: Kerr–McGee Corporation, Oklahoma City, Oklahoma 73125.

In the absence of oxygen, organic materials such as natural wastes are decomposed by anaerobic digestion. According to McCarty,[8-10] the process occurs in three stages which are performed by two different groups of bacteria acting as a coupled system. In the first step, complex organics such as fats, proteins, and carbohydrates are converted by enzymatic hydrolysis to simpler organic compounds. The second step involves fermenting these simpler compounds to mostly volatile fatty acids by a group of facultative anaerobic bacteria commonly called "acid formers." In the third step, the organic acids are converted to carbon dioxide and methane by a group of strictly anaerobic bacteria collectively called "methane formers." Biogas formed in this step is methane-rich fuel having a heating value of 600–700 Btu/ft^3.

Many facets of anaerobic digestion have been explored through a variety of disciplines. Topics involve microbiology,[8-14] biochemistry,[15-20] toxicology,[36,39,41] and enzymology, among others, and kinetics,[10,11,21-28] process control,[11,22,26,27,29] and thermodynamics[9] to define the operational parameters[19,25,29-40] of anaerobic digestion. Some of the more prominent investigations are included in the References.

2.2. Large-Scale Commercial Endeavors

Worldwide use of biomass including agricultural products and waste has grown steadily in efforts to ameliorate the energy crunch.[42-48] The largest and most advanced effort so far is a $1 billion bioconversion plant in Brazil to convert sugar cane to ethanol for transportation fuels.[49] The International Energy Agency (IEA), Paris, is also sponsoring a multinational research and development program featuring a clearinghouse on biogasification. In the U.S., biogasification research-and-development funding has been intensified by the Department of Energy (DOE), Washington, D.C. The Institute of Gas Technology (IGT), Chicago, is exploring several bioconversion routes for the anaerobic fermentation of biomass.[2,3,49] Calorific Recovery Anaerobic Process (CRAP), a division of Thermonetics, Inc., Oklahoma City, has successfully gone commercial with the startup of their bioconversion plant in Guymon, Oklahoma.[49] The plant has begun converting about 7,300 tons/yr of dry cow manure to 640 million ft^3/yr of methane in addition to 200 tons/day of sludge which is slated to be used as fertilizer. Biogas of Colorado, Inc., Denver, is also engaged in designing a similar large plant.

2.3. Potential for Small-Scale Facilities

Potential methane production by anaerobic digestion of the manure produced by the over 100 million head of cattle in the U.S. could supply only 2.5–3% of the nation's current annual demand of over 25 trillion ft^3/yr of natural gas.[50] Although bioconversion can offer little alleviation of the energy crisis on a national basis,[44,51] it would have significant ramifications for pollution control.[5,52-55] Small-scale biogasification systems do offer indirect aid to a faltering food economy due to energy shortages in the agricultural sector. There is more manure available, by far, for bioconversion at the many thousands of dairies and farms scattered across the country than in the large feedlots distributed mostly in the Southwest. These farms and dairies are in need of a system not as large as the aforementioned commercial plants but much larger than past laboratory models.[56,57] Collection and transportation to a central regional bioconversion plant will probably remain economically unjustified; therefore, biogas should be produced and used on-site at a farm or dairy.[58] Considering the farmer's current unrest due to escalating costs to produce crops, a self-sufficient energy and pollution-control plant would be a welcome relief to the farmer's financial burdens.[45,59]

3. Small-Scale Biogas-Processing Facilities

3.1. Biogas Plants for Dairies and Farms

Farms and dairies are generally very energy-intensive installations.[7,54,55,58] Figure 1 illustrates the distribution of energy on a typical dairy farm with 100 cows.[7,54,55] Although agriculture uses only 3% of the nation's total energy demand, spiraling costs of farm fuels have a direct impact on food costs and, consequently, inflation. An alternative means of energy supply such as a farm-installed anaerobic-digestion facility could alleviate dependence on conventional high-cost fuels. A flow scheme which shows the dispensation of energy when bioconversion is implemented at a farm or dairy is given in Figure 2.

Animal wastes are not the only substrate available for anaerobic fermentation. Energy is also available for crop residues such as corn silage, stover and grain, bedding, waste feed, and other farm refuse. Biomass of these types from a 100-cow farm can be used to supplement animal wastes providing a total of 1.6 billion kilocalories potentially available for bioconversion.[7,54,55,58,60] Due to the efficiency of the digestion process, only about 15-20% of this energy input can be converted to energy in the form of methane gas.[7] Nevertheless, anaerobic digestion is capable of generating a substantial portion of the energy needed to maintain dairy-farm operations.

3.2. Pollution-Control Advantages

Many other prospective benefits may be reaped with a biogasification system at a farm or dairy. In addition to valuable fuel production, anaerobic fermentation provides an effective means for managing animal wastes and controlling pollution associated with these wastes.[61-63] Organic solids are

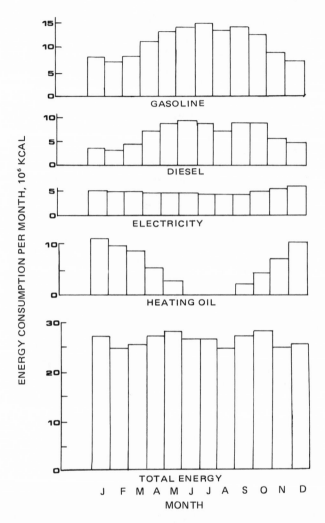

FIGURE 1. Distribution of energy on a 100-cow dairy farm.[7] Monthly energy consumption is given by type of fuel.

stabilized and reduced to a homogenized material less offensive in odor and less attractive to rodents and insects.[40,53] Also, pathogenic organisms are destroyed during the digestion process.[40] Nitrogen and other nutrients become more available for use by flora since the reduced carbon concentration resulting from carbon conversion to biogas increases the nitrogen concentration with respect to carbon.[37,38,40,64] It is further noteworthy that animal-feed supplements from, for example, single-cell protein production via bioconversion would

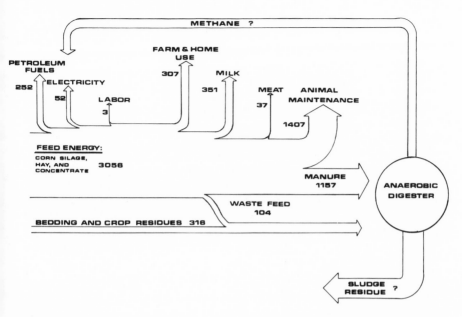

FIGURE 2. Dispensation of energy on a 100-cow dairy farm.[7] Each number is given in millions of kilocalories.

aid tremendously in that expensive category.[5] Indeed, "small systems are beautiful"[65] for biogas conversion when one considers the net impact on agricultural concerns.

3.3. Development of Small-Scale Biogas Plants

On the part of either government[66,67] or industry,[4] concerted efforts to build small-scale biofacilities have yet to be made. However, various innovative designs have emerged in mostly private ventures to implement biogasification on individual farms. A number of experienced biogas plant designers have published booklets concerning farm-power-from-farm-wastes or "dungas" production.[68–73] These entrepreneurs share the common objective of rural energy independence. As an example, one system features a displacement digestor which is a novel design concept whereby a cylindrical tank, open on one end, is inverted with the open end immersed in a fermentation vat.[69] Biogas produced under pressure in the fermentor displaces the biomass in the cylinder, which causes the cylinder to rise as it fills with methane. Thus, one basic unit serves as both the fermentor and gas-collection system.

Several academic institutions[1,5,53–56,60–62,74–78] and private research-and-development organizations[4,6,47,79,80–84] have conducted feasibility studies to determine the efficacy of small-scale biogasification units in agricultural operations. These amenability studies typically include parametric sensitivity analyses in attempts to define an optimum design.[4] Further, other technical aspects[23,24,25] concerning implementation of on-site biogas plants have been explored and well delineated. Many design proposals[5,61,85] have followed these studies, but the number of actual working units is relatively few.

4. Biogas Plant Design and Construction

A bioconversion system which generates methane from the anaerobic digestion of cow manure has been designed and constructed for a dairy by the University of Oklahoma School of Chemical Engineering and Materials Science (CEMS).[86−88] The primary objective has been to produce a "package" facility which is easy to install and operate, consisting of several modular components that can be assembled and put to use much like a "do-it-yourself" kit. Modular design is incorporated so that the plant can be retrofitted to a wide variety of existing farms and dairies.

4.1. Process Flow Description

The CEMS bioprocess incorporates a fairly straightforward method for converting raw manure to biogas and fertilizer as shown in Figure 3. The biosystem package shown in Figures 4–6 consists of a manure-loading sump, pump, hairpin double-pipe heat exchanger, fermentor, drying bed, floating-cover gas-storage-and-collection system, and a hydrogen sulfide stripper.[86−88] Fresh incoming manure is mixed with an equal amount of warm water in a 48-ft³ concrete sump. Concrete is fairly inexpensive and very durable. The sump is located in the milking area for two reasons. One is the ease with which manure can be collected and transferred to the sump. Since the typical cow spends about a third of her life in the milking area,[59] an ample amount of manure feedstock is concentrated in a comparatively small space. The other reason is that waste hot water used to clean the milk-barn floors can be drained into the manure sump. Use of this waste hot water minimizes heat-exchanger operations and thus partially defrays the cost for warming the manure to digestion temperatures.

Two concrete aprons encompass the site. Waste hot water used to wash off the milking-room floors is then added and mixed with a shovel to render the manure in a form more conducive to digestion. Approximately one part manure is mixed with one part water to insure a slurry content of about 7–9% dry solids by weight. Usually the farmer scrapes manure into a drain which is routed to an anaerobic lagoon; thus, no more time need be taken for loading the sump than is normally spent for disposing of the manure.

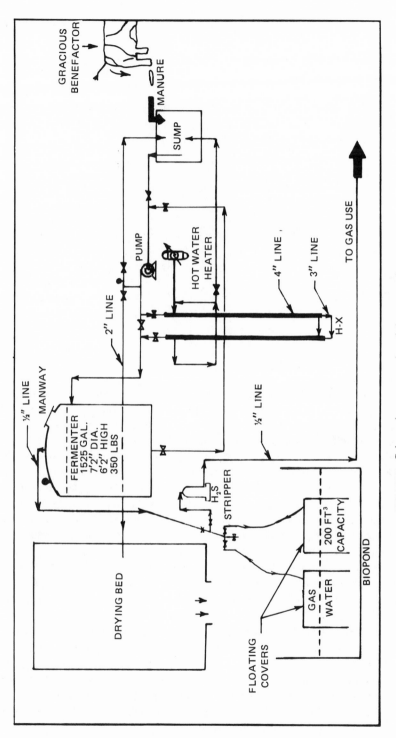

FIGURE 3. Schematic representation of the process flow.

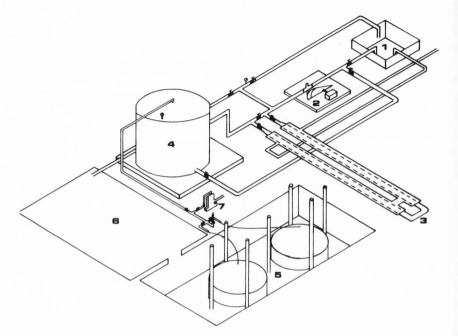

FIGURE 4. Isometric projection of site. 1 = sump; 2 = pump; 3 = heat exchanger; 4 = fermentor or anaerobic digestor; 5 = floating gas-storage covers; 6 = drying bed; 7 = H₂S stripper.

The manure–water mixture is pumped from the sump to the digestor by a suitable pump[88] which can pass some incompressible debris. A shell-and-tube, hairpin heat exchanger is used to control the temperature of the incoming manure. During cold weather, manure is warmed when passed through the heat exchanger; in the summer, the heat exchanger is bypassed and manure is routed directly from the sump to the fermentor. Simple, inexpensive, and fast construction of the heat exchanger, gas-collection system, as well as other plant equipment, is a paramount design consideration.

The use of lightweight plastic tanks proved to be quite advantageous to the construction of the CEMS biogas plant. These tanks are a product of recent technology and are much more convenient to work with than steel tanks. The polyolefin tanks weigh only 350 lb and can be easily moved by three people. In addition, these tanks have a fairly high impact resistance and can withstand working pressures of up to 15 psig. Bulkhead fittings were used to tie in all process lines to the fermentor. Fittings were installed with a small drill and sabre saw as compared with welding a steel tank. Digestor specifications and line sizes are given in Figure 3.

Each day, approximately 100 gallons of slurry, consisting of 50 gallons of manure and 50 gallons of water, can be pumped out of the fermentor and

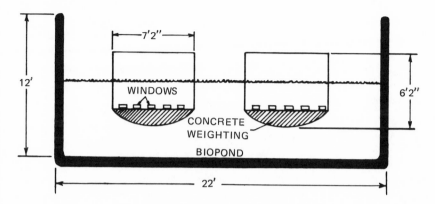

FIGURE 5. Floating-cover system.

FIGURE 6. The CEMS methane bioconversion system.[88] Biogas produced by the facility is to be used to fuel a hot-water heater located in the milk barn on the far right. The effluent on the drying bed shown on the left is suitable for spreading on pasture or cultivated land. The gracious benefactors in the background are admiring their new plant.

replaced with 100 gallons of fresh manure–water slurry. The outgoing digested manure is deposited on the drying bed. This method of charging the digestor on a daily basis versus a once-a-month batch loading has been found to optimize gas production.[88] Also, manure is not allowed to accumulate on the grounds,[89,93] making daily loadings favorable from a pollution-prevention standpoint.

4.2. Methane Storage System

Biogas produced in the digestor is routed to two storage covers floating in an anaerobic lagoon nearby. The floating covers are constructed from two 1525-gallon polyolefin tanks identical to the tank used to fabricate the digestor. The tops of the tanks are removed and weighted with about two tons of concrete to pressurize the stored gas. Each tank floats in an inverted position as shown in Figure 5. Open at the bottom, these tanks rise in the water as they fill with biogas. No compressors are required for this design making a very safe and inexpensive system to build and operate.

Additional gas generated by the biopond is also collected providing an even larger yield in gas production. The role of the biopond in the system design is twofold. First, it retains any liquid runoff from the surrounding milking area[91] and, second, it serves as a reservoir for the floating gas-storage vessels. Since anaerobic lagoons are present at most farms and dairies, no more space is needed for gas storage, not to mention compressors and gas-storage tanks that would also be needed. Again, the main objective for designing this system is economics. Polyolefin-tank use proves to be exceptionally innovative in that category as all three tanks used to fabricate the fermentor and gas-collection-and-storage system cost less than one steel tank that would have been used for the fermentor alone. Furthermore, gas-storage covers could be mass-produced complete with concrete weighting, which would eliminate any preparation before placing the covers in an anaerobic lagoon.

A hydrogen sulfide stripper has been installed to remove any H_2S in the gas before burning. The stripper is constructed from a 5-gallon plastic cylinder rated at 10 psig pressure. The cylinder is filled with wood chips impregnated with iron-(II) oxide. A drop-in cartridge filled with treated wood chips has been designed to make replacing expended chips easier. Very little hydrogen sulfide has been detected in the product gas, but the stripper should be installed to insure Environmental Protection Agency (EPA) standards have been met.

4.3. Fertilizer Production

Digested manure pumped out of the fermentor each day is deposited on a 25 ft × 30 ft drying bed. The bottom of the bed is lined with a 2-in. layer of concrete. A 6-in. layer of sand covers the concrete lining and acts as a medium on which the manure can percolate and thus become dewatered. Any excess liquid drains into the adjacent biopond. Once every two weeks the farmer can remove the dried manure for use as a fertilizer.

More detailed design descriptions, specifications, equipment costs, testing procedures, and recommended safety precautions are reported elsewhere.[87,88] The CEMS biogas plant, which was completed in 1975 at Norman, Oklahoma,[59] is shown in Figure 6.

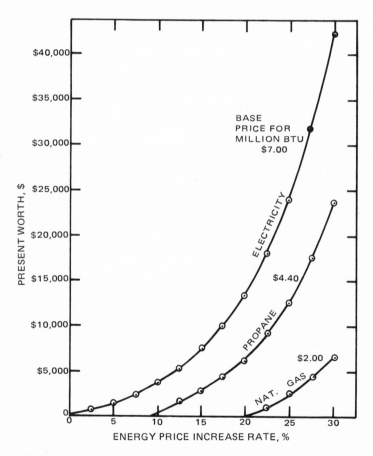

FIGURE 7. Present worth vs. annual energy-price-increase rate as a function of current energy price.[91,92]

5. Economic Feasibility and Marketability

The biggest deterrent to an on-site means for utilizing animal wastes has always been high cost; work has been aimed at designing and implementing a biogas facility which, when mass-produced, could be available at a low price. Such a unit may be built for an estimated 1975 cost of $4,000. This represents a significant decrease in the cost of biogas systems proposed heretofore,[56,61] putting these plants in a cost range attractive to farmers. Formerly, the installation of a biogasification system at a 100-cow dairy would encumber three to five years profit—a capital drain too large for any business.

Economic studies were conducted to evaluate the CEMS biogas system covering three cases: the first for dairies using natural gas, the second for those

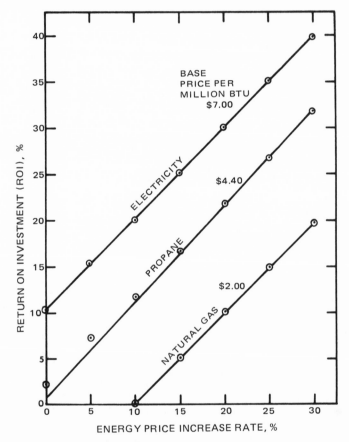

FIGURE 8. Rate of return on investment vs. annual energy-price-increase rate as a function of current energy price.[91,92]

using propane, and the third for dairies using electricity.[91] These studies assume a $4,000 capital investment with a 15-yr life and an annual interest rate of 10% and are based on the net energy produced, i.e., net biogas less parasitic power requirements for the pump and heat exchanger. The analyses in terms of present-worth and rate-of-return calculations are presented in graphical form in Figures 7 and 8. The results show the savings a farmer would realize compared to the cost of currently available forms of energy. For example, Figure 7 indicates that a biogas plant for a dairy farm using propane would have a present worth of almost $3,000, assuming an annual propane price rise of 15% per year[92] and a discount rate of 10%. Similarly, Figure 8 indicates that the methane plant will earn a 17% ROI for dairies using propane (15% annual price rise), which corroborates the results from the present-worth analysis of Figure 7. All cost estimates[92] are in 1975 dollars. The 1975 economic analyses are obviously conservative and favor the system more in the present market of the 1980s since energy costs are rising far more rapidly than equipment costs. A definite economic advantage exists for the CEMS system in that it is not very capital-cost sensitive to scale-up; a fifty-cow-size plant can be doubled in capacity for only a few hundred more dollars. A more detailed discussion of the economics and marketability of this biofacility is presented elsewhere.[87,88,91]

References

1. C. R. Engler, L. T. Fan, L. E. Erickson, and W. P. Walawender, Feedlot manure and other agricu'ture wastes as future material and energy resources, No. 45: Process Descriptions, Kansas State Univ., Manhattan, Kansas (1973).
2. Chementator, *Chemical Engineering,* (October 28, 1974).
3. Chementator, *Chemical Engineering,* (September 1, 1975).
4. E. Ashare, D. L. Wise, and R. L. Wentworth, Fuel Gas Production from Animal Residue, Dynatech Engineering Report No. 1551, Dynatech R/D Company, 99 Erie St., Cambridge, Massachusetts (January 14, 1977).
5. S. M. Barnett and D. J. Romano, Food, Fuel, and Fertilizers from Manure, Paper delivered at American Institute of Chemical Engineers National Meeting, Houston, Tx. (1975).
6. J. O. Burford, and F. T. Varani, Energy Potential through Bioconversion of Agricultural Wastes, final report to Four Corner Regional Commission, Grant FCRC 651-366-075 (1976).
7. W. J. Jewell, F. R. Morris, D. R. Price, W. W. Gunkel, D. W. Williams and R. C. Loehr, *Methane Generation from Agricultural Wastes: Review of Concept and Future Applications,* Cornell University, Ithaca, New York (1974).
8. P. L. McCarty, Anaerobic waste treatment fundamentals: I, Chemistry and microbiology; II, Environmental requirements and control; III, Toxic materials and their control; IV, Process design, *Public Works* **95,** 107–112 (September 1964), 123–126 (October 1964), 91–94 (November 1964), and 95–99 (December 1964).
9. P. L. McCarty, Thermodynamics of biological synthesis and growth, *Int. J. Air Water Pollut.* **9,** 621 (1965).

10. A. W. Lawrence and P. L. McCarty, Kinetics of methane fermentation in anaerobic treatment, *J. Water Pollut. Control Fed.* **41** (2), (1969).

11. J. Monod, The Growth of Bacterial Cultures, *Annu. Rev. Microbiol.* **3**, 371 (1949).

12. H. A. Barker, *Bacterial Fermentation,* John Wiley and Sons, Inc., New York (1956).

13. C. G. Golueke, W. J. Oswald, and H. B. Gotaas, *Appl. Microbiol.* **5**, 47 (1957).

14. H. J. Pelczar, Jr. and R. D. Reid, *Microbiology,* 2nd Edition, McGraw-Hill Book Company, New York (1965).

15. A. M. Buswell and H. F. Mueller, Mechanisms of methane fermentation, *Ind. Eng. Chem.* **44**, 550–552

16. J. F. Andrews and S. P. Graefe, *Adv. in Chem.* **105**, Anaerobic Biological Treatment Processes, American Chemical Soc. (1976) p. 126.

17. J. S. Jeris, and P. L. McCarty, The biochemistry of methane fermentation using C_{14} tracers, in: *Proceedings of the Seventeenth Industrial Waste Conference,* Purdue University, West Lafayette, Indiana (1963).

18. W. A. Pretorius, The effect of acetate utilizing methanogenic bacteria, *Water Res.* **6**, 1213–1217 (1972).

19. W. A. Pretorius, Principles of anaerobic digestion, in: *Water Pollut. Contr.* pp. 202–204 (1973).

20. P. G. Thiel, D. F. Tarieu, W. H. J. Hattingh, J. P. Kotzé, and M. L. Siebert, Interrelations between biological and chemical characteristics in anaerobic digestion, *Water Res.* **2**, 393–408 (1968).

21. E. S. Kirsch and R. M. Sykes, Anaerobic digestion in biological waste treatment, *Progr. Ind. Microbiol.* **9**, 155–236 (1971).

22. J. T. O'Rourke, Kinetics of anaerobic treatment at reduced temperatures, Doctoral dissertation, Stanford University, Stanford, California (1968).

23. A. E. Humphrey, S. Aiba, N. F. Millis, *Biochemical Engineering,* Academic Press, New York (1973).

24. A. E. Humphrey, Current developments in fermentation, *Chem. Eng.* (December 9, 1974).

25. B. Atkinson, *Biochemical Reactors,* Pion Ltd., London (1974).

26. J. L. Gaddy, E. L. Park, and E. B. Rapp, Kinetics and economics of anaerobic digestion of animal waste, *Water, Air, Soil Pollut.* No. 3. (1974).

27. S. Ghosh and F. G. Pohland, Kinetics of substrate assimilation and product formation in anaerobic digestion, *J. Water Pollut. Control Fed.* **46** (4) (1974).

28. S. Ghosh, J. R. Conrad, and D. L. Klass, *J. Water Pollut. Control Fed.* **47**, 30 (1975).

29. A. W. Lawrence, Application of process kinetics to design of anaerobic processes, in: *Advances in Chemistry Series 105* (F. G. Pohland, ed.), American Chemical Society, (1971).

30. A. M. Buswell, *Industrial Fermentations* **2**, Chemical Publishing Co., New York (1949). p. 518.

31. M. P. Bryant, S. F. Tzeng, I. M. Robinson, and A. E. Joiner, Nutrient requirements of methanogenic bacteria, in: *Advances in Chemistry, Series 105: Anaerobic Biological Treatment Processes,* American Chemical Society (1971).

32. K. Imhoff and G. H. Fair, *Sewage Treatment,* John Wiley and Sons, Inc., New York (1940).

33. A. W. Lawrence, Anaerobic biological waste treatment systems, in: *Agricultural Wastes: Principles and Guidelines for Practical Solutions, Proceedings, Conference of Agricultural Waste Management,* Cornell University, Ithaca, New York (1971).

34. L. G. Rich, *Environmental Systems Engineering,* McGraw-Hill, New York (1973).

35. J. Maly and H. Fadrus, Influence of temperature on anaerobic digestion, *J. Water Pollut. Control Fed.* **43**, 4 (1971).

36. H. R. Zablatsky and S. A. Peterson, Anaerobic digestion failure, *J. Water Pollut. Control Fed.* (40), (1968).

37. E. Rubins and F. Bear, Carbon–nitrogen ratios in organic fertilizer materials in relation to the availability of their nitrogen, *Soil Sci.* **54**, 411–423 (1942).
38. F. A. Sanders and D. Bloodgood, The effect of nitrogen to carbon ratio on anaerobic decomposition, *J. Water Pollut. Control Fed.* **37**, 1741 (1965).
39. C. N. Sawyer, Anaerobic units, in: *Proceedings, Symposium on Advances in Sewage Treatment Design*, New York (1961).
40. J. C. Converse and R. E. Graves, *Facts on Methane Production from Animal Manure*, University of Wisconsin, Madison, Wisconsin (1974).
41. I. J. Kugelman and P. L. McCarty, Cation toxicity and stimulation anaerobic waste treatment, *J. Water Pollut. Control Fed.* (37), (1965).
42. L. J. Ricci, Garbage routes to methane, *Chem. Eng.* (May 27, 1974).
43. L. J. Ricci, Scavenging of wastes promises a gas bonus, *Chem Eng.* (November 10, 1975). 10, 1975).
44. T. H. Maugh II, Fuel from wastes: A minor energy source, *Science*, **178**, 599–602 (1972).
45. T. M. McCalla, Think of manure as a resource, not a waste, *Feedlot Manage.* **14**, 10, 11, 68 (1972).
46. J. Jones, Converting solid wastes and residues to fuel, *Chem. Eng.* 87–94 (January 2, 1978).
47. J. L. Jones, Overview of solid waste and residue generation, disposition, and conversion technologies, presented at a symposium on Advanced Thermal Processes for Conversion of Solid Wastes and Residues, American Chemical Society, Anaheim, California (March 15–16, 1977).
48. J. C. Kuester and L. Lutes, Fuel and feedstock from refuse, *Environ. Sci. Technol.* **10**, 339–341 (1976).
49. P. M. Kohn, Biomass: A growing energy source, *Chem. Eng.* 58–62 (January 30, 1978).
50. L. L. Anderson, U.S. Bur. Mines Inf. Circ. No. 8549 (1972).
51. *Energy Alternatives: A Comparative Analysis*, University of Oklahoma Science and Public Policy Program, Norman, Oklahoma (May, 1975).
52. *Federal Register*, Effluent guidelines and standards, *Environ. Prot. Agency (U.S.) Publ.* **39**(32), Parts 1, 2, and 3 (1974).
53. T. L. Willrich and G. E. Smith, *Agricultural Practices and Water Quality*, Iowa State University Press, Iowa (1970).
54. W. J. Jewell, Anaerobic fermentation of agricultural wastes—Potential for improvement and implementation, Cornell University, Ithaca, New York (1976).
55. W. J. Jewell, H. R. Davis, W. W. Gunkel, D. J. Lathwell, J. H. Martin, T. R. McCarthy, G. R. Morris, D. R. Price, and D. W. Williams, *Bioconversion of Agricultural Wastes for Pollution Control and Energy Conservation*, New York State College, College of Agriculture and Life Sciences, Cornell University, Ithaca, New York (1976).
56. P. L. Silveston, *Methane Production from Manure in Small Scale Units*, University of Waterloo, Ontario, Canada (1975).
57. P. L. Silveston, Professor of Chemical Engineering, University of Waterloo, Canada, personal communication with M. R. Brulé (May 8, 1975).
58. W. W. Gunkel, W. J. Jewell, T. R. McCarty, D. R. Price, and D. W. Williams, *Analysis of Energy Utilization on Beef Feedlots and Dairy Farms*, Cornell University, Ithaca, New York (1975).
59. Herbert Kuhlman, owner of the site where the biogasification system has been installed, personal communication (1975).
60. G. L. Casler, W. W. Gunkel, and D. R. Price, *Accounting of Energy Inputs for Agricultural Production in New York State*, Cornell University, Ithaca, New York (1975).
61. G. R. Morris, W. J. Jewell, and G. L. Casler, *Alternative Animal Wastes; Anaerobic Fermentation Designs and Their Costs*, Cornell University, Ithaca, New York (1974).

62. R. J. Smith, *The Anaerobic Digestion of Livestock Wastes and the Prospects for Methane Production,* Iowa State University (November, 1973).

63. N. W. Snyder, Energy recovery and resources recycling, in: *Chemical Engineering/Deskbook Issue* McGraw-Hill, New York (October 21, 1974).

64. James C. Converse, Professor of Agricultural Engineering, University of Wisconsin, Madison, Wisconsin, personal interview with M. R. Brulé (1975).

65. E. F. Schumacher, *Small is Beautiful—Economics as if People Mattered,* Harper & Row, New York (1973).

66. *Proceedings of Conference on Capturing the Sun Through Bioconversion,* Washington Center for Metropolitan Studies, Washington, D.C. (1976).

67. H. W. Parker, D. M. Wells, and G. A. Whetstone, Study of Current and Proposed Practices in Animal Waste Management, *Environ. Prot. Agency U.S. Publ.,* Report No. 43019-74-003, Washington, D.C. (1974).

68. L. M. Auerbach, *A Homesite Power Unit: Methane Generator,* Alternative Energy Concepts Co., 242 Copse Rd., Madison, Connecticut (1974).

69. L. J. Fry, *Practical Building of Methane Power Plants for Rural Energy Independence,* available from L. J. Fry, 1223 North Nopal Street, Santa Barbara, California (1974).

70. L. J. Fry and R. Merrill, *Methane Digesters for Fuel Gas and Fertilizer with Instructions for Two Working Models,* New Alchemy Institute, Woods Hole, Massachusetts (1973).

71. R. B. Singh, The biogas plant, generating methane from organic wastes, *Compost Sci.* 13, 20–25 (1972).

72. R. B. Singh, *Bio-Gas Plant—Designs and Specifications,* Mother's Print Shop, P.O. Box 70, Hendersonville, North Carolina 28739 (1975).

73. R. B. Singh, *Some Experiments with Biogas,* Gobar Gas Research Station, Ajitmal, Etawah (U.P.) India. (1971).

74. T. P. Abeles, and P. Atkinson, *Economics and Energy Considerations for Anaerobic Digestion of Farm Waste,* College of Environmental Sciences, University of Wisconsin, Green Bay, Wisconsin (1975).

75. J. A. Alich, *Economic Assessment on Energy from Agricultural Residues, A Case Study,* Stanford Research Institute (SRI) Report (1977).

76. S. Donatiello, S. Graefe, L. Viamontes, J. Wright, and T. Yademec, *Cow Power (Production of Methane from Manure),* Washington University, St. Louis, Missouri (1973).

77. R. H. Shipman, D. P. Palmer, W. P. Walawender, L. T. Fan, Final project report on the production of energy for Decatur County dairy, Oberlin, Kansas, by the anaerobic digestion of livestock manures, Department of Chemical Engineering, Kansas State University, Manhattan, Kansas (December 31, 1975).

78. R. D. Wingo, Feedlot Waste and Methane Production from Oklahoma Dirt-Type Feedlots, Master's thesis, University of Oklahoma Norman, Oklahoma (1974).

79. T. H. Crane, Energy and resource recovery from manures and agricultural wastes, Resource Recovery Systems, Barker Colman Co., Irvine, California (1975).

80. C. N. Ifeadi and J. B. Brown, Jr., An assessment of technologies suitable for the recovery of energy from agricultural wastes, Waste Control and Process Technology Section, Battelle, Columbus, Ohio (1975).

81. L. C. Gramms, L. B. Pokowski, and S. A. Witzel, Anaerobic digestion of farm animal wastes, *Transactions of the American Society of Agricultural Engineers,* (July 11, 1971) p. 14.

82. R. E. Inman and J. A. Alich, Jr., Availability of agricultural residues as energy feedstocks, Stanford Research Institute, Menlo Park, California (1975).

83. D. L. Klass, Make SNG from waste and biomass, *Hydrocarbon Processing* 55 (4) 76–82 (April, 1976).

84. *Production of Power Fuel by Anaerobic Digestion of Feedlot Waste,* Hamilton Standard Division of the United Aircraft Corporation prepared for the USDA, Peoria, Illinois (1974).

85. C. C. Holloway, Use of Ruminant Animals in Refuse Disposal, Doctoral dissertation, University of Oklahoma Norman, Oklahoma (1975).

86. M. R. Brulé and S. S. Sofer, Small-scale biogas processing facilities, *Proceedings, The Gas Conditioning Conference 28:* B1–21 (1978).

87. M. R. Brulé and S. S. Sofer, A biogasification system at a dairy, *Proc. Okla. Acad. Sci.* **56**, 18–23 (1976).

88. M. R. Brulé, Process Design of an Anaerobic Digestion Facility, M.S. Thesis, Chemical Engineering and Materials Science, University of Oklahoma Norman, Oklahoma (1975).

89. O. H. Linguist, Environmental Protection Agency, Public Affairs Division, private communication with M. R. Brulé (April 14, 1975).

90. H. Eby, Design criterion for manure lagoons, American Society of Agricultural Engineers (ASAE) Paper 61–935 (1969).

91. C. F. Pomeroy, A study of the engineering economics of a biogasification system on a dairy farm, Chemical Engineering and Materials Science (unpublished manuscript), University of Oklahoma, Norman, Oklahoma (December 2, 1975).

92. E. D. Devero, representative of Oklahoma Natural Gas Company, private communication with M. R. Brulé (August 22, 1975).

93. U.S. Environmental Protection Agency, Sludge treatment and disposal, U.S. Government Printing Office, Washington, D.C. (October, 1974).

17
Anaerobic Digestion of Kelp

DAVID P. CHYNOWETH, SAMBHUNATH GHOSH, AND DONALD L. KLASS

1. Introduction

The search for new sources of energy is intensifying as the demand for energy increases and supplies of fossil fuels are becoming depleted. One possible long-term solution to this dilemma is the conversion of renewable sources of organic matter, such as wastes and biomass, to products that are suitable for use as fuels. Although organic wastes represent a minor potential supplemental energy resource,[1,2] land- and water-based biomass could be developed into major resources.[1,3-5] This has led to the concept of land- and water-based energy farms directed at production of biomass for conversion to synthetic fuels. Development of energy farms in the marine environment seems particularly attractive because large areas are available. In fact, it is generally thought that seaweed farms could be used both for energy production and to promote growth of fish and other marine fauna.

Giant brown kelp *(Macrocystis pyrifera)* has several properties that make it an attractive candidate species for ocean farming and conversion to fuel: (1) attachment and flotation structures, (2) high potential growth rate, and (3) suitable organic composition.[4,6] As illustrated in Figure 1, kelp is structurally well suited for ocean farming in deep offshore waters. A basal holdfast structure enables the plant to anchor itself to floating structures in deep open waters, and gas-filled bulbous structures orient the plant vertically toward the water's surface. The fronds possess numerous blades giving an overall large surface

DAVID P. CHYNOWETH, SAMBHUNATH GHOSH, AND DONALD L. KLASS • Institute of Gas Technology, 3424 South State Street, Chicago, Illinois 60616.

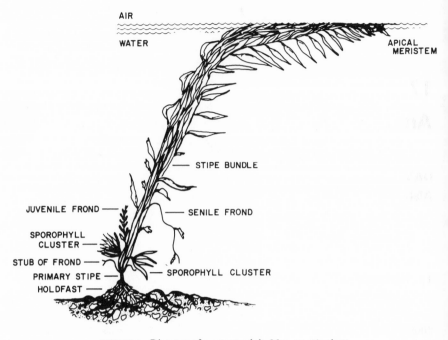

FIGURE 1. Diagram of a young adult *Macrocystis* plant.

area for collection of nutrients and solar energy. Because of the high growth rate and unlimited regenerative capacity of this plant, its upper portions can be effectively harvested on a regular basis with minimal disturbance to the artificial and main plant structures located 15–24 m below the surface.

Although natural growth rates of kelp can be limited under nutrient deficient conditions in surrounding waters, laboratory studies have shown that a growth rate as high as 18% per day in nutrient-rich deep water is possible.[5] For this reason the concept of kelp ocean farming includes upwelling of nutrient-rich deep water into the beds to stimulate kelp productivity. Although *Macrocystis pyrifera* is naturally a cold-water species, upwelling of cold deep water into the beds may even permit its growth in locations with warmer water temperatures.

The composition of kelp, discussed in more detail later, suggests that it should be readily biodegradable; the alga lacks the lignin that reduces biodegradability of many terrestrial biomass forms. Furthermore, the high potassium content of the inorganic fraction indicates a potentially high byproduct value of the effluent from the conversion process.

The concept of the kelp farm and conversion scheme (Figure 2), originally conceived by Wilcox[7] and currently in the concept-validation stage, envisions a large network of lines and buoyancy-control structures to serve as substrata for attachment of kelp plants. Nutrients required for stimulation of kelp productivity would be provided and supplemented by upwelling nutrient-rich deep water from depths of 150–300 m or by recycling conversion process effluent. Anaerobic digestion was selected for conversion of kelp to methane, since that process is effective in conversion of feedstocks with a high water content. Kelp will be harvested, chopped, and fed to batteries of digesters on a continuous basis. Product gases will be cleaned up and marketed. Digestion effluent may be recycled to the kelp farms as a source of nutrients or processed for recovery of a variety of products, including animal feed and fertilizer. A number of key details relating to the total system, such as farm size and design, and digester design and location relative to the farm, are dependent upon data from current and future experiments on kelp farming and processing. The result of a preliminary systems analysis of the farm and conversion system was reported by Bryce,[5] and is summarized in Table 1. The calculated gas costs ranged from $3.9–5.9 per GJ ($10^9$J), depending on farm size. These costs indicate that the overall system is economically attractive.

This chapter discusses the conversion of kelp to methane by anaerobic digestion. Preliminary studies at the Institute of Gas Technology (IGT)[8,9] established that kelp would support a methanogenic fermentation under conventional mesophilic conditions with reasonable performance and yields and

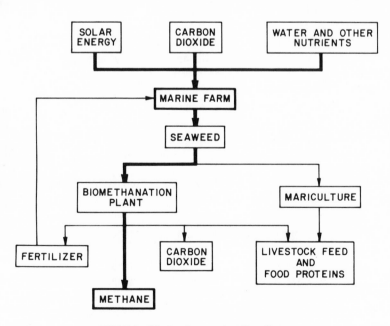

FIGURE 2. Marine farm project flow diagram.

TABLE 1. Feedstock and Gas Costs as a Function of Farm Yield[a,b]

Annual farm yield[b]		Feedstock cost	Gas costs
10^6 mg/km²	tons/acre	$/$10^9$ J	$/$10^9$ J
5.6	25	3.3	5.7
11.2	50	2.4	4.2
15.7	70	2.1	3.8
23.5	105	1.5	3.0

[a]Source: Reference 5.
[b]Costs in 1978 dollars.
[c]Dry ash-free.

without addition of external nutrients. The objectives of the recent work at IGT has been to identify the factors that limit methane yields and fermentation rates, and to optimize the methanation process for high energy recovery efficiencies. A summary and discussion of the research results follows.

2. Characteristics of Kelp Feeds

Two types of kelp preparations were considered as feeds for the anaerobic digestion process: raw kelp (RK) and baseline-treated kelp (BLTK). The RK was harvested 30 cm below the water surface, drained of physical water, chopped, ground with a hammer mill, and frozen prior to use. The BLTK started as raw kelp, which was then mixed with a 0.5 wt % $CaCl_2$ solution, heat treated at 95°C for 30 min, drained, pressed in a belt press, and frozen prior to use. The purpose of this treatment was to develop a feed with a lower salt and moisture content and a higher concentration of organic matter. In view of the good digester performance observed with RK and the high energy penalty associated with producing BLTK, most of the later digestion studies were conducted with RK.

Tables 2 and 3 compare several properties of typical lots of these two kelp preparations. Compared with RK, BLTK had higher levels of total solids, organic matter [expressed as volatile solids (VS)] and heating value; higher levels of carbon, nitrogen, and calcium; and lower concentrations of water, ash, sodium, potassium, and sulfur.

A typical organic-constituent analysis conducted only on RK is shown in Table 3. Protein, mannitol, algin, and cellulose accounted for 99% of the organic matter (as VS). Other minor constituents included laminarin and fucoidin. An amino acid analysis of the protein fraction is shown in Table 4. These types of analyses permit a preliminary evaluation of the suitability

TABLE 2. Characteristics of Kelp Feeds[a]

eed type	RK[b,c]	RK[b,c]	RK[b,c]	RK[c,d]	RK[b,c]	RK[b,c]	BLTK[b,e]
ot number	1	26	37	41	42	44	26
arvest date	2/19/76	9/30/76	10/25/77	6/6/78	9/13/78	11/1/78	2/19/77
oisture, %	89.7	88.8	88.2	86.9	88.1	89.7	75.8
otal solids (TS), %	10.3	11.2	11.8	13.1	11.9	10.3	24.2
olatile solids (VS), % TS	54.2	57.9	58.9	65.0	62.9	60.4	72.7
sh, % TS	45.8	42.1	41.1	35.0	37.1	39.6	27.3
lements, % TS							
C	26.1	27.80	28.00	28.90	28.30	27.9	36.1
H	3.69	3.73	3.92	4.00	3.61	3.51	4.61
N	2.56	1.63	1.86	1.23	1.18	1.89	3.36
P	0.49	0.29	0.33	0.21	0.22	0.29	0.46
S	1.09	1.05	1.09	1.06	1.35	1.62	0.89
Na	4.20	3.50	3.60	3.61	3.40	4.0	1.8
K	14.4	14.7	14.0	10.7	12.00	13	6.3
Ca	1.05	—	1.40	—	1.40	1.7	3.2
eating value, MJ/ kg	10.3	10.7	11.0	10.9	10.6	10.5	14.1

ource: References 8 and 9.
larvested and processed by U.S. Department of Agriculture, Western Regional Research Center, Albany, California, nd shipped to IGT in Trans-Temp containers at $-10°$ to $-15°C$.
K = Raw kelp: freshly harvested kelp drained of free water, chopped, ground in hammer mill equipped with a 4.78-nm screen, and frozen.
he same as b, but commercially harvested.
lLTK = Baseline-treated kelp: raw kelp mixed with 0.5% $CaCl_2$, heat treated at $95°C$ for 30 min, drained, pressed in belt press, and frozen.

TABLE 3. Composition of Organic Fraction
of Raw Kelp from Lot Number 1[a]

Organic group	Sample weight (% volatile solids)
Protein	29.5
Carbohydrates	
Mannitol	34.5
Algin	26.1
Cellulose	8.8
Laminarin	1.3
Fucoidin	0.4
Total	100.6

[a]Source: Reference 8.

TABLE 4. Analysis of Free and Bound Amino Acids in Raw Kelp Lot
No. 1[a]

Amino acid	Sample weight (% dry)
Glutamic acid	1.10
Alanine	0.96
Aspartic acid	0.80
Leucine	0.61
Glycine	0.50
Valine	0.47
Lysine	0.45
Phenylalanine	0.38
Serine	0.34
Threonine	0.34
Proline	0.33
Arginine	0.32
Isoleucine	0.31
Tyrosine	0.23
Methionine	0.16
Tryptophan	0.11
Cystine	0.04
Histidine	Trace
Total	7.45
Crude Protein (Kjeldahl N times 6.25)	8.56

[a]Source: Reference 8.

of the biomass species as a feedstock for various conversion processes. For example, the high water content and low heating value of RK is indicative of its poor suitability as a feedstock for direct combustion or thermochemical conversion processes. On the other hand, the high concentrations of water, nitrogen, and phosphorus and the lack of lignin indicate that it might be suitable for biomethanogenesis. The high salt content raises the uncertainty of potential salt toxicity. Furthermore, because of the high salt content and the fact that the major organic components are unique to marine environments, a marine inoculum might result in a more effective process. In addition to these general predictions, feed analyses are also useful for calculating theoretical yields and feed inputs and for conducting component and energy analyses for the conversion process. These uses are illustrated below.

A common error in evaluation of biomass feedstocks is the assumption that the composition of a given species is unaffected by variations in harvest season and location. The invalidity of this assumption is shown in the variation in nitrogen and phosphorus content for several kelp samples obtained from different locations at different harvest dates (Table 2). Similar fluctuations in nitrogen and phosphorus and also in specific organic constituents were described by Lindner et al.[10] and shown to be related to nutrient concentrations in the surrounding waters. Data presented below will illustrate the marked effect of nitrogen fluctuations on the kelp biomethanogenesis process.

3. Biomethanogenesis of Kelp

Biomethanogenesis of kelp or other substrates is a process occurring only under strict anaerobic conditions where mixed populations of bacteria decompose organic matter to carbon dioxide and methane, as illustrated in Figure 3. The first step carried out by nonmethanogenic bacteria is hydrolysis and involves cleavage of interpolymer bonds (which contributes to the plant's structure) and subsequent hydrolysis of polymers such as carbohydrates, proteins, and nucleic acids to small soluble molecules that can be transported into the bacterial cells for further metabolism.

These hydrolytic products are further metabolized to acetate, hydrogen, and carbon dioxide by a relatively hearty and fast-growing group of nonmethanogenic bacteria. Acetate and hydrogen (with carbon dioxide) are then converted to methane by a fastidious, slow-growing population of methano-

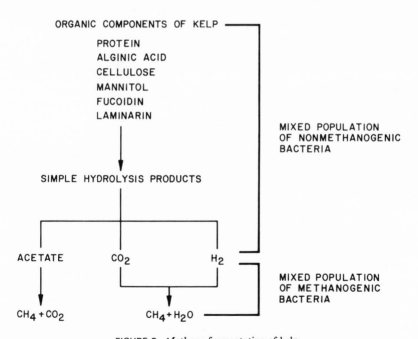

FIGURE 3. Methane fermentation of kelp.

bacteria. A balanced intermicrobial association between these two groups of bacteria is required for a stable high-performance fermentation.[11,12]

A generalized process scheme based on the above microbial process is illustrated in Figure 4. Upon harvest, biomass is chopped and ground for size reduction and possibly subjected to some kind of pretreatment to enhance biodegradation. The feed is slurried to 5–10% water and added to a primary digester having a typical volume of 14,150 m³ (500,000 ft³) at a loading of 1.6–6.4 kg VS/m³-day (0.1–0.4 lb VS/ft³-day) and a hydraulic retention time of 10–20 days. This unit is completely mixed and maintained anaerobic at a temperature of 35°C. Effluent is settled in an anaerobic secondary digester. Settled solids may be recycled (directly or following posttreatment), or dewatered and processed as fertilizer or animal feed. Supernatant from the secondary digester and dewatering processes can be recycled, used as fertilizer, or processed for extraction of inorganics or disposal. Process gas is often treated to remove carbon dioxide and hydrogen sulfide. A number of variations in this scheme are possible, including batch, continuous, and plug-flow operation, separation of nonmethanogenic and methanogenic phases, and maintenance of organisms on a fixed bed.

3.1. Performance Parameters

Several performance parameters are traditionally used to evaluate the stability of the biomethanogenesis process and its performance after it has

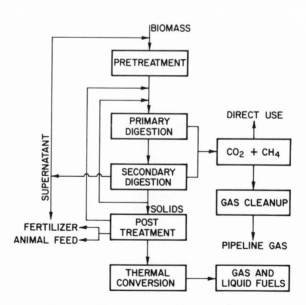

FIGURE 4. Biomethanogenesis process scheme.

achieved steady state. These include gas yield and methane content, methane yield, methane production rate, volatile fatty acids, pH, alkalinity, and conversion of organic matter.

Although anaerobic digesters receive feed additions once or twice per day, the overall biomethanogenesis process exhibits continuous culture kinetics because of the slow growth rates of the bacteria involved. With changes in feed conditions it can be shown by using Equation (1) of Figure 5 that 2.3 and 4.6 hydraulic retention times are required for the digester to achieve 90% and 99% replacement of the digester contents with the feed, respectively. Therefore digesters are not considered to be at steady state until: (1) four hydraulic retention times have elapsed and (2) the coefficient of variation of methane production and concentration of volatile acids is less than 12%. The latter requirement reflects the fact that biological steady state may require a longer period to achieve.

3.1.1. Gas Production

Since the ultimate products of biomethanogenesis are methane and carbon dioxide, gas production is the primary indicator of overall performance. Equation (2) of Figure 5 can be used to standardize gas measurements made at ambient conditions of temperature and pressure.

In our laboratory, gas measurements are made daily, and a mean daily value is used in Equation (3) of Figure 5 to calculate a mean gas yield for each week. The mean methane yield is then calculated from weekly methane content data, using Equation (4) of Figure 5. This parameter relates the quantity of methane produced to the amount of organic matter added and represents ultimate yield at long hydraulic retention times.

The methane production rate calculated from Equation (5) of Figure 5 is a measurement of the rate of the overall process and is expressed as methane volume per unit *culture volume* (not digester volume) per day.

3.1.2. Conversion of Organic Matter

The ash-free dry weight of volatile solids is used as a measurement of organic matter in determining the loading and the conversion efficiency of organic matter in the anaerobic digestion process. In view of errors associated with sampling feeds or digester effluent and with ashing technique, additional measurements of organic matter —such as chemical oxygen demand and carbon in the feed, effluent, and gas—and heating value determinations of feed and effluent are often made. We have found that the most reliable method for calculation of conversion of organic matter is by Equation (6) of Figure 5, which estimates organic reduction efficiencies by dividing the carbon in the product gas by the carbon in the feed.

1. Replacement of Digestor Contents as a Function of Time

$$\frac{C_t}{C_0} = e^{-t/\theta}$$

where C_t = concentration at time t, C_0 = initial concentration, t = time (in days), θ = hydraulic retention time (in days).

2. Gas Production

$$G = G_m \left(\frac{P - P_{H_2O}}{T + 273} \right) (0.380)$$

where G = gas production (in liters, 760 mm Hg, 15.7°C), G_m = measured gas production (in liters), P = atmospheric pressure (in mm Hg), P_{H_2O} = vapor pressure at T, T = temperature of gas (in °C), 0.380 = constant for correction to 760 mm Hg and 15.7°C.

3. Gas Yield

$$G_y = \frac{G}{VS_{added}}$$

where Gy = gas yield (in liters/g VS added), VS_{added} = volatile solids added [in g VS/liter-day (kg VS/m³-day)].

4. Methane Yield

$$M_y = G_y M_c$$

where M_y = methane yield [in liters/g VS added (m³/kg VS added)], M_c = methane content (in mol %).

5. Methane Production Rate

$$\text{MPR} = M_y L$$

where MPR = methane production rate (in m³/m³ culture-day), L = loading [in g VS/liter culture volume-day (kg VS/liter-day)].

6. Volatile Solids Reduction

$$VS_{red} = 50.8 \frac{G_y VS}{C}$$

where VS_{red} = volatile solids reduction (in %), VS = volatile solids content of feed (in % total solids), C = carbon content of feed (in % total solids), 50.8 = constant for conversion of gas yield to carbon.

FIGURE 5. Calculation of performance parameters.

3.2. Maximum Theoretical Yields and Heat of Reaction

Calculation of maximum theoretical yields of methane from biomethanogenesis of biomass feed stocks is useful in evaluation of digestion runs and provides a basis for establishing target yields for experimental work. Theoretical methane yields were calculated for several kelp lots as follows (illustrated in Figure 6). Composition data presented in Table 2 were used to calculate empirical formulas for each kelp lot. Stoichiometric equations for conversion of each feed to methane and carbon dioxide were determined. The resulting molar yields of methane were expressed as m^3/kg VS added. Because a fraction of organic matter in all bacterial fermentations is converted to bacteria, these yields were corrected for cell synthesis, using the data reported by McCarty[13] and by Speece and McCarty[14] for the anaerobic digestion of pure carbohydrate and protein. The final theoretical yields are reported in Table 5.

Theoretical yields corrected for bacterial growth ranged from 0.365 to 0.453 m^3/kg VS added for seven different kelp lots. These yields correspond to a theoretical VS reduction of 82.2%.

The theoretical heat of reaction for the biomethanogenesis of kelp was calculated to determine the contribution it would make to maintenance of the temperature of the reactor at $35\,^\circ$C. The estimate for conversion of raw kelp to CH_4 and CO_2 was 807 kJ/kg reacted. Corrected for actual VS reduction (about 50%) and microbial cell maintenance (estimated at about 20%), this would amount to 5.14 kJ/day in a 10-liter digester with a loading of 1.6 kg VS/m^3-day (0.1 lb VS/ft^3-day) and a hydraulic retention time of 10 days. That is equivalent to 8.2% of the energy required to heat the daily feed slurry (1 liter) from 20 to $35\,^\circ$C.

3.3. Bench-Scale Digestion Studies

Table 6 lists several factors that can cause experimental yields to be lower than calculated theoretical yields. Evaluation of these has been the basis for recent bench-scale digestion research aimed at evaluation and optimization of the kelp biomethanogenic process. Each of these factors will be discussed below in principle and in terms of established or potential influence on biological gasification of kelp.

3.3.1. Organic Composition of Feeds

As illustrated in the discussion above on theoretical yields, organic composition can influence the theoretical yield for biomethanogenesis of feedstocks. In addition, theoretical yields are rarely achieved because a fraction of the organic matter is refractive to hydrolysis and thus to decomposition under anaerobic conditions. The effect of organic composition on theoretical and

1. Raw kelp composition, % dry wt (Lot No. 26)

   $C = 27.8$ $VS = 57.9$

   $H = 3.73$ $Ash = 42.1$

   $N = 1.63$ Crude protein $= 10.2$[b]

   $S = 1.05$ Carbohydrate $= 49.7$

   $O = 23.7$

2. Raw kelp empirical formula (N and S neglected)

   $C_{2.32} H_{3.69} O_{1.48}$ 1 kmol $= 55.2$ kg (CHO)

                                      $= 57.9$ kg VS

3. Biomethanation of raw kelp (by stoichiometry)

   $$C_{2.32} H_{3.69} O_{1.48} + 0.66\ H_2O \rightarrow 1.25\ CH_4 + 1.07\ CO_2$$

4. Maximum theoretical methane yields

   $$1.25\ \text{kmol}\ CH_4 \times 23.6\ m^3\ CH_4/\text{kmol} = 29.6\ m^3\ CH_4$$

   $$1\ \text{kg kelp VS} \times \frac{1\ \text{kmol}}{57.9\ \text{kg kelp VS}} \times 29.5\ \frac{m^3\ CH_4}{\text{kmol}} = 0.511\ m^3\ CH_4$$

5. Estimation of VS converted to cells and therefore not to gaseous

   products (assumes 20% and 7% conversion, respectively, of carbohydrate

   and protein to bacterial cells)

   RK VS $= 83.0\%$ (CHO) and 17.0% protein

             $20 \times 0.83 = 16.6$

             $7 \times 0.17 = \underline{1.2}$

                    17.8% RK converted to bacteria

6. Maximum theoretical yield corrected for bacteria

   $$\text{Theoretical yield} \times 0.822 = 0.511\ \frac{m^3\ CH_4}{\text{kg VS}} \times 0.822$$

   $$= 0.420\ \frac{m^3\ CH_4}{\text{kg VS added}}$$

[a] By difference

[b] N x 6.25

FIGURE 6. Maximum theoretical yields of methane from anerobic digestion of raw kelp.

TABLE 5. Theoretical Methane Yields for Several Kelp Lots[a,b]

Kelp Lot	Empirical Formula[c]	Stoichiometric yield, SCM/kg VS added[d]	Corrected yield, SCM/kg VS added[e,f]
RK 1	$C_{2.17} H_{3.65} O_{1.27}$	0.533 (8.53)	0.438 (7.01)
RK 26	$C_{2.32} H_{3.69} O_{1.48}$	0.511 (8.18)	0.420 (6.72)
RK 37	$C_{2.33} H_{3.88} O_{1.48}$	0.515 (8.24)	0.423 (6.77)
RK 41	$C_{2.41} H_{3.96} O_{1.85}$	0.452 (7.23)	0.371 (5.94)
RK 42	$C_{2.36} H_{3.57} O_{1.77}$	0.444 (7.11)	0.365 (5.84)
RK 44	$C_{2.33} H_{3.48} O_{1.57}$	0.474 (7.59)	0.390 (6.24)
BLTK 26	$C_{3.01} H_{4.68} O_{1.73}$	0.542 (8.67)	0.453 (7.24)

[a]Source: References 8 and 9.
[b]760 mm Hg, 15.7°C.
[c]Excluding N, P, and S.
[d]Assumes that the reactants are kelp + H_2O and the products are CH_4 + CO_2.
[e]Numbers in parentheses have units of scf/lb VS added.
[f]Corrected for bacterial production based on data of McCarty[13] and Speece and McCarty[14] on enriched digesters receiving either soluble carbohydrate or soluble protein.

experimental yields is illustrated in Tables 5 and 7 for two different kelp preparations, raw kelp and desalted kelp. Although theoretical yields were higher for BLTK, experimental yields were lower under conventional operating conditions. The lower experimental yields for BLTK can be explained by the fact that significant amounts of biodegradable components of raw kelp (probably mannitol) are lost in the juice removed during the desalting treatment process. In view of the stable and high performance of RK-fed runs and the added expense and energy requirement associated with conversion of RK to BLTK, untreated RK was selected as the feed substrate for more intense study.

3.3.2. Potential Nutrient Limitation

Bacteria involved in the biomethanogenesis process require nitrogen and phosphorus for growth and metabolism. A carbon-to-nitrogen ratio of 11 and a carbon-to-phosphorus ratio of about 52 have been reported as optimal min-

TABLE 6. Factors That Could Cause Experimental Yields To Be Lower Than Theoretical Yields

Organic composition of feed	Feed concentration
Potential nutrient limitation	Particle size
Inoculum	Mixing
Temperature	Feeding frequency
Inhibitory substances	Catabolite repression
Hydraulic retention time	

TABLE 7. Comparison of Performances of Digesters Receiving RK and BLTK[a]

Run number	26	8	1	101
Feed	RK lot	RK lot	BLTK lot	BLTK lot
Retention times in progress	16	47	32	7
HRT, days	10	18	10	18
Methane yield, m³/kg VS added	0.215	0.278	0.207	0.212
Methane production rate, m³/ m³ culture-day	0.344	0.445	0.331	0.339
Methane content, mol % as acetic	58.4	58.2	59.3	59.9
Volatile solids reduction, %[b]	38.6	50.8	35.6	36.2
Total volatile acids, mg/liter as acetic	855	192	43.0	29.0
Alkalinity, mg/liter as CaCO₃	3,860	4,160	3,070	4,400
Conductivity, μmho/cm	17,500	20,600	8,830	12,000

[a] Loading = 1.6 kg/m³-day, temperature = 35°C.
[b] Carbon in product gas divided by carbon in feed.

ima for nonlimited digestion.[14,15] The use of these data and of experimental data on the composition of biomass feeds to predict nutrient requirements for biomethanogenesis should be interpreted with caution, as it is the ratio of *biodegradable* carbon and nutrients that actually dictates carbon and nutrient levels available for bacterial metabolism.

Nutrient levels in several lots of raw kelp are listed in Table 2. Data in Table 8, giving the performance of digester runs receiving supplements of nitrogen and phosphorus, show that these nutrients were not limiting in kelp Lot 26, which was used for most of the studies conducted thus far at IGT.

TABLE 8. Effect of Nutrient Concentrations on Kelp Biomethanation[a]

Run number	116	117	41	42	43	44
Kelp lot	26	26	42	42	42	42
Retention times in progress	12	4.5	4	4	4	4
HRT, days	12	12	15	15	15	15
Nutrients added	None	N, P	None	N, P	P	N
C-to-N ratio	17.0	11.6	24	15	24	15
C-to-P ratio	95.9	26.6	129	87.8	87.8	129
Methane yield, m³/kg VS added	0.238	0.234	0.0187	0.238	0.0312	0.222
Methane production rate, m³/ m³ culture-day	0.382	0.375	0.003	0.381	0.005	0.356
Total volatile acids, mg/liter as acetic	484	958	3510	802	2980	1233

[a] Temperature = 35°C, loading = 1.6 kg VS/m³-day.

Recent lots of kelp obtained in late summer contained lower levels of nitrogen and phosphorus (Table 8, Lots 41 and 42) and resulted in reduction of methane yields and production rates and, eventually, of digester imbalance. The data for Lot 42 in Table 8 show that the performance was restored to normal by supplementation of the nitrogen to a C-to-N ratio of 15. Phosphorus supplements did not affect the fermentation. Data reported by Lindner et al.[10] indicate that seasonal fluctuations in nutrient content of kelp occur and are related to nutrient concentrations in surrounding waters. With respect to the kelp farming concept discussed previously, some control over kelp nutrient levels might be exercised through variation in upwelling rates. Studies are currently in progress to define the critical nitrogen levels required for optimum digestion.

3.3.3. Inoculum

The lack of decomposition of certain components of feed substances may be related to the absence of organisms capable of degrading those compounds. It is generally thought that, for most types of organic wastes and cellulosic biomass feedstocks, a suitable inoculum can be developed from any environment under which anaerobic methanogenic decomposition of a mixture of organic compounds is occurring naturally, e.g., an anaerobic sewage digester, anaerobic lake sediments, or animal feces. Starting with such a source, an inoculum adapted to the feed under study can usually be developed within a reasonable period of time, i.e., two to four months.

Inoculum A used for most of the work on biomethanogenesis of kelp was derived from the effluent of digesters operated on domestic sewage sludge and municipal solid wastes. Because kelp contains substrates unique to marine algae (i.e., algin, fucoidin, and laminarin) and a high salt content, other inocula derived principally from anaerobic marine environments that decompose kelp naturally were investigated. Table 9 shows that the performance of Inoculum D, derived from anaerobic marine environments, was not significantly different from that of Inoculum A. Both inocula were evaluated at 35°C.

A second inoculum, E, was developed from a mixture of decaying kelp and anaerobic marine sediment. The culture was developed at room temperature (about 26°C) because that temperature more closely resembles the marine environment. The methane yield (Table 9) was approximately half that of the above two inocula developed and incubated at 35°C.

3.3.4. Temperature

Biomethanogenesis has been reported at temperatures ranging from 4°C (lake sediments) to 60°C (thermophilic dairy manure digesters).[16,17] Most digesters are operated at a mesophilic (35°C) or a thermophilic (55°C) temperature. Generally, it has been observed that higher temperatures give higher

TABLE 9. Effect of Inoculum and Temperature on Kelp Biomethanogenesis

Run number	8	122	123	29	SK-4
Feed	RK(26)	RK(26)	RK(26)	RK(26)	RK(26)
Retention times in progress	42	8	7	6	13
Inoculum	A[a]	D[b]	E[c]	A[a]	A[a]
Temperature, °C	35	35	26	55	55
HRT, days	18	18	18	18	7
Loading, kg VS/m³-day	1.6	1.6	1.6	1.6	3.2
Methane yield, m³/kg VS added	0.279	0.284	0.144	0.149	0.134
Methane production rate, m³/ m³ culture-day	0.446	0.459	0.230	0.238	0.430
Total volatile acids, mg/liter as acetic	1310	383	1590	4590	4900

[a]IGT's inoculum derived by mixing effluents from a digester receiving municipal solid waste-sewage sludge and from a municipal sewage sludge digester.
[b]Marine inoculum derived from mixing chopped decaying kelp, anaerobic marine sediment, and effluent from GE's marine-derived inoculum.
[c]Marine inoculum derived from mixing anaerobic marine sediment and kelp collected from a decaying kelp wrack pool (equivalent to Inoculum D).

reaction rates, thus permitting lower hydraulic retention times and high loadings without reduction in conversion efficiency.[17] However, most digesters are operated at 35°C because of high heating costs and high sensitivity of the bacteria in thermophilic digesters. The biomethanogenesis of kelp has been studied at 26, 35, and 55°C; methane production rates at 35°C were double those observed at 26°C (Table 9).

Several attempts have been made to develop thermophilic cultures that would give high yields and stable performance on RK feed. The performance data of two typical examples of thermophilic runs (29 and SK-4) are compared with mesophilic runs in Table 9. Performance was unstable, methane yields were low, and concentration of volatile acids was high for the thermophilic runs. These results are inconsistent with those obtained with other types of wastes and biomass.[17,18] Apparently kelp has properties (possibly high salt content) that prevent development of a healthy thermophilic population. The fact that the volatile acids concentration is high in the thermophilic runs suggests that the methanogenic bacteria are the limiting population. As is discussed later, the instability could be related to daily imbalance resulting from the fact that fermenters were fed on a daily rather than a continuous basis.

3.3.5. Inhibitory Substances in Feed

Biomass feeds can contain substances that are inhibitory to the biomethanogenesis process. These might be removed by pretreatment (such as that used to produce BLTK), or the effects can be minimized by dilution. Along these lines, studies were undertaken to evaluate potential salt inhibition related to the high salt content of raw kelp (typically 4.5%). These studies compared the performances of digesters that received undiluted and seawater-diluted kelp with a run that received kelp diluted with distilled water.

A profile of methane yield for each of these runs and the control is shown in Figure 7. Note that each of the three experimental runs began to show signs of inhibition after about two retention times (2θ). This inhibition was apparently related to the buildup of salts to critically high levels. Each of the experimental runs recovered after a period of about one month, indicating adaptation of the population to the inhibitory conditions. The conductivities of the two runs receiving direct kelp and seawater-diluted kelp were 45,000, 40,000, and 34,000 μmho/cm, compared to a value of 18,300 for the runs diluted with distilled water.

Table 10 compares other performance data for these runs. The performance of Run 119 receiving undiluted kelp was similar to that of Run 116 receiving kelp diluted with distilled water. The lower yield of Run 120, which received a higher loading of 3.2 kg/m³ culture-day must have been due to the higher loading (or associated lower HRT) rather than to the high salt concentration, because the salt concentration was equal to that in Run 119. (The latter run also received undiluted RK with no reduction in methane yields. The methane yield in Run 121, which received RK diluted with seawater, was lower

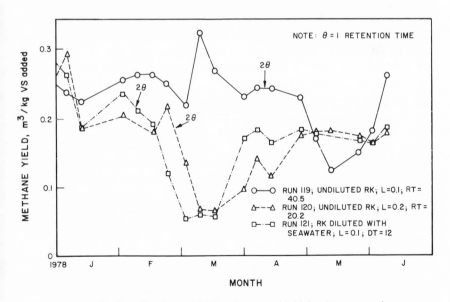

FIGURE 7. Profiles of methane yield data for runs with high salt concentrations.

TABLE 10. Comparison of Performances of Kelp Fermenters Receiving Freshwater-Diluted Kelp, Seawater-Diluted Kelp, and Undiluted Kelp[a]

Run number	116[b]	121[c]	119[d]	120[d]
Retention times in progress	12	18	5	11
HRT, days	12	12	40	20
Loading, kg VS/m³-day	1.6	1.6	1.6	3.2
Methane yield, m³/kg VS added	0.239	0.180	0.221	0.171
Methane production rate, m³/m³ culture-day	0.382	0.288	0.354	0.546
Methane content, mol %	57.7	52.6	54.3	48.2
Total volatile acids, mg/liter as acetic	484	2520	4110	8514
Conductivity, μmho/cm	17,200	34,000	45,000	40,000

[a]Temperature = 35°C, feed = RK Lot 37.
[b]Feed was diluted with distilled water.
[c]Feed was diluted with artificial seawater (Instant Ocean, Aquarium Systems, Eastlake, Ohio).
[d]Feed was undiluted.

than that of the control. The fact that the composition of salts in seawater is different from that in kelp might explain the inhibition in this high-salt run and the lack of inhibition in high-salt Run 119, which received undiluted kelp.

All three of these experimental runs had high concentrations of volatile acids in the effluent; the concentrations were highest in the two runs receiving undiluted RK, even though performance was stable. Although the high acids might explain the lower yields observed in Runs 120 and 121, the yield of Run 119 was not affected Conversion of these residual acids to methane would result in 11%, 31%, and 29% increases in methane yield for Runs 119, 120, and 121, respectively. This might be accomplished by passage of the effluent through a salt-adapted methane phase or by development of a salt-adapted methane phase in the mixed fermentation.

3.3.6. Hydraulic Retention Time (HRT)

Long hydraulic retention times (HRTs) should result in maximum VS reduction and methane yield, but slow decomposition and methane production rates. Compounds undecomposed at long retention times can be considered to be refractory to anaerobic decomposition. As the retention time is decreased, certain substrates and intermediates will wash out because of combinations of the following reasons: (1) organisms responsible for their degradation may have long generation times; (2) the solids retention time may not be long enough for degradation; and (3) easily metabolized substrates may prevent decomposition of more complex substrates via catabolite repression, a phenomenon to be discussed later. Recycling of settled digested solids should prevent the effects of Items (1) and (2), since the retention time of undigested solids and microorganisms would be increased by that process.

Although a systematic study is under way to evaluate the effect of HRT and effluent solids recycle on kelp biomethanogenesis, these experiments are not yet complete. However, data such as those presented in Table 7 for Runs 18 and 26 and Run 16 (Table 8) indicate that reduction in HRT from 18 to 12 and 10 days results in a corresponding reduction in methane yield.

3.3.7. Feed Concentration

Increases in feed concentration, independent of HRT, should result in higher fermentation rates and methane yields. At some upper concentration limit, further increases will saturate the population's ability to accommodate, and additional feed will remain unreacted. In fact, with some substrates, high concentrations may result in fermentation imbalance and digester failure. Recycle of solids often results in increases in the upper concentration limit without deterioration in performance. Studies are currently in progress to evaluate the effects of different loadings on kelp biomethanogenesis.

3.3.8. Particle Size

The particle size of feed for digesters usually has a significant effect on performance. Generally, smaller particles are more easily decomposed because of the larger surface area exposed to organisms or extracellular enzymes. Preliminary study showed that reduction of particle size from 2 to 4 mm to a puree consistency did not result in increased yields.[9] Since particle size requirements affect pretreatment process requirements, the effect of larger particle sizes on performance will be evaluated when large-scale digesters are available for study.

3.3.9. Mixing

The effects of mixing on anaerobic digestion have not been well documented. However, it seems reasonable to assume that, up to some point, mixing would enhance contact between organisms and substrates. Above that point, however, mixing could destroy organisms or obligate interspecies association[19] and thus decrease digester performance. Furthermore, effects of mixing could be sensitive to digester size and configuration, composition and particle size of feed, method of mixing, and a number of other factors.

3.3.10. Feeding Frequency

The ideal feeding frequency for the biomethanogenesis process is continuous, because that mode of operation enhances culture stability. However, it has been demonstrated that the kinetics of digesters fed daily under a conventional loading and hydraulic retention time resembles that of a continuous culture.[20] At some lower limit, culture instability would ensue because of the wide variation in growth rates of the organisms involved in the fermentation. This phenomenon becomes a problem particularly in small manually fed digesters. Full-scale digesters are automatically fed; thus the feeding frequency can be easily increased.

3.3.11. Catabolite Repression

Catabolite repression is a process common among organisms, whereby catabolites of easily decomposed substrates repress degradation of more complex substrates. Although this phenomenon has not been well studied in methanogenic fermentations, it probably exists; the effect would be that a simple substrate (such as mannitol found in kelp) would repress degradation of complex substrates at short retention times. This effect could be minimized by operating at long hydraulic retention times or by using two digesters in series.

3.4. Component and Energy Balance for Biomethanation of Raw Kelp

Data reported previously[8,9] were used to calculate a balance of the components, carbon, and energy for the biomethanation of raw kelp, and this is presented in Table 11. The validity of this balance is supported by the fact that 97.4%, 95.9%, and 102%, respectively, of feed organic components, carbon, and energy are accounted for in the products. Of primary importance is the fact that 100 kg of wet RK will yield 1.56 m^3 of methane. This yield is equivalent

TABLE 11. Component, Carbon, and Energy Balance for Run 8

	100 kg Kelp	35% C, $\dfrac{1.6 \text{ kg VS}/m^3\text{-day}}{47.5\% \text{ VS reduction}} \longrightarrow$	Digested Solids	+	Gas (60% CO_2 − 40% CO_2)
	89.7 kg H_2O		4.7 kg Ash		$0.465 \dfrac{m^3 \text{ Gas}}{\text{kg VS added}} \times 5.6 \text{ kg}$
	4.7 kg Ash		2.94 kg VS (calculated)		= 2.60 m^3 Gas
	5.6 kg VS				= 0.110 kg mol
Protein	1.65 kg (15.9%)		1.52 kg (20%)		
Carbohydrate					
Mannitol	1.93 kg		0.55 kg		
Algin	1.45 kg		0.22 kg		
Cellulose	0.49 kg		0.45 kg		
Laminarin	0.07 kg		0.01 kg		
Fucoidin	0.02 kg		0.11 kg		
Total protein and carbohydrate	5.61 kg		2.86 kg		
Carbon	2.68 kg		1.25 kg		1.32 kg
Energy	10,200 kJ		4696 kJ		5662 kJ

Accounted for: Feed organic components 97.3%; Feed carbon 95.9%; Feed energy 102%.

to 0.281 m³/kg VS added. The digester culture volume that would be required for this daily feed rate and yield would be 3.5 m³. Energy recovery in the product gas is 55.5%, and volatile solids reduction is 47.5%. A comparison of these experimental values for methane yield and volatile solids reduction with the maximum theoretical values reported in Table 5 shows that 64% of the methane yield and 57.8% of the volatile solids reduction have been achieved.

Examination of the organic components in the influent and effluent is useful in identifying the biodegradable and refractory components. For the kelp fermentation, mannitol and algin are the most biodegradable, and cellulose and protein the least biodegradable. It should be pointed out that some of the kelp protein was probably converted to bacterial protein. Laminarin and fucoidin are only minor components of kelp, and thus have minimal influence on the overall component balance.

A more detailed component and energy balance can be made for the biomethanation process when sufficient data have been obtained for formulation of the total process scheme, including preprocessing and pretreatment, anaerobic digestion, solids separation, processing of effluent supernatant and solids, and gas cleanup. It is estimated that for conventional digestion 20% of the product energy will be required for maintenance of the process.[2]

4. Summary

A stable fermentation was obtained for the mesophilic biomethanogenesis of giant brown sea kelp under conventional conditions in bench-scale fermenters (i.e., loading 1.6 kg VS/m³-day and retention time of 18 days). The methane yield obtained regularly for this feed substrate (0.28 m³/kg VS added) is high as compared with values reported for other wastes and biomass types composed primarily of carbohydrates, including municipal solid wastes, grass, feedlot cattle waste, and dairy manure (see Table 12). The references included in Table 12 all contain hard data to document the methane yields presented. (Note that high methane production rates are not always associated with high methane yields. Ultimately, the objective is to maximize both parameters.)

The studies reported show that the anaerobic fermentation of raw kelp is not limited by phosphorus or by nitrogen when the C-to-N ratio is 17 or less. Thermophilic digestion of kelp resulted in an unstable fermentation and lower methane yields as compared with the mesophilic fermentation. Direct addition of kelp without dilution water or diluted with seawater resulted in retarded digestion after about two retention times, apparently because of the resulting high salt concentrations. Performance gradually returned to normal following a period of adaptation. Two inocula derived from anaerobic marine environments did not show improved performance over IGT's inoculum developed from digesters receiving sewage sludge and municipal solid wastes. A search

TABLE 12. Performance Data for Anaerobic Digestion of Various Types of Biomass

Reference	Biomass type	MY^a	MPR^b	Special conditions^c
Chynoweth *et al.*[9]	Raw kelp	0.281	0.445	$T = 35$, $L = 1.60$, $RT = 18$
Pfeffer[21]	MSW-sludge^d	0.151	—	$T = 55$, $RT = 20$
Pfeffer[21]	MSW-sludge	0.099	—	$T = 35$, $RT = 20$
McCarty *et al.*[22]	MSW-sludge	0.204	0.750	$T = 35$, $L = 3.68$, $RT = 15$
Ghosh and Klass[15]	MSW-sludge	0.238	0.530	$T = 35$, $L = 2.24$, $RT = 12$
Klass *et al.*[23]	Grass mixture	0.184	0.370	$T = 35$, $L = 1.92$, $RT = 16$
Varel *et al.*[17]	Feedlot cattle waste	0.256	2.21	$T = 60$, $L = 8.65$, $RT = 9$
Varel *et al.*[17]	Feedlot cattle waste	0.171^e	4.44	$T = 60$, $L = 25.95$, $RT = 3$
Converse *et al.*[24]	Dairy manure	0.208	0.899	$T = 35$, $L = 4.32$, $RT = 15$

^{a}MY = methane yield, m^3/kg VS added.
^{b}MPR = methane production rate, m^3/m^3 culture-day.
^{c}T = temperature, °C; L = loading, kg VS/m^3-day; RT = retention time, days.
^{d}MSW-sludge = municipal solid waste-sewage sludge.
^{e}4.15 m^3 CH_4/m^3 culture-day (highest rate of methane production in the literature).

for an inoculum that will exhibit high performance at ambient temperatures (about 26 °C) is continuing. A materials balance presented for this fermentation showed that most of the methane is derived from decomposition of algin and mannitol, and that protein and cellulose represent the major refractory components of the effluent solids. It has not been determined whether the effluent solids protein is associated with kelp or bacteria produced by fermentation. Preliminary studies showed that heat alone, as well as acid and alkaline hydrolysis in the presence of heat, increased the biodegradability of the effluent solids.

Studies currently in progress on the biomethanation of kelp are aimed at decreasing the reactor size through optimization of the retention time and loading and increasing the methane yield through evaluation of various post-treatment processes, culture optimization, and phase separation. Eventually this information will be used for the design and operation of a large-scale fermentor.

ACKNOWLEDGMENTS

The authors wish to express appreciation for financial support by the Gas Research Institute and the General Electric Company. The technical assistance of John Conrad made this work possible. Special analyses conducted by the staffs of James Ingemanson and Robert Stotz are included in this work. The U.S. Department of Agriculture, Western Regional Research Center is acknowledged for providing the kelp used in these studies. Finally, the numerous suggestions and critical advice provided by Dr. Fred Pohland are greatly appreciated.

References

1. D. L. Klass, in: *Clean Fuels from Biomass, Sewage, Urban Refuse, and Agricultural Wastes,* Symposium Papers of the Institute of Gas Technology meeting at Orlando, Florida (January, 1976), pp. 21–58, Institute of Gas Technology, Chicago, Illinois.
2. J. Davidson, M. Ross, D. Chynoweth, A. Michaels, D. Dunnette, C. Griffis, J. Sterling, and D. Wang, in: *The Energy Conservation Papers* (R. H. Williams, ed.), Ballinger Publishing Co., (1975), pp. 303–373.
3. M. D. Fraser, in: *Clean Fuels from Biomass and Wastes,* Symposium Papers of the Institute of Gas Technology meeting at Orlando, Florida (January, 1977), pp. 425–439, Institute of Gas Technology, Chicago, Illinois.
4. T. M. Leese, in: *Clean Fuels from Biomass, Sewage, Urban Refuse, and Agricultural Wastes,* Symposium Papers of the Institute of Gas Technology meeting at Orlando, Florida (January, 1976), pp. 253–259, Institute of Gas Technology, Chicago, Illinois.
5. Armond J. Bryce, in: *Energy from Biomass and Wastes,* Symposium Papers of the Institute of Gas Technology meeting at Washington, D.C. (August, 1978), pp. 353–377, Institute of Gas Technology, Chicago, Illinois.
6. W. J. North, *The Biology of Giant Kelp Beds (Macrocystis) in California,* J. Cramer, Lehrte, Germany (1971).
7. H. A. Wilcox, The ocean food and energy farm, paper presented to The American Association for the Advancement of Science, New York (January, 1975).
8. D. L. Klass and S. Ghosh, in: *Clean Fuels from Biomass and Wastes,* Symposium Papers of the Institute of Gas Technology meeting at Orlando, Florida (January, 1977), pp. 323–351, Institute of Gas Technology, Chicago, Illinois.
9. D. P. Chynoweth, D. L. Klass, and S. Ghosh, in: *Energy from Biomass and Wastes,* Symposium Papers of the Institute of Gas Technology meeting at Washington, D.C. (August 1978), pp. 229–251, Institute of Gas Technology, Chicago, Illinois.
10. E. Lindner, C. A. Dooley, and R. H. Wade, Chemical variation of chemical constituents in *Macrocystis pyrifera,* Ocean Food and Energy Farm Project Final Report, Naval Undersea Center, San Diego (January, 1977).
11. R. A. Mah, D. M. Ward, L. Baresi, and T. L. Glass, *Annu. Rev. Microbiol.* **31,** 309–341 (1977).
12. J. G. Zeikus, *Bacteriol. Rev.* **41,** 514–541 (1977).
13. P. L. McCarty, in: *Principles and Applications of Aquatic Microbiology,* (H. Heukelekian and N. Dondero, eds.), John Wiley, New York (1964), pp. 314–343.
14. R. E. Speece and P. L. McCarty in: *Advances in Water Pollution Research. Proceedings of the [First] International Conference* Vol. 2, Pergamon Press, New York (1964), pp. 305–322.
15. S. Ghosh and D. L. Klass in: *Clean Fuels from Biomass, Sewage, Urban Refuse, and Agricultural Wastes,* Symposium Papers of the Institute of Gas Technology meeting at Orlando, Florida (January, 1976), pp. 123–182, Institute of Gas Technology, Chicago, Illinois.
16. C. A. Kelly and D. P. Chynoweth, in: *Proceedings of the ASTM Symposium on Native*

Aquatic Bacteria: Enumeration, Activity, and Ecology (J. W. Costerton and R. R. Colwell, eds.) Published by American Society for Testing Metals, (1979), pp. 164–179.

17. V. H. Varel, H. R. Isaacson, and M. P. Bryant, *Appl. Environ. Microbiol.* **33**, 298–307 (1977).

18. J. T. Pfeffer, *Biotechnol. Bioeng.* **16**, 771–786 (1974).

19. J. G. Ferry and R. S. Wolfe, *Arch. Microbiol.* **107**, 33–40 (1976).

20. S. Ghosh, J. R. Sedzielarz, K. H. Griswold, M. P. Henry, S. J. Bortz, and D. L. Klass, Research study to determine the feasibility of producing methane gas from sea kelp, Final Report for U.S. Navy Contract No. N00123-76-C-0271, Institute of Gas Technology, Chicago (1976).

21. J. T. Pfeffer, in: *Microbial Energy Conversion* (H. G. Schlegel and J. Barnea, eds.), Pergamon Press, New York (1977), pp. 139–155.

22. P. L. McCarty, L. Y. Young, D. C. Stuckey, and J. B. Healy, Jr., in: *Microbial Energy Conversion* (H. G. Schlegel and J. Barnea, eds.), Pergamon Press, New York (1977), pp. 179–199.

23. D. L. Klass, S. Ghosh, and J. R. Conrad, in: *Clean Fuels from Biomass, Sewage, Urban Refuse, and Agricultural Wastes*, Symposium Papers of the Institute of Gas Technology meeting at Orlando, Florida (January, 1976), pp. 229–252, Institute of Gas Technology, Chicago, Illinois.

24. J. C. Converse, J. G. Zeikus, R. E. Graves, and G. W. Evans, Dairy manure degradation under mesophilic and thermophilic temperatures, paper presented at the Winter Meeting of the American Society of Agricultural Engineers, Chicago (1975).

18

Basic Principles of Ethanol Fermentation

DOUGLAS M. MUNNECKE

1. Introduction

The microbial production of ethyl alcohol from agricultural commodities for blending into gasoline to decrease our dependence on imported crude oil has recently received much regional, national, and international attention.[1-5] The microbial conversion of agricultural substrates into ethanol is, however, an ancient practice that certainly predates the science of microbiology, the chemistry of the distillation process, and the engineering of an ethanol fermentation plant. Only in the mid-1900s was microbial production of ethanol replaced by synthetic ethanol derived from petroleum as the major source of this chemical for our society. The development of synthetic ethanol occurred at approximately the same time that microbiologists were beginning to investigate the potential of microbial processes for the production of industrial feedstocks or solvents such as glycerol, lactic acid, citric acid, acetone, and butanol for the rapidly expanding chemical industries.[6] The development of the necessary support science and technology for this young fermentation industry was therefore repressed when chemical based technology proved so successful. With the recognition in the 1970s that our petroleum reserves are being rapidly depleted, researchers began to update the fermentation technology for industrial ethanol production. Fortunately, the biochemistry of this fermentation process was continually examined over the last 100 years. Pasteur's research with French wines in the 1860s defined the basic concepts of the fermentation process and com-

DOUGLAS M. MUNNECKE • University of Oklahoma, Department of Botany and Microbiology, Norman, Oklahoma 73019.

mercial interests in beer, wine, and hard liquor production promoted continual interest in understanding the biochemistry of ethanol fermentations. Thus, the basic, biochemical aspects of microbial ethanol production are relatively well understood, but technology for industrial ethanol production by microbial processes has, until recently, been neglected.

Central to the operation of an efficient ethanol fermentation process is the proper understanding of the biochemical mechanism by which the desired product, ethanol, is produced. Only then can the many parameters affecting the fermentation process be kept optimal for ethanol production and allow for maximal conversion of substrate into ethanol as well as lower the production costs due to efficient use of fermentation equipment. This chapter reviews the basic biochemical aspects of the ethanol fermentation process, lists which substrates can be fermented, and discusses how they are metabolized by various microorganisms. The significance of substrate preparation to the efficiency of the fermentation process and the importance of certain parameters which must be controlled for successful ethanol production is also reviewed. Finally, the fermentation technology being considered for ethanol production is examined.

2. Why Microbes Produce Ethanol?

When microorganisms are grown on sugars in the presence of oxygen, they obtain cellular material and energy by oxidizing these organic compounds. As a result of this oxidation, carbon dioxide and water are produced as metabolic waste products. The excess electrons from the oxidation of sugars are carried by an electron transport system to oxygen, the final electron acceptor, and water is formed. Certain microorganisms are able to grow on organic compounds such as sugars in the absence of oxygen, utilizing organic compounds as electron acceptors instead of oxygen. During this anaerobic growth, sugars or other fermentable carbon sources are oxidized and excess electrons are transferred to organic acceptor molecules and ethanol is produced as a waste product of the fermentation process instead of water. Microorganisms responsible for ethanol production are facultative, i.e., they can grow with or without oxygen. If air is allowed to enter the fermentation process in sufficient quantities, then microbial metabolism will switch from an anaerobic, ethanol-producing process to the more efficient, aerobic process (Krebs cycle, electron transport system), and no further ethanol will be produced. The previously produced ethanol may actually be utilized (glycolytic pathway) and oxidized to carbon dioxide and cell material. Thus, microbes produce ethanol when growth parameters do not support an oxidative metabolic process, thereby requiring these facultative microorganisms to employ a less efficient pathway which produces ethanol as a metabolic waste product.

3. Substrates for Ethanol Production

Microorganisms can metabolize anaerobically a wide array of organic compounds and produce ethanol. These substrates may be derived from agricultural commodities such as corn, potato, sugar cane, sugar beet, grains, or any other crop high in starch or sugar content. Fermentable substrates can also be derived from cellulosic materials such as wood, straw, newspaper, wastes from the lumber industry, or manure.

Sugars in agricultural commodities are found in two basic forms: dimers of simple hexose sugars such as sucrose (glucose—fructose) as in the case of sugar cane and sugar beets, or in the form of starch (glucose polymer) derived from corn, potatoes, and grains. If only agricultural crops were used to produce both liquid (ethanol) and gaseous (methane, hydrogen) fuels to supplement our diminishing fossil fuel supply, our agricultural productivity would not be great enough to simultaneously satisfy both our energy and food requirements.[3,4] Therefore, other sources of fermentable sugar or non-sugar substrates are being examined.

Cellulosic materials are certainly the largest and currently the most extensively investigated carbohydrate reserve which can be used for ethanol production. Wood contains cellulose (a glucose polymer), hemicellulose (mixed hexoses and pentoses), and xylan (a xylose polymer), all of which can serve as substrates for fermentation if properly pretreated. Cellulosic "waste products," such as straw, rice hulls, timber cuttings, and newspapers, can also be utilized as sources of fermentable sugars for ethanol production.[7] Other agricultural wastes which may be utilizable for ethanol production are whey from the cheese industry[8] and other food processing wastes.

4. Substrate Preparation

The importance of substrate preparation cannot be over emphasized in regard to its effect on substrate fermentability, overall process kinetics, and efficiency of substrate conversion into ethanol.[9] Since substrate cost strongly influences the price of ethanol, pretreatment procedures are vitally important for an optimized process and minimum product production costs.

Starch can be degraded by some microorganisms into the sugars maltose and glucose which can then be further metabolized to yield ethanol. However, species of *Saccharomyces,* the yeast culture used in most fermentation processes, is not able to convert starch into fermentable sugars. Therefore, starch must first be hydrolyzed into maltose and glucose molecules before ethanol production can begin.[2,10] In beer production, barley starch is hydrolyzed by plant enzymes (amylases) produced by the germinating seeds during the malting process.[9] In other processes, starch liquification is accomplished by fungal amylases which are added to starch. In most ethanol production processes using starch (corn, potato, or grains), there must be a pretreatment to reduce the high molecular weight starch polymers to maltose (a glucose dimer), glucose, and other fermentable sugars (Figure 1). The concentration of intermediate-molecular-weight polymers, dextrins, should be minimized since they are not metabolized by yeast.

Sugar cane and sugar beets contain sucrose (a fructose–glucose dimer), which can be directly metabolized by the organisms used in industrial ethanol production. Thus, for these substrates, pretreatment is a physical process and does not require sucrose hydrolysis. Molasses, a byproduct of sugar refining, is also an excellent substrate for ethanol fermentations.

The technology needed for utilization of starch or sucrose in yeast fermentations has developed over the years and is well optimized since these substrates are important in the beer, wine, and spirits industries. The technology for cellulose utilization is not as advanced, even though investigators have been concerned with cellulose utilization since the nineteenth century. Various physical, chemical, and biological processes have been examined for releasing cellulose from wood and from agricultural and industrial wastes. Chemical and physical processes involving heat, hydrolysis by strong acid or base, and milling have been examined for use as pretreatment processes. If microbial pretreatment processes are used, cellulose is hydrolyzed to cellobiose (a glucose dimer). This compound is further cleaved to glucose (Fig. 1).[11,12]

Liquid wastes derived from the food processing industry require another type of pretreatment. Whey from the dairy industry contains readily fermentable substrates, casein and lactose,[8] but their concentrations are too low for industrial ethanol production. Here, concentrating the waste stream is necessary to bring the fermentable substrate concentration up to significant levels so that the final yield of ethanol will be high enough for economical distillation.

The substrate pretreatment process is not only critical in determining the yield of fermentable sugars from a given starting substrate, but the process can also affect the type of sugars formed and their respective ratios. When two or more sugars are simultaneously available for microbial metabolism, one sugar may inhibit the utilization of the others. This inhibition, called catabolite repression, can strongly affect the yield of ethanol in some fermentation systems and is discussed in detail later.

5. Substrate Metabolism

The primary substrates involved in ethanol fermentations discussed in this chapter are glucose, sucrose, lactose, cellulose, and xylan. All substrates except xylan are hexose-based sugars and can be metabolized through three basic

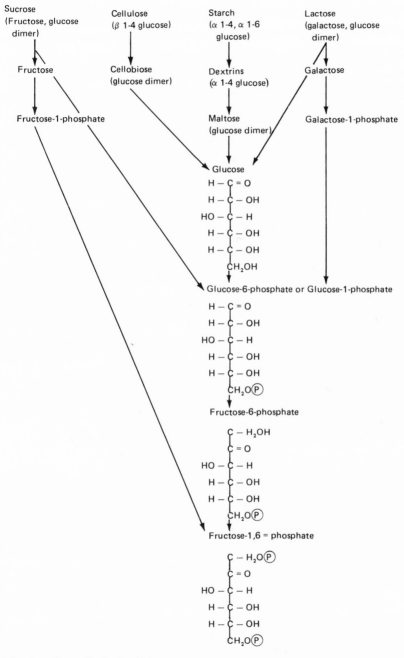

FIGURE 1. Generalized microbial pathways for the initial metabolism of hexose sugars from dimers and polymers.

metabolic pathways. Where each substrate will enter into a metabolic pathway is strongly governed by the pretreatment process used and the type of microbial culture employed. Figure 1 outlines the primary steps involved in the metabolism of these substrates as they enter into catabolic hexose sugar pathways.[13] For instance, sucrose is hydrolyzed by the enzyme sucrose phosphorylase to glucose-1-P and fructose which, in turn, is phosphorylated. Cellulose is degraded enzymatically in two steps—first to cellobiose (cellulose), then to glucose (cellobiose, or cellobiose phosphorylase). Starch, during pretreatment, is hydrolyzed into primarily two fermentable substrates, glucose and maltose. Maltose, a glucose dimer, is then degraded to two glucose molecules. In poor pretreatment processes, large amounts of dextrins are left unhydrolyzed and may or may not be fermented. Lactose is metabolized to glucose and galactose which eventually enter the glycolytic pathway.

Xylan is composed of pentose sugar monomers and does not enter the same pathways as do hexose sugars. Xylan is degraded to xylose and xyulose-5-phosphate which then enters a five-carbon sugar assimilation pathway. One such pathway leads to glyceraldehyde-3-P and acetyl-P, resulting in one ethanol and one pyruvate molecule from each xylose. Other compounds such as lipids and organic acids are metabolized through pathways associated with the glycolytic, dicarboxycylic acid, and glyoxylate pathways.

Since most of the substrates currently being examined for ethanol production are hexoses, it is important to understand how yeast or bacterial cultures metabolize these compounds and produce ethanol. Once the sugars reach a basic entrance point in the cellular metabolic machinery (Figure 1), there are three basic catabolic pathways which the various microorganisms can use. The Embden–Meyerhof pathway (Figure 2) is probably the most common pathway for glucose metabolism and proceeds through a series of transformations involving phosphorylation reactions which are important in energy production for cellular growth. The glucose molecule is sequentially metabolized to yield two moles of pyruvate per mole of hexose. The hexose-monophosphate shunt (Figure 2) is slightly different from the Embden–Meyerhof pathway and results in one mole of ethanol and one mole of pyruvate from each mole of glucose or equivalent sugar. The Entner–Doudoroff pathway (Figure 2) produces two moles of pyruvate from each mole of glucose but by a different mechanism than the Embden–Meyerhof pathway. Yeast, as a class of organisms, utilize both the Embden–Meyerhof and the hexose monophosphate shunt pathways; various bacteria employ all three.

The metabolic pathways shown in Figures 1 and 2 are not unique to anaerobic fermentation processes. Cultures growing with or without oxygen would metabolize the sugars by the same pathways. In aerobic processes, compounds reduced in these metabolic steps (NAD^+ to $NADH$) would be reoxidized through an electron transport system with oxygen serving as the electron acceptor and water being formed. In fermentations, however, oxygen is not

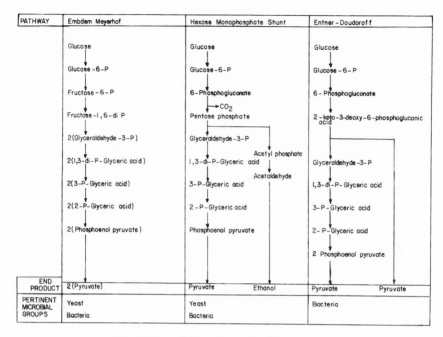

FIGURE 2. Microbial metabolic pathways leading to pyruvate formation.

present and so the microbe has to use an organic compound as an electron acceptor for recycling NADH. Numerous systems have evolved with pyruvate serving as a key branching point in fermentations. In an ethanolic type of fermentation, pyruvate is metabolized to acetaldehyde and carbon dioxide, and acetaldehyde is then reduced to ethanol while NADH is reoxidized to NAD^+. In this process, it is theoretically possible to achieve 2 moles of ethanol for each mole of glucose converted, or on a weight basis, a 51% conversion of glucose into ethanol. In practice only 90–95% of the theoretical amount of ethanol is produced because some of the pyruvate is consumed for cellular material during culture growth and is not available to serve as an electron acceptor. Also, low levels of higher alcohols are produced as metabolic waste products.

The yeast ethanolic fermentation is the most efficient pathway for ethanol production, but it is not the only pathway leading to ethanol accumulation (Table 1).[14] In a mixed-acid fermentation (*Clostridium* sp.), pyruvate is transformed into lactic acid, formic acid, carbon dioxide, hydrogen, acetic acid, and ethanol. The butanediol fermentation (*Bacillus* sp.) produces ethanol as well as four other major organic waste products while the butanol/acetone fermentation transforms pyruvate into five major organic end products.

One pathway listed in Table 1 is called the homolactic acid fermentation. No ethanol is produced in this process; only two moles of lactic acid per mole

TABLE 1. Anaerobic Metabolism of Pyruvate

Type of fermentation	End products	Microorganisms
Ethanolic	Ethanol	Yeast
	Carbon dioxide	*Zymomonas*
Mixed acid	Lactic acid	*Clostridium* and many
	Formic acid	enteric bacteria
	Acetic acid	
	Carbon dioxide	
	Hydrogen	
	Ethanol	
Butanediol	As in mixed acid plus	*Bacillus* and other
	2,3-butanediol	bacteria
Acetone/butanol	Acetic acid	*Clostridium*
	Butyric acid	
	Ethanol	
	Butanol	
	Acetone	
	Isopropanol	
	Carbon dioxide	
	Hydrogen	
Homolactic	Lactic Acid	*Lactobacillus*
		Streptococcus

of glucose are produced. It is mentioned in association with ethanol fermentations because if a culture having this metabolic machinery (e.g., *Lactobacillus* sp.) contaminates the ethanol fermentation, the yield of ethanol can be strongly reduced. Under some fermentation conditions, *Lactobacillus* or *Streptococcus* species are able to compete with ethanol-producing organisms and degrade hexoses to lactic acid.

6. Effect of Microorganisms on Ethanol Production

Although numerous microorganisms are capable of producing ethanol, not all are suitable for industrial processes. Also, no one culture is ideal for efficient conversion or high attenuation of all the aforementioned substrates. Yeast cultures,[15-17] (in particular *Saccharomyces* sp.) have been most extensively examined. Relatively recently, bacterial cultures[18] of *Bacillus* and *Clostridium* species have been explored for high-temperature ethanol fermentation processes.

The alcoholic beverage industry basically uses two types of yeast cultures, both *Saccharomyces,* in their alcoholic production processes. One yeast type is a top fermenter, so called because the majority of the yeast cells remain at the top of the fermentation broth during active fermentation, while the second yeast type is a bottom fermenter since it settles to the bottom of the fermenter during active fermentation. Various "wild type" yeasts are occasionally involved in beverage production since many different genera of true yeasts are capable of good ethanol fermentations. Generally, however, various species of *Saccharomyces* are examined for ethanol production processes because they are very efficient in converting sugars into ethanol[17] and are not as strongly inhibited by high ethanol concentrations as are other microbes. Table 2 indicates that theoretically two moles of ethanol can be produced from one mole of glucose (511 kg of ethanol from 1000 kg of glucose).

Bacterial cultures are being examined for use in ethanol production processes because of their higher temperature tolerance. Thus, selected species of *Clostridium* and *Bacillus* are able to grow as thermophilic microorganisms (i.e., they have a growth temperature optimum above 50°C) and may therefore reduce the cost of the fermentation and distillation processes. However, the yield of ethanol by bacterial cultures is not as high as in yeast fermentations (Table 2). For instance, *Clostridium* cultures can utilize either the mixed-acid or acetone–butanol fermentation process. Typical yields of organic waste products for cultures not selected for high ethanol production are listed in Table 2. More lactic acid or acetone and butanol are formed in these examples than ethanol, and only 5–25% of the theoretical maximal amount of ethanol from the yeast ethanol process is produced. Species of *Bacillus* employ a butanediol type of fermentation which produces 35% of the theoretical maximum ethanol yield.

TABLE 2. End Products of Various Anaerobic Fermentations

Culture/type of fermentation	CO$_2$	H$_2$	Formic acid	Acetic acid	Lactic acid	2,3-Butanediol	Ethanol
Yeast/ethanol	200	—	—	—	—	—	200
	(49)						(51)
Bacteria/butanediol	172	36	18	0.5	3	66	70
	(42)	(0.4)	(4.6)	(0.2)	(1.2)	(33)	(18)
Bacteria/homolactic	—	—	—	—	200	—	0
					(82)		(0)
Bacteria/mixed acid	88	75	2	37	80	—	50
	(21)	(0.8)	(0.5)	(12.3)	(33)		(13)
Bacteria/acetone–butanol	219	139	+	13	—	—	10
	(54)	(1.5)		(4.3)			(2.5)

Based on expected yields from 100 mmoles glucose. Figures expressed in mmole (wt %).

7. Effect of Fermentation Parameters

7.1. Temperature

Temperature has an important influence on the growth rate of the microorganism and the rate of ethanol production.[19-23] Wine and beer fermentations are generally conducted below 20°C, whereas higher temperatures (30–38°C) are being examined for industrial alcohol production by yeast cultures. The optimal temperature for cellular growth and ethanol production may decrease as the concentration of ethanol increases. At high broth concentrations of ethanol, its inhibitory effect can be more pronounced at higher temperatures than at lower temperatures due to greater instability of the cellular membrane structure at higher temperatures.

Many investigators are attempting to isolate thermophilic bacterial cultures which can grow at temperatures above 50°C.[18,20] The major advantage is not a higher metabolic rate of ethanol production, but possible lower costs in regard to cooling the fermenter and distilling the fermented broth. Temperature may also influence the ratio of chemicals produced by a bacterial, branched-pathway type of pyruvate metabolism. Thus, shifts in temperature may vary the amount of pyruvate going into ethanol, organic acids, and other alcohols.

7.2. pH

A very important factor for cellular growth is external pH.[23,24] Most alcoholic yeast fermentations are conducted below pH 4.5, although this may not be the optimal pH for growth or ethanol production. Yeast cultures can grow over a wide pH range from 3 to 8 with an optimum for growth generally in the slightly acid range. Shifts in pH can also affect the final ratio of organic waste products produced by yeast cultures. Thus, the optimal pH for a fermentation process must support a balance among ethanol production, cellular growth, and physiological effects on waste product pathways. Low pH values in yeast fermentations help to inhibit growth of contaminating bacterial cultures. Bacterial cultures generally have a pH optimum around 7–7.5, with less tolerance than yeast to acid conditions. Therefore, fermentations utilizing *Clostridium* or *Bacillus* species are conducted in higher pH buffered media.

7.3. Oxygen

The microorganisms involved in alcoholic fermentations are facultative microbes since they are able to grow with or without the utilization of oxygen. Thus, two types of drastically different pathways of pyruvate metabolism are available. With oxygen present, more cell mass will be produced from the initial substrate (5 to 10 times more than in anaerobic growth) and the growth rate will be increased. Therefore, for inoculum formation, aeration improves the yield of cell mass and its rate of production. However, for ethanol production, oxygen must be restricted from entering the fermenter and zero-to-very-low levels of oxygen in the broth should be maintained. The oxygen concentration which triggers aerobic or anaerobic growth processes varies from culture to culture[25-27] and is dependent on substrate concentration and cell density. Oxygen at very low concentrations may be needed by "anaerobically" growing yeast cultures for synthesis of lipids required for cell-wall synthesis.[28-30] If low levels of oxygen are not provided, certain lipid precursors must be added to the fermentation broth for cellular growth to occur. Ethanol production, however, can occur without cell growth and therefore does not require oxygen.

7.4. Ethanol

The concentration of ethanol in the fermentation broth can directly affect the growth rate of the culture and its ability to convert sugars into ethanol.[22,31] Inhibitory and toxic levels of ethanol vary from culture to culture. With some yeast cultures, 50% inhibition of growth occurs at 4–6% ethanol, while others are more ethanol tolerant. Generally, maximal ethanol concentrations produced microbiologically range from 11–14%. However, some investigators report that final ethanol concentrations of 20% are achievable.[32] If the ethanol is produced at low rates so that the intracellular ethanol concentration remains low, then higher levels can be tolerated. However, high production rates result in intracellular buildup of ethanol and in an increased sensitivity.[32] Certain other higher alcohols produced as minor waste products can be more inhibitory than ethanol. Thus, even though they are produced in much smaller quantities, they may have a significant inhibitory effect.[33]

For efficient ethanol production, the microbial and engineering processes require two opposing situations. Microbial processes such as growth and conversion of sugars into ethanol proceed best at 0% ethanol and are increasingly inhibited as the alcohol concentration rises. Yet, for efficient distillation, high concentrations (above 6%) are desired. Attempts to surmount this problem have involved screening programs for the isolation of ethanol tolerant microorganisms[34,35] and genetic studies to improve ethanol tolerance. Also, different fermentation processes have been examined for solving the problem of ethanol inhibition. In one process, a vacuum fermentation process, the fermentation is conducted under vacuum and ethanol is continuously drawn out of the broth as it is produced. The condensate has a much higher ethanol concentration and is then distilled.

7.5. Sugar

The concentration of sugar can affect the microbial ethanol fermentation in various ways.[36] The amount of alcohol produced is proportional to the amount of sugar added; thus, high sugar concentrations are desired. However, sugar concentrations which are too high can inhibit metabolism due to the increased osmotic pressure. Very low levels of sugar, on the other hand, as might be experienced in continuous-flow fermentation processes, may limit the rate of ethanol production. Hence, each fermentation process will have an optimal glucose or equivalent sugar concentration.

Secondly, the concentration of various sugars in a mixed substrate fermentation can affect their metabolism. For instance, if glucose is present in a mixture of sugars, some microbes will initially metabolize glucose, and this metabolism represses or blocks the metabolism of the other sugars (catabolite repression). Only when glucose is completely degraded will, for instance, the metabolism of maltose and other sugars derived from starch hydrolysis begin. This sequential sugar metabolism[15,24] may not present problems in batch fermentations, but may result in lower efficiency of ethanol production in continuous fermentations since one sugar in the feed may delay or repress the metabolism of a second sugar.

7.6.   Nutrients and Vitamins

Salts are critical for both the growth of ethanol-producing microbes and the production of ethanol. For growth, microorganisms need a proper balance of macronutrients such as nitrogen, phosphorus, potassium, sodium, and sulfur, as well as micronutrients such as zinc, copper, iron, magnesium, and manganese.[37] The macronutrients are required primarily for synthesis of cellular material while the micronutrients are required for coenzymes and as cofactors in enzymatic reactions. The proper amount of each macro- and micronutrient is dependent upon the type of ethanol fermentation process desired. For instance, by maintaining low levels of nitrogen, less cell mass will be produced and higher levels of ethanol will occur. In some processes, no cellular growth is desired. Therefore, only micronutrient addition is required for glucose conversion into ethanol. The various sugar feedstocks will contain different amounts of inorganic nutrients and the technical problem is how to balance the nutrients for optimum process kinetics. In many cases, only minor nutrient addition is required.

For some microorganisms, specific organic compounds (vitamins) are required. Biotin is usually a requirement for yeast growth, while lipid precursors are required if yeast are grown anaerobically. Bacterial cultures may or may not have specific organic nutrient requirements.

Buffering the medium is also required to maintain the pH of the broth within a range of 3.5 to 5.0. Complex natural substrates such as malt have an inherent buffering ability; however, a buffer is needed if sugar solutions are to be fermented. Potassium or sodium mono- and dibasic phosphate buffer is good for neutral-to-slightly acid ranges, whereas carbonate buffer is too alkaline for ethanolic fermentations.

8. Fermentation Systems

8.1. Batch

Conventional ethanol fermentations are conducted as batch processes where the reactor is charged with substrate, the microbial inoculum is added, and the process allowed to run to completion (4–10 days). The fermentation tank can be mechanically agitated by impellers to decrease diffusional limitations, or the natural agitation created by escaping carbon dioxide may be sufficient. In batch processes, the sugar or fermentable carbon source is added batchwise at decreasing intervals to the growing culture, or continuously at an increasing rate as the microbial population expands. After the fermentation is complete, the cells are removed before distillation.

8.2. Continuous

The same type of fermenter used in batch processes can also be used with slight modification for continuous-flow operation.[36,38,39] Here, the sugar and nutrient medium are continually added to the reactor, and the effluent, which contains ethanol and cell material is continuously treated for cell separation and product recovery. Since the concentration of sugar in the fermenter remains close to zero, there is no direct problem of high sugar concentrations adversely affecting cellular growth or ethanol production. The rate of sugar addition has to be regulated so that inhibitory levels of ethanol do not occur and cause decreased growth rates. The continuous fermenter for best efficiency should be operated near, but below, the maximal cellular growth rates (= maximum dilution rates). Too much initial sugar in the influent can cause a high ethanol concentration in the fermenter which will inhibit growth. If the microbial population grows slower than the dilution rate due to ethanol inhibition, the culture will be diluted and "washed out" of the fermenter.

A modification of the continuous fermentation process involves conducting the fermentation under a vacuum.[28,40,41] Operating under vacuum, ethanol can be continuously removed from the broth as it is produced and its inhibitory effects on cell growth are reduced. This modification allows for higher rates of ethanol production per liter of fermentation broth and creates a condensate containing a higher ethanol concentration for better distillation efficiency. With ethanol inhibition decreased, higher cell density is desired and this can be achieved by recycling the cells removed from the effluent back into the fermenter.[42]

8.3. Immobilized Cells

The immobilization of microorganisms in carrier gels for ethanol production has been examined.[43,44] Since ethanol production is not dependent on cellular growth, nongrowing cells can be immobilized in gels and placed into continuous-flow reactors. By maintaining nongrowth conditions, glucose conversion to ethanol can be expected to be above 95%.[44] Another advantage of this process is that high cell densities can be maintained, even higher than with cell recycling methods, and it does not require costly continual cell centrifugation and recycling. The efficiency of ethanol production by immobilized cells on a gram dry weight basis is reported to range upward from 80% in comparison to the productivity of free cell suspensions. Finally, this process produces a product stream which can be directly distilled without first removing the cells by centrifugation.

9. Conclusions

The production of alcohol by microorganisms for use as a fuel is currently shrouded in heated debate about whether the energy contained in a gallon of ethanol is worth the energy expended in its production.[3,4] Aside from this major debate, there are several other major technical problems involved in ethanol production. One involves the distillation process, and the other involves high substrate costs. For ethanol to be competitive with the price of gasoline, a cheap substrate must be available, and very efficient conversion of substrate into ethanol must occur. Currently, only yeast-type fermentations are capable of giving this efficient conversion. Research efforts need to be concentrated on upgrading the bacterial conversion process. In bacterial fermentations, ethanol is one of many end products. Its overall yield is greatly reduced if the substrate carbons are split among various acids, alcohols, and carbon dioxide. The bacterial metabolic pathways producing these waste products are under genetic and physiological control. By understanding these control mechanisms, the ratio of end products can be manipulated.[18] If experiments are successful, then thermophilic bacterial fermentations will be efficient enough for consideration in industrial processes. Secondly, for each substrate fermented into ethanol, strain selection experiments should be conducted to find the best microbial culture for the particular substrate being fermented. For instance, *Saccharomyces cerevisiae* may be the best microorganism for sucrose or glucose fermentations, but it is useless in lactose fermentations. Many industrial processes may be accepting more than one type of raw substrate, i.e., potato, corn, and grains. The effect of these mixed substrates on ethanol production are not completely understood, and it may be that instead of pure cultures, mixed yeast cultures will effect a more efficient fermentation. Another unknown is whether or not agricultural products contaminated with aflatoxins or pesticides, and therefore no longer food or feed-grade quality, can be blended into ethanolic fermentations. These contaminants may end up in the single cell protein fraction and thus restrict its use. Also, the toxins may inhibit the kinetics of sugar conversion to ethanol.

With regard to improving the distillation process, two approaches seem promising. Genetic engineering of ethanol-tolerant cultures would allow for the creation of microbes able to grow in high-ethanol solutions, and this would lower the cost of conventional distillations. However, there are no real indications that this technique will create a significant increase in ethanol tolerance. Therefore, advances in the field of chemical engineering for ethanol recovery will be most significant. The vacuum distillation process is one step in this direction, and, perhaps if combined with new fermentation processes such as immobilized whole cells, it may significantly lower the cost of the distillation process. Innovative solvent extractions may also be able to lower the cost of ethanol recovery.

References

1. D. A. O'Sullivan, UN workshop urges wider use of ethanol. *Chem. Eng. News* **57**(17), 11–12 (1979).
2. Novo Industry, Novo enzymes for alcohol production: Conversion of starch, Novo Industry, Mamaroneck, New York (1977).
3. E. V. Anderson, Gasohol: energy mountain or molehill, *Chem. Eng. News* **56**(31), 8–15 (1978).
4. D. Holzman, Alcohol fuels: A critical analysis, *People and Energy* **5**(2), 3–9 (1979).
5. V. Yand, and S. C. Trindade, Brazil's gasohol program, *Chem. Eng. Prog.* **75**(4), 11–19 (1979).
6. S. C. Prescott and C. G. Dunn, *Industrial Microbiology,* Third Edition, McGraw-Hill, New York (1959), pp. 208–218, 250–299.
7. R. D. Tyagi and T. K. Ghose, Production of ethyl alcohol from cellulose hydrolysate, in: *Proceedings of the Bioconversion Symposium,* Indian Institute of Technology, Delhi (1977), pp. 585–597.
8. S. Bernstein, C. H. Tzeng, and D. Sisson, The commercial fermentation of cheese whey for the production of protein and/or alcohol, *Biotech. Bioeng. Sym.* **7**, 1–9 (1977).
9. B. S. Eneroldsen, Dextrins in brewing, in: *European Brewery Convention, Proceedings,* 12th Congress, Interlaken, Switzerland (1969), pp. 205–223.
10. S. C. Prescott and C. G. Dunn. *Industrial Microbiology,* Third Edition, McGraw-Hill, New York, (1959), pp. 836–886 and p. 945.
11. S. G. Meyers, Ethanolic fermentation during enzymatic hydrolysis of cellulose. Presented at Second Pacific Chemical Engineering Congress, Denver (August 30, 1977).
12. G. Cysewski and C. R. Wilke, Utilization of cellulosic materials through enzymatic hydrolysis, I: Fermentation of hydrolysate to ethanol and single cell protein, *Biotech. Bioeng.* **18**, 1297–1313 (1976).
13. G. Gottschalk, *Bacterial Metabolism,* Springer Verlag, New York (1979), p. 283.
14. R. Y. Stanier, M. Doudoroff, and E. A. Adelberg, *The Microbial World,* Third Edition, Prentice-Hall, Engelwood Cliffs, New Jersey (1970) p. 873.
15. V. S. O'Leary, R. Green, B. C. Sullivan, and V. H. Holsinger. Alcohol production by selected yeast strains in lactase hydrolyzed acid whey. *Biotech. Bioeng.* **19**, 1019–1035 (1977).
16. R. E. Kunkee and M. A. Amerine, Yeasts in wine making, in: *The Yeasts,* Vol. 3, (A. H. Rose and J. S. Harrison eds.), Academic Press, London (1979), pp. 6–71.
17. R. B. Gilliland, Yeast strain and attenuation limit, in: *European Brewery Convention, Proceedings,* 12th Congress, Interlaken, Switzerland (1969), pp. 303–313.

18. R. Lamed, and J. G. Zeikus, Catabolic pathways of two ethanol producing thermophilic anaerobes, Abstract 026, Annual American Society of Microbiology meeting (May, 1979).

19. R. M. Walsh and P. A. Martin, Growth of *Saccharomyces cerevisiae* and *Saccharomyces uvarum* in a temperature gradient incubator, *J. Inst. Brew.* **83**, 169–172 (1977).

20. A. Atkinson, D. C. Ellwood, C. G. T. Evans, and R. G. Yeo, Production of alcohol by *Bacillus stearothermophilus, Biotech. Bioeng.* **17**, 1375–1378 (1975).

21. L. G. Loginova and E. P. Guzheva, Dehydrogenase activity in thermotolerant yeasts, *Mikrobiologiya* **30**(5), 917–920 (1961).

22. S. Aiba, M. Shoda, and M. Nagatani, Kinetics of product inhibition in alcohol fermentation, *Biotech. Bioeng.* **10**, 845–864 (1968).

23. J. S. Harrison and J. C. J. Graham, Yeasts in distillery practice, in: *The Yeasts,* Vol. 3 (A. H. Rose and S. S. Harrison, eds.), Academic Press, London (1970).

24. T. Wainwright, Biochemistry of brewing, in: *Modern Brewing Technology* (W. P. Finlay, ed.), MacMillan, New York (1971).

25. C. M. Brown and B. Johnson. Influence of oxygen tension on the physiology of *Saccharomyces cerevisiae* in continuous culture. *Antonie van Leeuwenhoek; J. Microbiol. Serol.* **37**, 477–487 (1971).

26. E. Oura, Effect of aeration intensity on the biochemical composition of Baker's yeast, I: Factors affecting the type of metabolism, *Biotech. Bioeng.* **16**, 1197–1212 (1974).

27. P. J. Rogers and P. R. Stewart, Mitochondrial and peroxisomal contributions to the energy metabolism of *Saccharomyces cerevisiae* in continuous culture, *J. Gen. Microbiol.* **79**, 205–217 (1973).

28. G. R. Cysewski, and C. Wilke, Rapid ethanol fermentations using vacuum and cell recycle, *Biotech. Bioeng.* **19**, 1125–1143 (1977).

29. A. D. Haukeli and S. Lie, The effects of oxygen and unsaturated lipids on the physiological conditions of Brewer's yeast, Fifth International Fermentation Symposium, Berlin, Abstract Y.01 (July, 1976).

30. M. H. David, and B. H. Kirsop, Yeast growth in relation to the dissolved oxygen and sterol content of work, *J. Inst. Brew.* **79**, 20–25 (1973).

31. C. D. Bazua and C. R. Wilke, Ethanol effects on the kinetics of a continuous fermentation with *Saccharomyces cerevisiae. Biotech. Bioeng. Symp.* **7**, 105–118 (1977).

32. S. Hayashida and M. Hongo, The mechanism of formation of high concentration alcohol in saki brewing, Fifth International Fermentation Symposium, Berlin, Abstract 20.12 (July, 1976).

33. T. W. Nagodawithano and K. H. Steinkraus, Influence of the rate of ethanol production and accumulation on the viability of *Saccharomyces cerevisiae* in rapid fermentations, *Appl. Environ. Microbiol.* **31**(2), 158–162 (1976).

34. A. A. Ismail and A. M. M. Ali, Selection of high ethanol yielding *Saccharomyces. Folia Microbiol.* **16**, 346–349 (1971).

35. A. A. Ismail and A. M. M. Ali, Selection of high ethanol yielding *Saccharomyces. Folia Microbiologica* **16**, 349–354 (1971).

36. H. G. W. Leuenberger, Cultivation of *Saccharomyces cerevisiae* in continuous culture, *Arch. Mikrobiol.* **83**, 347–358 (1972).

37. V. V. Zhirova, L. A. Ivanova, and Y. P. Gracher, Influence of trace elements on the formation of acids and ethanol by the yeast *Saccharomyces carlsbergensis, Mikrobiologiya* **46**(3), 423–427 (1977).

38. K. Rosen, Continuous methods for production of alcohol from molasses, Fifth International Fermentation Symposium, Berlin, Abstract 20.03 (July, 1976).

39. H. Schatzmann and A. Fiechter, Anaerobic chemostat experiments with *Saccharomyces cerevisiae,* Fifth International Fermentation Symposium, Berlin, Abstract 6.06 (July, 1976).

40. A. Ramalingham and R. K. Finn, The vacferm process: A new approach to fermentation alcohol, *Biotech. Bioeng.* **19**, 583–589 (1977).

41. R. K. Finn and R. A. Boyajian, Preliminary economic evaluation of the low temperature distillation of alcohol during fermentation. Fifth International Fermentation Symposium, Berlin, Abstract 3.10 (July, 1976).

42. W. Englebart and H. Dellweg, Continuous alcoholic fermentation by circulating agglomerated yeast, Fifth International Fermentation Symposium, Berlin, Abstract 20.04 (July, 1976).

43. M. Kierstan and C. Bucke, The immobilization of microbial cells, subcellular organelles, and enzymes in calcium alginate gels. *Biotech. Bioeng.* **19**, 387–397 (1977).

44. D. Williams and D. M. Munnecke, The production of ethanol by immobilized yeast cells, *Biotech. Bioeng.* **23**(7) (1981).

19
Ethanol Production by Fermentation

D. BRANDT

1. Introduction

Conversion of biomass to ethanol is an attractive route for biomass utilization because ethanol can be easily assimilated by the liquid fuel and chemical markets. Ethanol is somewhat unique as a fermentation product because it can be part of both markets. This market diversity gives long-range stability to ethanol fermentation projects by uncoupling them from the vagaries of the liquid fuels market. This discussion will treat ethanol as a liquid fuel; however, its value as a chemical feedstock could be an even greater incentive for its production from biomass.

Conversion of biomass to ethanol is comprised of several process elements. These generalized elements define plant sections for which major conceptual design decisions must be made early in the project's schedule.

A substrate supply sytem must extract the required amount of fermentable carbohydrates from the raw feedstock source. This system can extend from the collection of raw feedstock to the delivery of fermentable solids to the fermentation. Its extent and complexity depend upon how close the raw feed solids are to being directly fermentable.

The fermentation system is open to some process options which are determined by the feedstock and capital investment parameters. This system includes the fermentation vessels and all auxiliary equipment required for their operation. A choice of batch or continuous fermentation can significantly affect

D. BRANDT • Stone and Webster Engineering Corporation, 245 Summer Street, Boston, Massachusetts 02107.

the quantity and design of equipment. This choice must be made, however, in coordination with the substrate supply system which could limit the options.

The product recovery system is greatly influenced by the end-use specifications. This system will probably be based on distillation as no more efficient process appears ready for large-scale use in the near future. The product specifications determine the number of separations and thereby set the system's design. Distillation is energy intensive, so the number of distillations has an impact on the energy efficiency and the operating cost. Byproduct recovery may be the deciding factor in an ethanol project's economic viability. Byproducts can be coproducts of fermentation or feedstock components which are not used by the fermentation but which have their own value. The recovery operation usually consists of removing water from solid byproducts by various solid–liquid separators and evaporators. The feedstock choice influences byproduct values, because all the feedstock's solid residue must be sold or disposed with the byproduct.

Treatment of the effluent water and vapor streams from these plants may also be a significant economic factor. The cost of treating or containing these effluents may be high if they cannot return a byproduct credit. The waste treatment systems required are not exotic, but the borderline economics of many ethanol fermentation projects cannot tolerate much "nonproductive" capital investment.

This chapter will describe the design schemes typically used for the process elements and the factors which impact on their choice. The viability of ethanol production for biomass fuel conversion will also be discussed in terms of the operating and economic efficiencies which these designs can achieve.

2. Feedstock Selection

The choice of feedstock has a significant effect on the process design and operating economics. Fermentable solids can be obtained from several sources. The extremes in choice extend from biomass production devoted to ethanol fermentation to fermentable waste streams. The economic impact of this choice is basic to the project's viability.

2.1. Pretreatment Design

The substrate supply system is unique to the type of feedstock. For example, cereal grains can be fed directly to a plant with no attempt at prior recovery of byproducts. These feedstocks require dry milling to remove the grain's kernel structure and allow access to the fermentable solids. Storage and solids-handling equipment to feed the dry mills are required. This equipment is common for devoted feedstocks such as cereal grains. Dry milling is employed to avoid unnecessary dilution and water consumption early in the process. The milled grains are slurried in preparation for the degradation operations. No separation of fermentable solids from the rest of the grain is made, and the entire mass passes through the fermentation.

Separation of fermentable solids, starch, and other byproducts from cereal grains before fermentation can be done by wet-milling techniques. Sets of equipment to remove the fiber, recover protein and oil, and produce a pure starch stream are used. The resulting investment for such "pretreatment" is, of course, greater than for dry milling. This increased investment must be justified by increased byproduct values which are greater than those from dry-milling. This investment is usually not justified unless the operator has access to the wet-milling byproduct markets or other special incentives exist.

Waste streams containing fermentable solids may require sterilization and concentration. These streams are good substrates if they are available on a regular basis. Effluents from food or other agricultural product processing plants are excellent examples because many of the solids are directly fermentable. These streams may be quite dilute and require concentration to the proper level for fermentation. If no "cooking" is required for degradation and no evaporation or other heating is employed for concentrating, sterilization is needed. Sterilization inactivates microorganisms which compete for substrate with those producing ethanol. It is the minimum substrate preparation required.

The extent of preparation for other biomass feedstocks falls within the extremes of wet-milling and simple sterilization. Sugar cane juice, for instance, requires destruction of the cane plant and recovery of the juice, but not extensive separation or refining. Cellulose requires destruction of its structure and hydrolysis to glucose, but again, no separation or refining steps. Two likely candidates for ethanol feedstocks are dry-milled cereal grains and waste streams containing sugars. The following discussion compares the designs for these two feedstocks.

The fermentation solids of most biomass resources are locked into natural polymers with fermentable sugars as monomers. These polymers are resistant to physical destruction so this degradation is usually accomplished by enzymes or acid. The resistance of the polymers to degradation has direct bearing on the size, complexity, and, ultimately, cost of the equipment for this step. The investment for equipment varies directly with the processing required to convert feedstock to fermentable sugars. The investment for equipment to convey, dry-mill, and degrade grain is much greater than for a pump and sterilizer to deliver a fermentable waste stream of proper concentration. The difference in investment may require an entirely different economic approach for each feedstock.

2.2. Pretreatment Costs

Operating expenses also vary with the processing required by different feedstocks. Operating expenses are not only related to the quantity of equipment required but can be affected by such things as the intensity of the energy or the quantity of enzyme required. Again, these costs are usually directly related to the extent to which the feedstock solids are directly fermentable.

2.3. Feedstock Price

The feedstock's direct cost to the process is obviously important. Current estimates indicate that the raw feedstock accounts for as much as two-thirds of the production cost for fermentation ethanol.[1] Unfortunately, many of the easily fermented substrates are attractive sources for other products and are priced accordingly. Less dependable feedstocks are available at lower costs but usually require more pretreatment and yield less ethanol for the total mass fed. Cellulose is a good example of a very resistant, but low-cost abundant material which thereby sustains many enzyme or acid degradation research efforts.

2.4. By-product Credit

The byproduct credit potential of a feedstock currently plays a large role in venture economics. A feature which makes cereal grains, particularly corn, attractive feedstocks is the value of their unfermentable components. Credit for the residue after fermentation of dry-milled corn, known as distiller's grains, can recover as much as 58% of the production costs.[2]

2.5. Waste Treatment

Some feedstocks affect capital and operating costs by the treatment which their waste streams require. These feedstocks can produce high biological oxygen demand (BOD) and high salts streams with little value as animal feed. Molasses is such a feedstock.[3]

It is necessary to consider, not only the supply cost, but the full economic impact of a feedstock in order to evaluate it realistically. The factors which have been discussed are an indication of what should be considered.

3. Process Description

The process design can be made in proper detail once the feedstock is selected. Each process system will consist of several pieces of equipment. This discussion is illustrated in Figures 1, 2, and 3, and in Table 1.

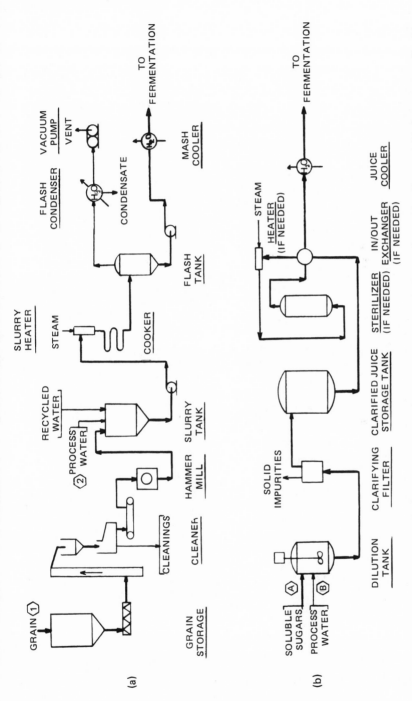

FIGURE 1. Process designs for preparation of cereal grain or soluble sugar feedstocks. Numbers or letters identify material balance streams listed in Table 1. (a) Cereal grain preparation; (b) soluble sugars preparation.

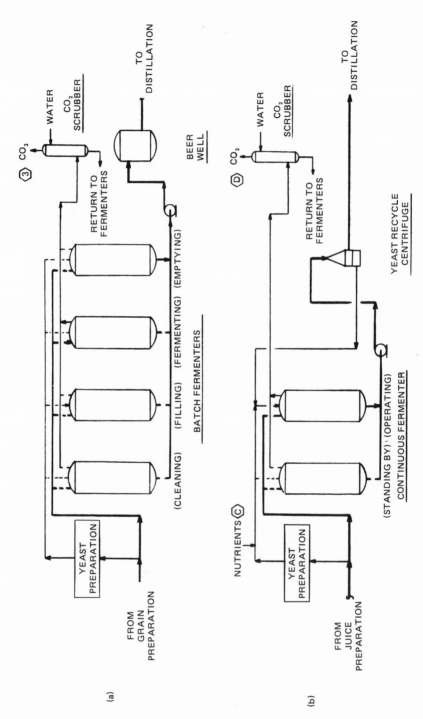

FIGURE 2. Process designs for batch or continuous fermentation of ethanol. Numbers or letters identify material balance streams listed in Table 1. (a) Batch fermentation; (b) continuous fermentation.

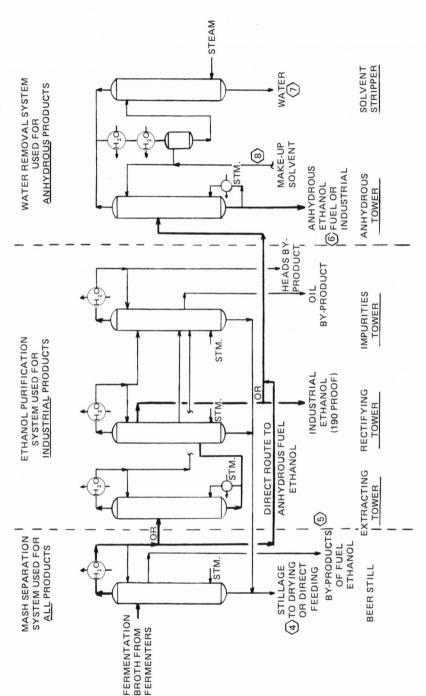

FIGURE 3. Process designs for distillation of various ethanol products. Numbers identify material balance streams listed in Table 1.

TABLE 1. Overall Material Balances and Ethanol Fermentation Mass
Balances for 38 Million Liters Per Year of Fuel Ethanol[a]

I. Batch Grain Fermentation

Component	Stream 1	Stream 2	Stream 3	Stream 4
Starch	6,999	—	—	—
Protein	1,669	—	—	1,869
Oil/fat	263	—	—	263
Other solids	2,096	—	—	2,295
Carbon dioxide	—	—	3,579	—
Water	1,520	182,266	1,433	182,351
Total	12,547	182,266	5,012	186,778

II. Continuous Molasses Fermentation

Component	Stream A	Stream B	Stream D	Stream E
Sugars	7,720	—	—	—
Protein	728	—	—	767
Oil/fat	146	—	—	146
Other solubles	3,204	—	—	3,551
Carbon dioxide	—	—	3,579	—
Water	2,768	200,682	1,433	195,134
Total	14,566	200,682	5,012	199,598

III. Ethanol Distillation

Component	Stream 5	Stream 6	Stream 7	Stream 8
Ethanol	—	3,441	—	—
Secondary products	290	—	—	—
Benzene	—	—	—	small
Water	—	17	77	—
Total	290	3,458	77	—

[a]Values are in kg/hr.
[b]Stream letter C (nutrients) not needed in molasses feed.

3.1. Feedstock Preparation

3.1.1. Physical Reduction

Physical reduction by milling is necessary for most solid materials. Disk mills consisting of rotating flat plates with protruding teeth to tear the kernels are typically used for grains. Sugar cane juice would require the first steps in sugar refining to recover juice from the cane. Cellulose requires extensive physical degradation to obtain reasonable yields of fermentable materials.[4,5] Ball milling or extensive treatment with solvents is required by current technology.

3.1.2. Substrate Hydrolysis

Physical size reduction is followed by hydrolysis to form fermentable sugars from the carbohydrate polymers. The most fermentable portion of grain is starch, which is fully liquefied by high-temperature enzyme treatment in a "cooking" step. The hydrolysis is completed to glucose with a second enzyme. Cellulose can be hydrolyzed by enzymes or acids in processes that are currently more complex than the hydrolysis of starch. Cellulose also leaves a resistant solid residue.[6] Sugar cane juice or molasses would not require a hydrolysis step because their solids are already fermentable.

3.1.3. Feedstock Sterilization

The reduced and slurried or soluble substrate must be free of competing microorganisms. If high-temperature hydrolysis or concentrating steps have not sterilized the substrate solution, a separate sterilization is required. This can be accomplished in batch fermenters by putting the broth through a heating cycle before fermentation. This procedure wastes fermenter capacity in large plants and is not possible in continuous fermentations, so prior sterilization is necessary. Standard heat exchangers and drums with adequate residence times can be used in continuous modes for soluble substrates. Solid substrates which form slurries require more extensive equipment to assure that every crevice and inner part of the particles attain the necessary temperature.

3.1.4. Concentration Adjustment

Adjustment of the concentration of fermentable solids to an optimal level for fermentation is the final preparation step. It is usually an in-line dilution step. The concentration is determined by the capacity of the microorganism to tolerate the ethanol produced from it and by the operating objectives of the fermentation. These objectives are set by balancing high production rates at high concentrations against high yields at low concentrations.

3.2. Ethanol Fermentation

Fermentation of ethanol suffers, perhaps, from being one of the oldest commercial-scale chemical production processes. It has not benefited fully from recent advances in equipment and technology design because, except for potable alcohol, recently most U.S. ethanol has been synthesized from ethylene.[7] It is likely that installation of new ethanol fermentation facilities will prove the feasibility of new methods and add to the options presently available.

3.2.1. Batch Fermentation

Traditional batch fermentation is straight forward and well tested. It may be the only reasonable approach for some feedstocks. Dry-milled cereal grain feedstocks usually require that the nonfermentable solids pass through the fermentation. This makes recovery of yeast cells from the solids in the fermentation broth for continuous recycle inefficient and uneconomical. Slurry substrates are susceptible to contamination which can be more easily isolated in batch operation when it occurs.

Batch operation requires a bank of fermenters to be operated in parallel on a sequenced schedule. Typically, a full batch fermentation cycle for one vessel will extend over 2 or 3 days. Holding capacity is provided for fermentation feed and effluent so that the remainder of the plant can achieve the design and operating advantages of continuous processing. Batch operation is characterized by large fermenter volumes that are not continuously active.

3.2.2. Continuous Fermentation

Continuous fermentation is possible for soluble feedstocks and may reduce the capital investment because less fermentation volume is needed. Soluble feedstocks make the recovery of yeast cells for recycling relatively easy because no other solids are present. Sterilization of the soluble substrates is also easier, making contamination less likely. Continuous fermentation will require more sophisticated control and yeast recovery equipment which may counteract the reduced fermenter investment in smaller installations. Larger installations, conversely, may reach sizes where the reduced investment can justify prior separation of nonfermentable solids from a grain mash and a secondary fermentation of the sugars left in the solids so that the primary fermentation can be continuous.

3.2.3 Yeast Supply

The availability of the proper strain of yeast is important to sustain regular ethanol production. Yeast may be purchased externally or produced on-site. On-site propagation of a strain on the plant's substrate may be economically advantageous. The strain will acclimate to the plant's substrate and environment as well as become regularly available. Smaller fermenters are used to produce inoculating volumes with sufficient yeast cells for immediate rapid fermentation in the larger vessels. The yeast propagation system contains sets of these seed fermenters and transfer equipment designed to produce and transfer the inoculum without contamination. Batch operation requires an inoculation for every fermentation.

Continuous fermentation requires that most of the yeast cells be recovered from the broth and recycled to sustain the mature population in the fermenter. The recovery is usually accomplished by centrifugation designed to minimize loss of cells to the beer still from which they will exit with the bottoms stream. The overflow cells are replaced with new cells from a propagation system which is much smaller than would be required for a batch fermentation of similar capacity. This recycle also reduces the consumption of substrate for yeast propagation.

3.3. Ethanol Recovery

Ethanol recovery has been traditionally accomplished by distillation, and this appears likely to continue for the immediate future. A train of towers operating in series is employed, each accomplishing one or two separations of the ethanol from components of the fermentation broth.

3.3.1. Stillage Separation

The first tower removes fermentation products (mainly ethanol) and some water from the solid nonfermentables in the broth or "beer." This tower is traditionally referred to as the "beer still." It is desired that all ethanol be stripped from the broth in this tower. Rectification to increase the ethanol concentration in the overhead is a secondary purpose included in some designs. The stripping section must be designed to accommodate the solids in the broth which will move down through the stages in the tower from the feed point and ultimately exit with the bottoms stream.

3.3.2. Anhydrous Distillation

The distillation sequence after the beer still will vary with the type of ethanol product. Potable ethanol requires refining to the specifications of a buyer or for the product in which it is used. Industrial ethanol requires removal of impurities, including fusel oils, which are byproducts of the fermentation. In addition, anhydrous industrial ethanol requires that an entrainer be added to break the water–ethanol azeotrope in a separate tower. The ethanol and entrainer are then separated in yet another tower. Anhydrous industrial ethanol requires at least four distillations in a standard design.

Fuel-grade ethanol should be anhydrous to prevent separation of water from gasoline–ethanol mixes which may become saturated at winter temperatures.[8] It is anticipated, however, that the fusel oil impurities need not be removed. A fuel-grade ethanol distillation train would therefore consist of a beer still with rectification and the anhydrous distillation's entraining and stripping towers.

3.3.3.   Product Specifications

The energy consumed in these refining steps is directly related to the number of towers. Each tower requires sufficient vapor boil-up from its bottom to sustain the equilibrium vapor–liquid separations on the trays above. This boil-up requires energy, so it is important to select the product specifications carefully. Decisions to produce over-specification material or to install flexibility to produce all the products, can result in more expensive operation.

3.4.   By-product Recovery

Recovery of byproducts can result in a significant credit if there is an established market for them. They must be provided at buyer's standards and may require extension of the ethanol plant's systems to integrate with a particular user's operation.

High-purity carbon dioxide can be recovered from the fermentation. This gas will require purification and pressurization to bottle it for commercial distribution. It has excellent byproduct value, but the market for it is highly localized and the product from this size unit probably could not support a large distribution operation. Installation of a carbon dioxide recovery unit requires case-by-case evaluation of its economic potential. The fermenter vent should be scrubbed free of entrained broth to control air emissions even if no recovery is made.

The beer still bottoms can be a pollution problem or a valued byproduct, depending upon feedstock and/or local circumstances. Streams with little animal feed value can be used as fertilizer, for anaerobic methane production or for aerobic single cell protein production.[3] These uses attempt to achieve a return for treating this high BOD waste. Its direct use is preferred but usually requires integration with another local operation such as irrigation of fields or direct consumption in animal feed lots. Such outlets are not common and, again, require a case-by-case evaluation.

A protein-rich distiller's grains product may be recovered at some drying cost and marketed when grains are used as feedstock. The plant protein and other solids from the grain are enriched by protein from the entrained yeast cells. It is premium-priced animal feed additive which has been available from limited sources to date. Drying this material for distribution is costly and is only done to lower the long distance transportation charges per unit of protein mass. Local, immediate use of this stream would require less drying. It is conceivable that a cattle feed lot of sufficient size adjacent to a fermentation plant could feed this stream directly in properly drained troughs as an appropriate portion of the livestock's diet. The credit from this byproduct can contribute greatly to a project's economic viability if the distiller's grains from such a plant do not overwhelm the local market.

4. Conversion Efficiency

The overall efficiency of converting biomass to ethanol for fuel is affected not only by the process, but also by definition of the boundaries for the calculation. It is imperative that a design appropriate to fuel ethanol be the basis. This process design should incorporate energy-saving technology which may not yet have been applied to ethanol but which has been proven in other applications. It is not appropriate to evaluate the efficiency of converting biomass to ethanol on the basis of historical practice for which there had been little incentive to save energy.

The boundaries for the efficiency calculation are significant. Biomass conversion plants are not isolated and the impact of their energy efficiency extends beyond their battery limits. They interact with their surroundings in many ways, and the resulting effects should be included in any evaluation.

These interactions assign values to the types of energy used or produced which go beyond pure thermodynamics. This assignment of value is not a superficial cultural or economic phenomena, but is realistically related to the fuel's versatility in producing usable energy. A form of energy which has restricted uses or which would require extensive new supply systems for its use cannot claim the same value as a versatile source which fits into established systems. The value of the form of energy produced should therefore be a part of an efficiency evaluation. Proper weighting of such values may show that the plant's efficiency within its battery limits is not an absolute indication of a biomass conversion project's overall effectiveness.

4.1. Independent Plant

An independent or "grass roots" ethanol plant has several efficiency disadvantages which must be overcome to make a viable project. The grass roots plant considered here is limited to converting a biomass grown specifically to provide the plant's feedstock ("devoted" biomass), cereal grain, to ethanol with no outside integration of feedstock supply or product distribution systems.

Feedstocks grown to supply the grass roots plant require prices that return the full cost of their production plus a profit for the producer. Market prices reflect the energy used in grain production and collection, so they are indicators of the efficiency of grain production and delivery. Relative to other sources, the cost of using cereal grains feedstock is high, and efficiency suffers when devoted feedstocks are used.

Installation and operation of a power plant for a grass roots facility to consume imported fuel is costly. Just as for the feedstock, devoted power production requires that the price for power reflect the full cost of delivery which includes payment for fuel and for power plant investment. Use of a low-cost, indigenous fuel such as agricultural wastes, wood, or waste paper, can reduce fuel costs and improve the energy efficiency. A large amount of the energy used to grow grains can be recovered, if a portion of their harvest residue is used as fuel. These savings are countered by the increased cost for equipment to accept these fuels, but the overall conversion efficiency for a devoted biomass is improved.

Devoted production of process energy is costly whatever its source. It is one of the largest ethanol production costs in a grass roots plant, second only to the feedstock cost. It will require that the prices for ethanol and byproducts absorb its cost.

Distiller's grains from a grass roots plant are typical of many byproducts in that some processing costs are necessary before their value can be recovered. This byproduct is commonly dried to reduce its weight for market distribution. Producers of potable alcohol find this profitable even with the energy-intensive drying operation. Under the constraints of the efficiency requirement for fuel ethanol, however, drying these grains is an energy and cost detriment. Byproducts must have premium prices and/or meet market specifications directly before they can contribute to a fuel ethanol project's profitability.

A grass roots plant can be successful if the net effect of feedstock availability, power generation, and return on the product and byproducts is favorable. This author believes it will be difficult for a grass roots fuel ethanol plant to achieve a favorable result by these criteria. This will likely not be achieved until fuel ethanol commands a price which can absorb the economic and energy disadvantages of a grass roots plant.

4.2. Integrated Plant

Integration of ethanol production with other operations can improve its overall efficiency. Integration with other operations allows the resource consumed by the ethanol facility to be discounted by sharing its delivery or production cost. Waste streams are good examples because the cost of their production is borne by their product's price. Similar examples of the advantages of integration can be found in all of the process steps.

Waste streams from carbohydrate processing plants provide feedstocks which usually require little pretreatment. Molasses, cheese whey, and corn wet-mill wastes all contain solids that are much closer to being fermentable than in the raw form in which they are harvested. The production of sugar, cheese, or corn starch absorbs the delivery costs of the waste streams, and additionally, pretreatment costs are less. A disadvantage of using such streams is that ethanol capacity is restricted by the waste stream's volume. This may restrict the economics of scale which can be obtained per unit of ethanol produced.

Waste energy from electric generating plants, process plants, or waste incinerators has great potential for application in ethanol distillations. The heat required for ethanol production is of relatively low quality. Exhaust steam from power plant turbines, for example, is hot enough to reboil atmospheric ethanol distillation towers. Production of ethanol for addition to gasoline suggests use of waste heat from a refinery. Many refining processes use higher temperatures than those for ethanol production, so heat energy from the refinery could be reused at lower temperatures in an adjacent ethanol plant. Waste incinerators recover heating value from material whose production has been written off against their original service. These sources of heat will usually be large enough for most ethanol plants. The sources must also be consistent, so a base-load power plant or close coordination with a process or waste-burning plant are required.

Compatibility of the byproducts with adjacent feedlots or process plants can reduce the energy needed to prepare them for shipment or disposal. Direct feeding of undried distiller's grains eliminates a large energy-consuming step. The water would evaporate to the atmosphere or be absorbed in feed-mixing operations at the feedlot. Coordination is required to assure that the feedlot is large enough to accept the stillage produced. Some return of energy from a feedlot may be possible by using the manure to produce low-Btu gas for fuel.

Stillages from some feedstocks are not suitable as distiller's grains but can be used as feedstocks for further processing. Using them as substrates to produce protein in another fermentation for aquaculture or farm feeding is a possibility. When there are no options for downstream integration, disposal is necessary, but even then, a properly designed system, such as incineration, may recover some energy.

5. Energy Efficiency

An ethanol production complex which makes full use of its feedstock can result in a positive energy ratio. Figure 4 illustrates how such a complex could achieve a positive result using wheat grain as a feedstock. The complex includes an ethanol plant and a cattle feedlot. The energy to produce wheat and the energy to produce steam and power enter the balance. Contingency and power generation losses are covered by additional energy inputs. Fuel-grade ethanol is an energy product as well as the distiller's grains which are fed undried to the cattle. The metabolic energy of the cattle feed can be claimed in full because virtually no energy is expended to prepare and transfer it to the feedlot within the complex. The ratio of energy input-to-output is greater than one. This ratio is absolute and does not take into account any weighting of energy forms based on versatility.

6. Conclusion

Ethanol can be an economical and efficient method of converting biomass to liquid fuel where the net result of feedstock availability, operational integration, and market proximity produce a favorable situation. Such an economically favorable situation becomes more probable as the price of fuels from

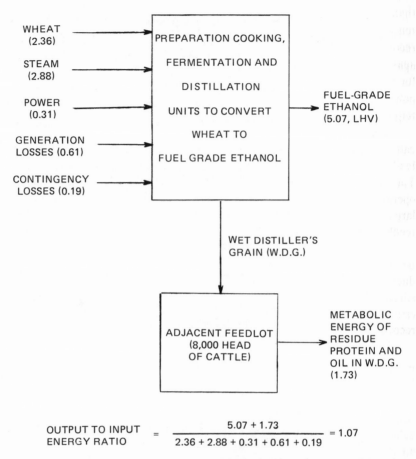

FIGURE 4. Energy balance for fuel-grade ethanol production integrated with adjacent cattle feedlot.

petroleum increases. The number of these situations should grow, making the conversion of biomass to ethanol an alternative source of liquid fuels.

ACKNOWLEDGMENT

The authors acknowledge the able assistance of Ms. L. E. Fournier in preparing the figures.

References

1. Stone & Webster Canada Limited, Internal Fermentation Ethyl Alcohol Study, Toronto, Canada (October, 1977).

2. D. L. Miller, Fermentation ethyl alcohol, in: *Biotechnology and Bioengineering Symposium* (E. Gaden *et al.* eds.), John Wiley and Sons Inc., New York (1976), pp. 307–312.
3. V. Yang and S. C. Trindade, Brazil's gasohol program, *Chem. Eng. Progr.* **75**, 11–19 (1979).
4. M. Mandels, L. Hontz, and J. Nystrom, Enzymatic hydrolysis of waste cellulose, *Biotech. Bioeng.* **16**, 1471 (1974).
5. H. E. Grethlein, Comparison of the economics of acid and enzymatic hydrolysis of newsprint, *Biotech. Bioeng.* **20**, 503 (1978).
6. D. Brandt, L. Hontz, and M. Mandels, Engineering aspects of the enzymatic conversion of waste cellulose to glucose, *American Institute of Chemical Engineers Symposium Series* **69**(133), 127–133 (1973).
7. D. Brandt, Long-term changes are possible in ethylene feedstocks, *Oil Gas J.,* **77**(6), p. 51–56 (February 5, 1979).
8. W. A. Scheller, Ethanol as a renewable energy source in automotive fuel, Nebraska GASOHOL, presented at the Stone & Webster International Biochemical Symposium, Toronto, Canada (October, 1977).

PART III
TECHNICAL AND ECONOMIC
CONSIDERATIONS

20

Technical Considerations of Biomass Conversion Processes

JOHN M. RADOVICH

1. Material Balances

Material balances are a restatement of the law of conservation of mass, "mass is neither created nor destroyed," as applied to processes involving chemical and physical changes. A material balance is a balance on mass not on volume nor on moles. It is an accounting of all the material entering and leaving a system. The balance is made with respect to clearly defined, but often arbitrarily chosen, system boundaries. The boundaries of the system must be stated precisely in order for the balance to be useful. The material balance becomes essential in process evaluations because it permits the engineer to (1) determine the flow rates or amounts of unknown streams, (2) recognize incomplete or incorrect data and subsequently obtain usable data from additional measurements or accurate estimates, and (3) design equipment for the process.

1.1. Basic Equations and Guidelines

The generalized material balance equation is written as:

$$\begin{array}{l}\text{accumulation of mass} \\ \text{within the system}\end{array} = \begin{array}{l}\text{mass input across} \\ \text{system boundaries}\end{array} - \begin{array}{l}\text{mass output across} \\ \text{system boundaries}\end{array}$$
$$+ \begin{array}{l}\text{mass generation} \\ \text{within the system}\end{array} - \begin{array}{l}\text{mass consumption} \\ \text{within the system}\end{array} \quad (1)$$

JOHN M. RADOVICH • School of Chemical Engineering and Materials Science, University of Oklahoma, 202 W. Boyd, Room 23, Norman, Oklahoma 73019.

Processes which are often encountered when converting biomass to energy are operated at steady state; thus, the accumulation term is zero. For processes which do not involve a chemical reaction (e.g., drying or mixing), the generation and consumption terms are zero.

Space limitations do not permit a detailed discussion of techniques for setting up and solving material balances. Details and examples of material balance, and solution techniques, are given by Himmelblau,[1] Henley and Rosen, [2] and Hougen et al. [3] Some general guidelines for developing or evaluating material balances for the types of processes frequently encountered in biomass conversions systems are discussed below.

1.1.1. System Boundaries; Choosing a Basis; Composition Data

The system boundaries are *arbitrarily* defined by the design engineer. For a biomass conversion process for energy and fuels, we are interested in evaluating the appropriateness of the design of the facilities for converting biomass into energy and fuels. Consistent with this purpose, the system should encompass all processing steps from receipt of the biomass, as delivered to the "plant gate," to the energy or fuel product which will be delivered to a user. It is important that the extent of these boundaries by *explicitly* defined.

Many material balances can be written for the defined system, such as:

1. Total mass balance (all components) for the entire system.

2. Mass balance on a single chemical component or atomic species for the entire system. (Note that the mass and moles of an atomic species, e.g., oxygen, will be conserved.) Total mass balances may be substituted for one of the component balances.

3. Total mass balance for each piece of process equipment, for groups of equipment (subsystems), and each mixing point within the system. The overall mass balance for the system may be substituted for a subsystem balance, and subsystem balances may be substituted for an equipment or mixing point balance.

4. Component balances for the equipment, mixing points, and subsystems. Total mass balances may be substituted for the component balances in these cases.

Each material balance should have a consistent basis—the amount of reference material. When analyzing subsystems, the basis chosen for each calculation should be consistent with the basis for the entire system. Subsystem calculations can be done on a separate basis for convenience, but the results *must* be converted to the same basis as the entire system before proceeding to the balance for the next subsystem.

The nature of the composition data used in these calculations should be carefully scrutinized. For gases at relatively low pressures (less than ten atmospheres), volume percent is equivalent to mole percent, but not to weight percent. The composition analysis may be on a water-vapor-included basis or on a dry basis (Orsat Analysis). It is preferrable to use composition data which reflect the amounts of *all* components since it is easy to convert these data to an "*x*-component-free basis" by

percent composition on an "*x*-free" basis

$$= \frac{\% \text{ composition including all components}}{100\% - \% \ x \text{ composition}} \quad (2)$$

Similar comments apply to compositions of solid or liquid streams. For solids, it is then rather simple to convert composition data from an as-received analysis to a moisture and ash-free basis or some other basis.

1.1.2. Chemical Reactions and Yields

If a chemical reaction occurs in the system, a chemical equation for each reaction can be written. A balanced chemical equation provides the qualitative and quantitative information necessary for calculating the weights of reactants and products. Stoichiometry gives the relative ratio of the combining moles of elements and molecules. The numerical coefficients in the chemical equation are the stoichiometric ratios relating the moles of reactants to the moles of products. The amounts must be converted to a mass basis before proceeding with a material balance.

In industrial processes, the exact stoichiometric amounts of reactants are rarely used. Also, the reaction may not go to completion, or there may be side reactions. In both these cases, the material balances must account for all the mass in the process. In order to evaluate existing data on processes involving chemical reactions, the following terms should be familiar to the design engineer:

1. Limiting reactant: the reactant is present in the smallest stoichiometric amount. This is easily determined by examining the ratio of reactant molar feed rate-to-stoichiometric ratio. The reactant with the lowest value is the limiting reactant.

2. Excess reactant: the reactant is present in excess of the limiting reactant. The percent excess of a reactant is the amount above that required to react with the limiting reactant according to the chemical equation. This is based on complete reaction of all the limiting reactant.

3. Incomplete reaction: the limiting reactant is not completely converted into products.

4. Conversion: the fraction of a key reactant (usually the limiting reactant) in the feed stream which is consumed to generate the reaction products.

$$x_c = \text{moles reactant consumed}/\text{moles of reactant in feed}$$

$$5. \text{ Yield} = \begin{cases} \text{moles of a particular product/initial moles of key reactant} \\ or \\ \text{moles of product/moles of key reactant consumed} \end{cases}$$

The limiting reactant should be the key reactant mentioned above.

6. Selectivity = moles of product A generated/moles of product B generated

7. Inert substances are chemical components that do not react in any way, e.g., N_2 in an air-blown gasifier.

Because of the complexity and multiplicity of the reactions involved in the conversion of biomass to fuel, it may not be possible to write the chemical equation or equations. However, evaluations and design calculations can be made based on the conversion and yield data for the reactants and products. In many cases, the only information available is the amount of material in and out of the system. In that case, a valid overall material balance for the system with a chemical reaction can be calculated if the amounts of all reactants entering and all the products leaving are known. Examples of different types of material balances are presented later.

1.1.3. Units

Mass balances can be calculated in any convenient units. The International System of Units (SI) is recommended. However, calculations, or at least results, should be presented in those units which are familiar to the industry (e.g., barrels, standard cubic feet, etc.). Tables 1 and 2 contain basic information and selected conversion factors for use with engineering and SI units.

1.2. Applications to Conversion Processes

1.2.1. Air-Blown Gasifier

The first example is from a report published by the Electric Power Research Institute.[5] The material balance, for a steady-state operation, is of the form "total mass in = total mass out." The system boundaries are drawn around the gasifier and condenser as shown in Figure 1. The mass balance data are given in Table 3. This material balance contains only limited information. The compositions of the inlet and outlet streams are unknown. Because a chemical reaction occurs in which water vapor is formed, we cannot be sure that the water leaving the condenser is due to the condensation of input steam or product water vapor. Although a chemical reaction occurs within the system, it is not necessary to know how much mass has been generated or consumed when writing this type of material balance. This is still a valid balance because the total mass in includes all reactants and inerts, while the total mass out includes

TABLE 1. Basic and Derived SI Units[a]

Basic units		
Quantity	SI unit	In terms of other SI units
Length	meter (m)	—
Mass	kilogram (kg)	—
Time	second (sec)	—
Thermodynamic temperature	kelvin ($^\circ$K)	—
Amount of substance	mole (mol)	—
Derived units		
Force	Newton (N)	kg-m/sec^2
Pressure	Pascal (Pa)	N/m^2
Energy, work, quantity of heat	joule (J)	N-m
Power	watt (W)	J/sec
Heat flux		W/m^2
Heat capacity, entropy		J/$^\circ$K
Specific heat capacity, specific entropy		J/kg-$^\circ$K

[a]Source: Reference 4.

TABLE 2. Selected Conversion Factors to SI Units[a]

To convert from	To	Multiply by
barrel (42 gallons for petroleum)	m^3	1.5898729×10^{-1}
British thermal unit (Btu, International table)	J	1.0550559×10^3
Btu/lb$_m$-$^\circ$F	J/kg-$^\circ$K	4.186800×10^3
Btu/hr	W	2.9307107×10^3
Btu/sec	W	1.0550559×10^3
Btu/ft^2-hr-$^\circ$F	J/m^2-sec-$^\circ$K	5.678263
Btu/ft^2-hr	J/m^2-sec	3.1545907
calorie	J	4.1868000
cal/g-$^\circ$C	J/kg-$^\circ$K	4.186800×10^3
ft^3	m^3	2.8316847×10^2
gallon	m^3	3.7854118×10^{-3}
horsepower (550 ft-lb$_F$/s)	W	7.4569987×10^2
psi	Pa	6.8947573×10^3
watt-hour	J	3.6000000×10^3

[a]Source: Reference 4.

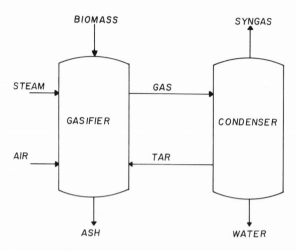

FIGURE 1. Simplified diagram of air-blown gasifier process.

all inerts plus unreacted material, plus reaction-generated products minus consumed reactants. If we know the precise nature of the reaction and the yield of products, then we are able to calculate the composition of all outlet streams.

1.2.2. Pyrolysis – Gasification Reactor

The material balance data in Table 4 for a combined pyrolysis–gasification process were obtained from a detailed analysis by Brink and Thomas.[6] The process flow diagram is presented in Figure 2. The water from the wet wood is first evaporated and then both are sent to a reactor for conversion to gas. This is a more detailed material balance but it is still in the form "total mass in = total mass out." However, component material balances have also

TABLE 3. Material Balance for Air-Blown
Gasification Process[a,b]

In		Out	
Component	(kg)	Component	(kg)
Wood	100	Gas	263
Air	164	Ash	3
Steam	17	Water	15
Totals	281		281

[a]Source: Reference 5.
[b]Basis: 100 kg of wood to gasifier.

TABLE 4. Material Balance for Pyrolysis–Gasification Reactor[a,b]

Component		In (kg)	Out (kg)
First-stage reactor			
Wood, organic		997.0	997.0
ash		3.0	3.0
water		1103.0	1103.0
	Total	2103.0	2103.0
Pyrolysis–gasification reactor			
Wood, organic		997.0	—
ash		3.0	3.0
water vapor		1103.0	—
	Subtotal	2103.0	3.0
Air			
oxygen		482.8	0.0
nitrogen		1575.5	1575.5
argon		26.9	26.9
carbon dioxide		1.1	938.0
hydrogen		0.0	65.3
carbon monoxide		0.0	505.8
methane		0.0	26.5
water vapor		11.3	1059.6
	Subtotal	2097.6	4197.6
	Total	4200.6	4200.6

[a]Source: Reference 6. Used with permission.

been presented. Thus, we know the amount of material generated or consumed by the chemical reaction and the amount of the inerts for each stream. In order to calculate these balances the following information is required is available in the original publication.[6]

1. The composition of the wood, i.e., the weight percent carbon, hydrogen, oxygen, and ash.

2. Information on the reactions and the yields of hydrogen, carbon monoxide, carbon dioxide, and methane.

Had the wood composition and reaction yields been presented here, a check on the component balances could have been made by calculating balances on the atomic species.

1.2.3. Pyrolysis of Wood

The final example is taken from an SRI report[7] for the pyrolysis process shown in Figure 3. Detailed material balances for the components and atomic

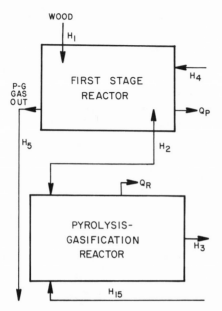

FIGURE 2. Simplified diagram of combined pyrolysis–gasification process.

species for the process are given in Table 5. The yields based on the moisture- and ash-free wood feed used in these calculations were 30 wt% char (moisture- and ash-free); 25 wt% oil (with 12 wt% water), and 13 wt% water. There are sufficient data in Table 5 to check the balance on atomic species (e.g. carbon). We can also check the total material balance:

$$
\begin{aligned}
\text{Total mass in} &= \text{Streams } (1 + 2 + 7 + 9) \\
&= 166.7 + 28.8 + 809.3 + 34.1 = 1033.9 \times 10^3 \text{ lb/hr} \\
\text{Total mass out} &= \text{Streams } (8 + 10 + \text{oil} + \text{char}) \\
&= 937.9 + 50.0 + 20.8 + 25.2 = 1033.9 \times 10^3 \text{ lb/hr}
\end{aligned}
$$

The amount and type of data presented in this example are the most useful when evaluating a biomass conversion process.

2. Energy Balances

This section discusses energy balances and information needed to apply them correctly. As we did with the mass balances, we shall invoke the law of conservation of energy.

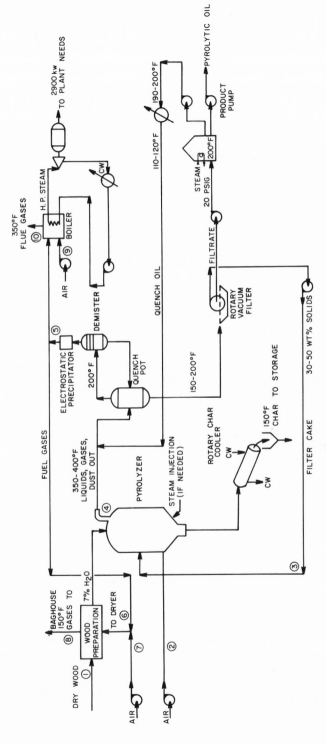

FIGURE 3. Detailed diagram of wood pyrolysis process.

TABLE 5. Material Balance for Pyrolysis of Wood[a,b]

Component	Stream number										Oil	Char
	1	2	3	4	5	6	7	8	9	10		
C	44.88	—	—	—	—	—	—	—	—	—	(58.8)[a]	(78.5)
H	4.74	—	—	—	—	—	—	—	—	—	(4.6)	(5.6)
O	31.795	—	—	—	—	—	—	—	—	—	(23.4)	(8.8)
N	0.165	—	—	—	—	—	—	—	—	—	(0.2)	(0.5)
S	0.08	—	—	—	—	—	—	—	—	—	(0.1)	(0.1)
Ash	1.67	—	—	—	—	—	—	—	—	—	(0.04)	(6.6)
H_2O	83.33	—	—	13.91	13.91	9.87	—	95.03	—	7.21	(12.0)	(1.0)
H_2	—	—	—	0.57	0.57	0.40	—	—	—	—	—	—
CO	—	—	—	11.40	11.40	8.07	—	—	—	—	—	—
CO_2	—	—	—	21.08	21.08	14.92	—	33.54	—	13.78	—	—
N_2	—	22.10	—	22.13	22.13	15.69	620.78	636.47	22.33	28.77	—	—
O_2	—	6.72	—	—	—	—	188.48	172.83	6.78	0.32	—	—
CH_4	—	—	—	2.08	2.08	1.48	—	—	—	—	—	—
C_2H_4	—	—	—	0.61	0.61	0.42	—	—	—	—	—	—
C_3H_6	—	—	—	0.23	0.23	0.16	—	—	—	—	—	—
Char	—	—	2.56	2.57	trace	—	—	—	—	—	—	25.16
Oil	—	—	2.56	23.40	trace	—	—	—	—	—	20.83	—
Total	166.66	28.82	5.12	97.98	72.01	51.01	809.26	937.87	29.11	50.08	20.83	25.16

[a]Source: Reference 7. Used with permission.
[b]Basis: 1000 dry tons/day wood feed rate.

2.1. Basic Equations and Guidelines

The first law of thermodynamics, which is a generalized statement of the law of conservation of energy, for a steady-state flow process is

$$\Delta H = Q - W_s \tag{3}$$

where ΔH is the change in enthalpy of the stream from inlet to outlet conditions; Q is the heat crossing the system boundaries; W_s is the "shaft" work done on the fluid by a piece of equipment. Heat transferred to the system is positive; heat removed from the system is negative. Work is positive when it is transferred from the system to the surroundings. The kinetic and potential energy terms are negligible compared to the others. Detailed discussions of the derivation of this equation and its application can be found in standard textbooks on thermodynamics.[8,9]

Enthalpy has no absolute value; only changes in enthalpy with respect to a reference state can be calculated. It must be emphasized that enthalpy is a state function, which means that its value only depends on the state of the system, i.e., its temperature, pressure, and volume. The change in enthalpy depends only on the initial and final states of the system; it is independent of the path or method of accomplishing the change in state.

2.1.1. Calculation of Enthalpy Changes

Evaluation of the ΔH term for a given process usually involves determining the enthalpy changes for a change in physical state, change in temperature, change in pressure (which is usually negligible), or chemical reaction. A brief discussion of these ΔH calculations follows.

2.1.1a. Phase Change. The most common phase change in biomass conversion processes is vaporization or condensation. For this change, ΔH is just the difference in enthalpy at constant temperature and pressure of the substances in the liquid and vapor state. This is commonly called the "latent heat of vaporization."

2.1.1b. Temperature Changes. This enthalpy change is easily calculated from

$$\Delta H = \int_{T_1}^{T_2} C_p \, dT \tag{4}$$

where ΔH is the enthalpy per unit mass (or per mole) and C_p is the heat capacity per unit mass (mole). Heat capacities of gases, liquids, and solids have been tabulated (e.g., see Perry[10]).

2.1.1c. Chemical Reactions and Heats of Combustion. Most of the energy changes that occur in thermochemical conversion processes are a result of the liberation or absorption of energy by the chemical reaction. The heat of reaction is the difference in enthalpy between the products of the reaction and the reactants. For an *exothermic* reaction, the enthalpy of the products is less than the enthalpy of the reactants and heat is released to the surroundings during the reaction. For an *endothermic* reaction, the enthalpy of the products is greater than that of the reactants, and heat will be absorbed during the reaction. The heat of reaction is expressed by Hess' Law of heat summation (on a per-mole basis) as given by the following:

$$\Delta H \text{ (reaction)} = \Sigma \Delta H_f^\circ \text{ (products)} - \Sigma \Delta H_f^\circ \text{ (reactants)} \qquad (5)$$

where $\Sigma \Delta H_f^\circ$ = sum of the standard heats of formation of the products or reactants. Standard heats of formation are the enthalpy changes accompanying a reaction which forms 1 mole of a single compound from the elements which make it up. The reactants and products are in their standard states at 1 atm and 298°K.

Heats of reaction can also be calculated from standard heats of combustion (ΔH_c°) which are the enthalpy changes (at 298°K and 1 atm) associated with reactions in which compounds combine with O_2 to form specified products of combustion (e.g., H_2O and CO_2). When evaluating a biomass conversion process, the heats of combustion are most often used to compute the enthalpy changes for the process or to compare the fuel values of the feed and product streams. The term "heating value" is used interchangeably with the standard heat of combustion. The "gross" or "higher heating value" (HHV) is ΔH_c° for the reaction which produces carbon dioxide and liquid water. The net or lower heating value is ΔH_c° when the water produced is in the vapor phase. The difference between the two is the latent heat of vaporization of the water formed. Note that according to our sign convention, the heat of combustion will be negative since combustion is an exothermic reaction and heat is released to the surroundings.

Because of the complex, multicomponent nature of biomass, the usual tabulated heat of combustion[8,10] are of limited usefulness for determining heat effects associated with biomass conversion processes. Heating values for selected crops and forest species are given in a report by Radovich et al.[5]; heating values for different types of wood are presented by Tillman,[11] Shafizadeh,[12] and de Groot.[13] When experimental data are not available, Tillman suggested the following formula for calculating HHV on a dry weight, ash-free basis:

$$\text{HHV (Btu/lb)} = (7527)\chi_c + (1 - \chi_c)11479 \qquad (6)$$

where χ_c is the weight fraction of cellulose plus hemicellulose. The predicted value is within 5% of the experimental HHV. Shafizadeh[12] and de Groot[13] presented the following formula for determining the heat of combustion in calories/gram (multiply by 1.8 to get Btu/lb):

$$\Delta H_c^\circ = 94.19 \, (\% \ C) + 55.01 \qquad (7)$$

where % C is the percent carbon obtained from chemical analysis of the biomass fuel. This is the equation for the data obtained for combustion of biomass and for char and volatiles derived from pyrolysis of the biomass. Since many conversion process reactions do not occur at 25°C, we must use the enthalpy change of the reaction at the specified temperature when making energy balances. A formal procedure for determining the enthalpy change is presented later.

Recall that the enthalpy change of any process or transformation depends only on the initial and final states of the process and not on the actual path of the process. Consider the following process:

$$\text{Reactants at } T_r, \, P_r \xrightarrow{\Delta H \ = \ ?} \text{Products at } T_p, \, P_p \qquad (8)$$

Then,

$$\Delta H = H \, (\text{products at } T_p, \, P_p) - H(\text{reactants at } T_r, \, P_r)$$

An alternate path utilizes the standard enthalpy changes mentioned above:

$$
\begin{array}{ccc}
\text{Reactants to } T_r, \, P_r & \xrightarrow{\ \Delta H\ } & \text{Products } T_p \ P_p \\
\Big\downarrow \Delta H_1 & & \Big\uparrow \Delta H_3 \\
\text{Reactants at } 298°\text{K} & \xrightarrow{\ \Delta H_2\ } & \text{Products at } 298°\text{K} \\
(1 \text{ atm}) & & (1 \text{ atm})
\end{array}
$$

Now, $\Delta H = \Delta H_1 + \Delta H_2 + \Delta H_3$. The quantity ΔH_1 is the enthalpy change for taking the reactants at their standard states from T_r to 298°K. Likewise, ΔH_3 is the enthalpy change for taking the products in their standard states from 298°K to T_p. The standard heat of formation or combustion for the generation of the specified products is given by ΔH_2. The ΔH at a higher temperature is much less than that determined at 298°K especially when an exothermic reaction such as combustion, is involved.[13]

2.1.2. Simplified Forms of the Energy Balance Equation

Energy balance calculations based on the First Law can be simplified for a number of special cases. In each of these cases the system consists of the biomass in the process unit. The special cases are characterized as follows.

1. No work added or extracted from the system: $\Delta H = Q$. Any net change in enthalpy of the inlet and outlet streams must equal the heat exchanged between the system and surroundings.

2. No work and adiabatic ($Q = 0$) operation: $\Delta H = 0$. The total enthalpy of the entering streams must equal the total enthalpy of the existing streams. This is the most common situation for energy balances for processes involving chemical reactions.

3. No work and constant temperature: $\Delta H = Q$. This case is different from Case 1 in that the only enthalpy changes will be associated with phase changes (e.g., drying) or chemical reaction ($T_p = T_r$).

If a work term is involved in the energy balance, it must be evaluated in terms of the shaft work exchanged between the system and the surroundings *at the point of exchange,* the system boundary. For example, if a gas stream (the system) is compressed, the work term is the shaft work added to the gas at the compressor blades. The efficiency of the compressor (or any other mechanical device) must be used to calculate this shaft work.

2.2.   Applications to Conversion Processes

2.2.1.   Pyrolysis of Wood

The overall energy balance for the pyrolysis process described in Figure 3 is given in Table 6.[7] This type of energy balance corresponds to application of the First Law to an isothermal system with $W_s = 0$. The system boundary encloses the entire process. The inlet energy is calculated from the higher heating value of the wood, which is 19.1×10^6 Btu/dry ton (9.55×10^6 Btu/lb).

TABLE 6.   Overall Energy Balance for Pyrolysis of Wood[a,b]

Input	
Wet wood	796.7×10^6 Btu/hr
Output	
Oil	255.7×10^6 Btu/hr
Char	332.1
Dryer stock	87.4
Boiler stock	11.2
Heat rejected to cooling	42.9
Insulation losses	28.4
Miscellaneous losses	39.0
Total output	796.7×10^6 Btu/hr

[a]Source: Reference 7. Used with permission.
[b]Basis: 1000 tons/day dry wood feed.

TABLE 7. Energy Balance for Pyrolysis–Gasification[6]a

Energy-flow stream	In (10^6 cal)	Out
First Step Reactor		
H_1 (298°K)	0.0	—
H_2 (598°K)	—	936.3
(ΔH_c = 4730)		
H_4 (699°K)	1218.9	—
H_5 (325.7°K)	—	270.4
Q_p	—	12.2
Pyrolysis–Gasification Reactor		
H_{15} (879.6°K)	302.8	—
H_3 (1273°K)	—	2151.4
(ΔH_c = 3796)		
Q_r	—	21.7

aSource: Reference 6. Used with permission.
bBasis: 1000 kg dry wood, datum temperature = 298°K.

The exit energy is calculated from the higher heating value of the oil (12,2͡ꞈ Btu/lb) and char (13,200 Btu/lb), and from the sensible heat for the exit gases and cooling water which are referred to the datum temperature, 77°F. The heat crossing the system boundaries are the insulation and mechanical losses. The First Law can thus be written as

$$H_{in} = H_{out} + Q \qquad (9)$$

The process power requirements, the W_s term, are excluded from the energy balance.

Most "energy balances" presented in the literature are incomplete balances because they account only for the inlet and outlet energy based on the higher heating value of the biomass feed and fuel product streams. The differences between these values must be the enthalpy changes of the nonbiomass and nonfuel product streams plus the Q terms. However, as will be seen later, even these incomplete energy balances can be used to evaluate the efficiency of the process.

2.2.2. Pyrolysis–Gasification Reactor

The energy balance calculations for the pyrolysis–gasification reactor described in Section 1.2.2 and shown in Figure 2 are given in Table 7.[6]

The energy balance around the first-stage reactor is

$$H_1 + H_4 = H_2 + H_5 + Q_p$$
$$0.0 + 1218.9 = (936.3 + 270.4 + 12.2) \quad (10^6 \text{ cal}) \qquad (10)$$
$$1218.9 = 1218.9 \quad (10^6 \text{ cal})$$

The enthalpy of Stream 1 is zero because it is at the datum temperature; the enthalpy of Stream 2 includes the sensible heat of the wood at $1000°C$ plus the sensible heat and latent heat of vaporization for the water at that temperature; Q_p is the heat loss from the reactor which was calculated as 1% of the total heat input to the reactor; H_5 is the sensible heat of the gas out of the reactor heating jacket.

The energy balance around the pyrolysis gasification reactor is

$$H_3 + Q_R = H_2 + H_{15}$$
$$2151.4 + 21.7 = 936.3 + 302.8 \qquad (11)$$
$$2173.1 \neq 1239.1 \ (10^6 \text{ cal})$$

As written, the energy balance is incorrect because it does not include the heat of reaction for the formation of the gaseous products in H_3. The heat of reaction is the difference in $\Delta H_c°$ for Streams H_2 and H_3:

$$\Delta H_{\text{reaction}} = H_2(\Delta H_c°) - H_3(\Delta H_c°) \qquad (12)$$
$$= 4730 - 3796 = 934(10^6 \text{ cal})$$

The energy balance is complete when $\Delta H_{\text{reaction}}$ is added to the input energy flows ($H_2 + H_{15}$). As for the first reactor, the stream enthalpies were calculated from the sensible heats at the given temperature, and Q_r was assumed to be 1% of the inlet energy.

The data presented in this example are most nearly in the form of the original First Law energy balance for a process in which no work is done.

3. Evaluation of Process Efficiency

Biomass is a raw material that can be used as an energy source. What type of energy should be produced? A difficult question, whose complete answer goes beyond the scope of this chapter. An answer which can serve as a framework for our consideration was given by Miles,[14] who stated that energy resource utilization should maximize the recovery of energy and byproducts by judiciously combining the appropriate technology with the available resource. Only those energy products which are most scarce or most economic should be produced.

The purpose of this section is to determine which process or combination of processes gives the maximum energy from biomass.

3.1. Criteria

In order to make valid comparisons among conversion processes, both system boundaries and a basis must be clearly and consistently defined:

1. System boundaries: Selection of the boundaries for comparison of biomass conversion processes has been very arbitrary. It is suggested that the system boundaries for a plant which produces fuel encompass all processes from receipt of the biomass at the plant gate to the fuel product as it leaves the plant to be converted into work or another form of energy. If the process is to produce energy (steam, electricity) then the boundaries should cross this energy form as it exits the plant. The points at which mass or energy cross this boundary must be clearly identified.

2. Basis: The material balances should be based on a suitable unit mass of biomass (of known composition) as received at the plant gate or as fed to the first processing unit if significant changes in composition occur during storage. Any convenient time period is acceptable for a flow process. A reference state is also needed for the enthalpy calculations required in the energy balances. This should be 298°K with all components in their standard states.

3.2. Thermal or Energy Efficiency

The efficiency of a process is a function of system boundaries, and more importantly, the engineer's preference. When evaluating a given conversion process, we must compare "technical" merits and "economic" merits. We would like to assign an economic value to the product streams. For a process producing fuel or energy, some energy output streams, such as heat rejected to cooling water, have no economic value and are thus neglected in our considerations. Sliepcevich et al.[15] have defined a general energy efficiency, which is useful for the comparisons we must make:

$$\eta_e = \text{(net useful output)}/\text{(valuable input)} \qquad (13)$$

The net useful output is defined as the desired energy output produced by the conversion process. The valuable input is the energy that is required (the "energy that costs")[16] to produce the useful output. This is a so-called first law efficiency[17] since it is only concerned with heat effects and does not consider the quality of the energy terms directly. The determination of "useful output" and "valuable input" will depend on the system as discussed later.

Consider the simplified biomass conversion process shown in Figure 4,

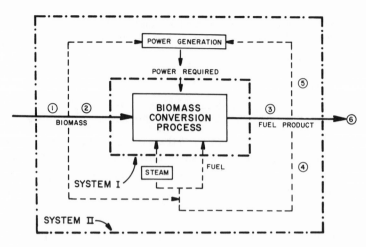

FIGURE 4. Biomass conversion process for efficiency calculations.

with the energy balance streams given in Table 8. The energy efficiency for System I, at first glance, is given by

$$\eta_e = \frac{3}{2 + \alpha + \beta + \gamma} \tag{14}$$

The net useful output is the HHV of the fuel product. The net useful input is the HHV of the biomass plus all process energy requirements, regardless of form, which cross the system boundaries. We are only concerned with the actual biomass entering the conversion process. This calculation also neglects

TABLE 8. Energy Balance Streams for General Conversion Process

Stream	Stream Number[a]	Amount (energy/time)
Power required		α (electrical)
Fuel required		β
Steam		γ
Biomass	1	(HHV)
Biomass	2	(HHV)
Fuel products	3	(HHV)
Fuel products	4	(HHV)
Fuel products	5	(HHV)
Fuel products	6	(HHV)

[a]Refer to Figure 4.

the quality of input energy to the process, i.e., one Btu of electrical energy is equivalent to one Btu of the heating value of raw biomass. It should not be used. The recommended calculation of η_e is given by

$$\eta_e = \frac{3}{2 + (\alpha/\eta_p) + (\beta/\eta_s) + \gamma} \tag{15}$$

where η_p and η_s are the actual conversion efficiencies for the use of either the fuel product or raw biomass to generate power and steam respectively. Note that new biomass or fuel product may or may not be used for the process energy requirements.

The energy efficiency for System II is:

$$\eta_e = 6/1 \tag{16}$$

Note that *all* the energy requirements are within the system boundaries. This may not always be the case.

Other efficiencies have been defined,[5,17,18] many in the form

$$\eta_e = \frac{\text{HHV of product} - \text{operating energy}}{\text{HHV of feed}} \tag{17}$$

These efficiencies are commonly referred to as thermal efficiencies because of the comparison made with the heating value of the fuel. It is also similar in form to boiler efficiency based on the indirect method of calculation.[16] However, this equation for efficiency is not recommended because we are logically interested in comparing energy outputs to inputs. This is done more explicitly by using η_e.

It is important to note that η_e is based on arbitrarily selected streams from the energy balance. The basis for selection was our ability to assign an economic value to the energy stream. The ratio, total energy out/total energy in, is always 1.0 by definition of the energy balance given by the First Law.

The major fault with using the energy efficiency is the arbitrary choices which are made concerning the definition of useful output and energy input.. When all the choices are consistent among a group of conversion processes producing fuel, the comparisons among energy efficiencies are probably meaningful. However, it is probably not valid to use η_e to compare processes which produce energy or work by different means. A more valid comparison can be made by evaluating the thermodynamic efficiency[15]:

$$\eta_t = \frac{\text{performance of an actual process}}{\text{performance of an equivalent ideal process}} \tag{18}$$

The data required for calculating η_t are generally not available. A brief discussion of thermodynamic efficiency is presented in the next section.

3.3. Thermodynamic Efficiency

The definition of thermodynamic efficiency, η_t, given in the preceeding section, is general and ambiguous. Its value for comparing various processes depends on decisions concerning[15]:

1. The definition of performance. This can usually be decided on by answering the question: What is the primary purpose of the process?

2. The ideal process to be used for comparison: There are many ideal devices which could accomplish the desired purpose or change in the process without entropy production ($S_p = 0$).

3. The basis for establishing the size equivalence.

These items have been agreed upon for a number of simple processes.[15] For a heat engine, $\eta_t = W_{actual}/W_{ideal}$ for equal heat transfer at the elevated temperature. The ideal work is obtained from application of the mass, energy, and entropy balance to the ideal process, under the restriction that $S_p = 0$. The value of η_t will depend on the choice of the ideal process. For a heat engine cycle, the Carnot cycle is the usual basis for comparison. Examples of η_t calculations for other simple processes are given by Sliepcevich et al.[15]

There have been relatively few applications of the thermodynamic efficiency to evaluate complex processes, particularly biomass conversion processes. Powers et al.[19] illustrated its use in evaluating the production of electricity by fluidized-bed combustion of coal. They showed that the thermodynamic efficiency is given by Equation (19):

$$\eta_t = \frac{W_{actual}}{W_{ideal}} = (W_{actual}) / \left[\Sigma M_o(H_o - T_D S_o) - \Sigma M_i(H_i - T_D S_i) \right] \quad (19)$$

where W_{actual} is the electricity generated, M is the mass flow rate, H and S are enthalpy and entropy, respectively, T_D is the datum temperature, and the subscripts i and o denote inlet and outlet streams. The term $(H - T_D S)$ is known as the availability or available energy. It is defined as the maximum amount of work that can be extracted from a unit mass in steady-state flow without any chemical reactions or any change in kinetic or potential energy. The heat energy is transferred at the datum temperature.

Evaluation of the terms of this equation requires the application of material, energy, and entropy balances* to the system. Powers et al.[19] also showed

---

*Development of the entropy balance and the second law of thermodynamics is given in the previously cited References 8, 15, and 17.

the similarity between η_t and a thermal efficiency defined by

$$\eta_t = (W_A)/(\text{heating value of fuels}) = W_A/(M_o \Delta H'_{c_o} - M_i \Delta H'_{c_i}) \quad (20)$$

where $\Delta H'_c$ are the *net* heating values (water vapor rather than liquid is a product of combustion) of the mass-flow streams.

Bailie[17] has defined another efficiency factor, the "figure of merit":

$$\text{Figure of merit} = \frac{\text{Maximum work obtainable from all outlet streams}}{\text{Maximum work obtainable from all inlet streams}} \quad (21)$$

All streams, mass, heat, and work, are considered, and none is arbitrarily neglected. This efficiency is related to the thermodynamic efficiency in that the maximum work is evaluated by considering ideal processes ($S_p = 0$) operating between the actual conditions and the datum conditions (298°K, 1 atm). Bailie gives numerous examples of calculating the figure of merit for fossil fuel combustion and coal conversion processes. All calculations depend on a knowledge of H, S, or $H - T_D S$ for the mass components, data which Bailie has tabulated.

3.4. Recommendations

We strongly recommend explicitness and consistency in defining the boundaries of the system under consideration. When evaluating the efficiency of a biomass conversion process to produce fuels, the energy efficiency will give a valid comparison. However, should the biomass conversion process include steps for the production of work (i.e., steam or electrical energy), a more meaningful comparison will be obtained by considering the thermodynamic efficiencies. The thermodynamic efficiency will indicate the best process for producing the work/energy as compared to an ideal process. Also one should be wary of inferring overall system efficiencies from individual process or subsystem efficiencies, regardless of the type of efficiency considered. The only quantity of importance in a complex system is the *overall* efficiency and not the efficiency of a single process or combination of processes.

Development of the entropy balance and the second law of thermodynamics is given in the previously cited References 8, 15, and 17.

References

1. D. M. Himmelblau, *Basic Principles and Calculation in Chemical Engineering,* 3rd edition, Prentice-Hall, Englewood Cliffs, New Jersey (1974).

2. O. A. Hougen, K. M. Watson, and R. A. Ragatz, *Chemical Process Principles, Part 1: Material and Energy Balances*, John Wiley & Sons, New York (1954).

3. E. J. Henley and E. M. Rosen, *Material and Energy Balance Computations*, John Wiley & Sons, New York (1969).

4. J. Y. Oldshue, AICHE goes Metric, *Chem. Eng. Prog.* **73**(8), 135–138 (1977).

5. J. M. Radovich, P. G. Risser, T. G. Shannon, C. F. Pomeroy, S. S. Sofer, and C. M. Sliepcevich, Evaluation of the Potential for Producing Liquid Fuels from Biomass, prepared for Electric Power Research Institute, EPRI AF-974, University of Oklahoma, Norman, Ok. (January, 1979).

6. D. L. Brink and J. F. Thomas, The Pyrolysis Gasification–Combustion Process: Energy effectiveness using oxygen vs. air with wood fuel systems, in: *Fuels and Energy from Renewable Resources* (D. A. Tillman, K. V. Sarkanen, and L. L. Anderson, eds.), Academic Press, New York (1977), pp. 141–168.

7. S. M. Kohan and P. M. Barkhordar, Mission Analysis for The Federal Fuels from Biomass Program, Vol. IV: Thermochemical Conversion of Biomass to Fuels and Chemicals, prepared for Department of Energy by Stanford Research Institute International, Menlo Park, California (January, 1979).

8. J. M. Smith and H. C. Van Ness, *Introduction to Chemical Engineering Thermodynamics*, 3rd Edition, McGraw-Hill, New York (1975).

9. G. J. VanWylen and R. E. Sonntag, *Fundamentals of Classical Thermodynamics*, John Wiley & Sons, New York (1976).

10. R. H. Perry and C. H. Chilton, *Chemical Engineer's Handbook*, 5th Edition, McGraw-Hill, New York (1973).

11. D. A. Tillman, *Wood as an Energy Source*, Academic Press, New York (1978).

12. F. Shafizadeh, American Institute of Chemical Engineers Symposium Series **74**(177), 76 (1978).

13. F. Shafizadeh and W. F. De Groot, Thermal analysis of forest fuels, in: *Fuels and Energy from Renewable Resources* (D. A. Tillman, K. V. Sarkanen, and L. L. Anderson, eds.), Academic Press, New York (1977), pp. 93–113.

14. T. R. Miles, Logistics of energy resources and residues, in: *Fuels and Energy from Renewable Resources* (D. A. Tillman, K. V. Sarkanen, L. L. Anderson, eds.), Academic Press, New York (1977), pp. 225–248.

15. C. M. Sliepcevich, J. E. Powers, and W. J. Ewbank, *Foundations of Thermodynamic Analysis, Part 2*, McGraw-Hill, New York (1971).

16. A. W. Culp, Jr., *Principles of Energy Conversion*, McGraw-Hill, New York, (1979).

17. R. C. Bailie, *Energy Conversion Engineering*, Addison-Wesley, Reading, Massachusetts (1978).

18. J. K. McCartney, The value of energy from waste at the production interface, in: *Clean Fuels from Biomass and Wastes*, Symposium Papers of the Institute of Gas Technology meeting in Orlando, Florida (January, 1977), pp. 291–301, Institute of Gas Technology, Chicago, Illinois.

19. J. E. Powers, D. L. Katz, D. E. Briggs, E. R. Lady, M. Rasin Tek, and B. Williams, Evaluation of Coal Conversion Process to Provide Clean Fuels, Part III, prepared for Electric Power Research Institute, EPRI 206-0-0, University of Michigan, Ann Arbor, Mi. (February 1979).

21

Economic Considerations of Biomass Conversion Processes

1. Introduction

Earlier chapters have described various biomass conversion processes and processing procedures. This chapter provides a systematic method of estimating biomass process economics and determining the revenue required to produce useful fuels and chemicals from biomass feedstocks. Included is a discussion of capital investment economics as well as operating and maintenance costs, feedstock costs, and financial considerations.

1.1. Level of Estimating the Desired Precision

Numerous levels of estimating precision have been employed to evaluate biomass conversion process economics. The evaluation level of accuracy has varied with such factors as the purpose of the analysis, the state of the biomass conversion technology, the time and funds available for the effort, and the skill and experience of the analysts. In general, the purpose of the analysis and the benefits derived from the estimating procedure are the most important factors. For example, a far greater level of estimating effort is required for a study whose purpose is to justify facility construction than for a study designed to establish a budget or to roughly compare alternatives. The first requires accounting levels of detail including material quantities and costs, and the sec-

FRED A. SCHOOLEY • California Institute of Technology, Jet Propulsion Laboratory 4800 Oak Grove Drive, Pasadena, California 91103 (Mail Stop 502–307).

ond may employ only overall measures such as facility dollar-per-square-foot
or dollar-per-cubic-foot factors.

1.2. Assumed Study Purpose and Conditions

In our analysis it is assumed the study purpose is to realistically compare
alternative biomass conversion processes and to derive a realistic price (revenue
requirement) for a fuel or chemical end product. As was described in earlier
chapters, this procedure requires the development of detailed process descriptions, process flow charts, material balance calculations, and energy balance
calculations.

In addition, assumptions must be made with regard to site-specific
resource availabilities such as biomass feedstocks, water, land, and labor. Also,
assumptions concerning environmental quality and pollution control requirements at the conversion site are required. In a study which does not involve a
specific site location or general proposed plant area, the analyst must use
"averaging assumptions" with regard to these factors and many others of a
similar nature. Specific averaging assumptions for biomass conversion options
are discussed later. The sections that follow describe general economic guidelines and financial considerations applicable to biomass process economics.

2. General Guidelines

Economics is far from being an exact science and, in many ways, may
more closely resemble an art rather than a well defined discipline allowing precise application of established rules and principles. However, there are several
general engineering economic guidelines which should be applied to the analysis of industrial processes to provide confidence in the end results of the procedure. The guidelines which follow are among the most important when evaluating biomass options.

1. Establish a uniform economic analysis procedure and consistent chart
of accounts which will allow a fair comparison of process alternatives and
resulting end product prices (revenue requirements).

2. Utilize a "life cycle costing procedure" which provides consideration
for all major cost elements over the useful life of the conversion facility. Include
all initial investment requirements as well as operating and maintenance cost
elements.

3. Consider the time value of money in the analysis by utilizing a procedure which provides for the minimum rate of return expected by conversion
facility ownership and which considers the time periods, amounts, and duration
of major investment and operating expenditures.

Each of these three engineering economic guidelines will be discussed in the context of the assumed study purpose and conditions.

2.1. Analysis Uniformity

When comparing biomass alternatives, the establishment of uniform procedures and assumptions are essential to the development of a usable endproduct. Assumptions concerning income taxes, investment tax credits, and the nature of the conversion organization can have major impacts on endproduct costs.

In most cases, investment tax credits are not taken in order to assure conservative study results. Federal taxes apply to taxable income at a current rate of 48% and 2–6% is typically added to represent an estimate of the average net state tax impact.

The decision to consider the venture as either regulated or nonregulated is usually dependent on the likely markets to be served by the products of the biomass conversion facilities. Products such as electric power, utility fuel oil, and synthetic natural gas are usually produced by regulated organizations. Products such as process steam, ammonia, ethanol, intermediate Btu gas, and other fuel oils are usually produced by nonregulated organizations. Regulated industries are normally assumed to partially finance the total capital investment of proposed biomass conversion facilities with debt capital. A ratio of 65% debt to 35% equity, a 9–12% interest rate on debt, and a 15–18% return on equity capital were typical for many biomass conversion studies in 1979.[1]

Other important ground rules required to assure estimate uniformity involve adjustments for dollar depreciation or inflation and assumptions concerning the stage of technology commercialization and product transportation. In the typical case, all monetary figures are given in constant U.S. dollars (e.g., mid-1979 dollars) and the "mature plant concept" is followed as opposed to a pioneering venture. Typically product transportation facilities are not included in the plant facilities investment and product prices are on a "plant gate" basis.

2.2. Life Cycle Costing

A thorough economic analysis must consider both investment and operating costs. In the past, studies have often emphasized only one aspect of system design and have either totally ignored or only casually addressed other aspects. This procedure has been typical for a company whose main product is conversion equipment or for a user organization whose main interest is in system performance. To prevent this emphasis on one phase of system operations at the expense of another, a life-cycle costing procedure should be adopted. Life-cycle costing requires the establishment of a standard chart of accounts

TABLE 1. Standard Chart of Accounts

Capital investment	$^{a} ....
Plant facilities	....
Equipment	....
Land	....
Organization and startup expenses	....
Working capital requirements	....
Initial catalysts and chemicals	....
Interest during construction	
Total capital investment	$ ....
Annual operating costs	
Biomass feedstock	....
Raw materials (other)	....
Maintenance materials	....
Water	....
Catalysts and chemicals	....
Labor	
Operating	....
Supervision	....
Maintenance	....
Administrative and support labor	....
Purchased electric power	....
General administrative expenses	....
Property taxes and insurance	
Total annual operating costs	$ ....

^{a}Monetary values vary by conversion process type, time period, and estimating assumptions and procedure.

which allows the estimation of costs over the entire life of the facility. Table 1 provides one example of a standard chart of accounts for a biomass conversion facility.

2.3. Money Time Value

Alternative capital investment and operating expenditure can be equated by considering the time value of the future funding stream. This equality can be calculated by several procedures including (1) discounting expenditures to a present value; (2) by determining alternative rates of return; and (3) by calculating alternative uniform annual costs. The procedure is explained in detail in Reference 2, Part II. However, basically it involves discounting distant expenditures in consideration for the minimum rate of return expected by conversion facility ownership. Typically, this rate is 9–12% on debt funds and 15–18% on equity capital, as previously indicated. The specific formula used to determine average annual costs is given in Section 6 on the calculation of product revenue requirements.

3. Capital Investment Economics

This section includes a definition of the typical boundaries (facilities normally included and excluded), as well as plant capacity limitations, construction periods, plant utilities, investment land costs, working capital requirements, startup costs and investment depreciation guidelines.

The typical plant facilities investment includes the total cost of the plant after final construction and ready for startup. It consists of all process facilities, necessary utilities, and general plant facilities, including equipment and all direct and indirect costs of installation.

3.1. Plant General Facilities

The plant general facilities investment includes the following buildings, shops, and installations:
- Office buildings
- Laboratory
- Flare and relief systems
- Maintenance shops
- Supplies warehouse
- Control laboratory
- Cafeteria, dispensary, and change house
- Plant roads, parking areas, walks, and storm sewers
- Plant communications system
- Plant fences, guard houses, and lighting system
- Fire protection facilities

The utilities and general facilities costs used do not include provision for site-specific offsites such as dams, water pipelines, power generation, stream diversion, access roads, railroads, air strips, bridges, tunnels, product pipelines, or townsite development. Many or most of these items may be necessary for the completion of a project in a remote area. Since it is assumed that in most cases electric power and makeup fresh water are purchased and delivered to the plant, the utilities investment does not include investment for water supply development and delivery.

Estimated construction periods appropriate to the biomass technologies under consideration vary depending on the size of the plant. These periods range, for example, from about two years for small biomass-fired steam electric power plants to about four years for large ammonia or methanol production facilities using biomass feedstock. When considering debt-financed ventures (as typified by regulated utility plants), interest during construction is included in the total capital investment.

3.2. Plant Utilities

The plant utilities investment typically includes the following facilities:
- Raw water filtering and softening facilities
- Cooling water system (includes either air-cooling or wet-cooling towers, distribution system, and blowdown treatment)
- Boiler feedwater demineralizers and deaerators
- Water distribution
- Wastewater collection and treatment
- Emission control equipment
- Plant fuel system
- Steam generation, distribution system, condensate lines, and boiler blowdown treatment facilities, if not separately identified
- Electric power substation and distribution system
- Instrument air and inert gas systems
- Ash disposal
- Fuel-gas handling

3.3. Land Investment

Land required for a conversion plant site is typically valued at $5,000–10,000/acre ($12,000–24,000/hectare) in rural U.S. locations and $25,000–50,000/acre ($60,000–120,000/hectare) in urban locations. These land costs include rough grading and site preparation expenses in 1979.

3.4. Working Capital

Working capital is provided for payroll and other cash operating expenses plus funds for feedstock purchase until receipt of accounts receivable. It typically includes the following items:
- Three months' total labor expense
- Two months' other cash operating expense
- One month's feedstock supply

3.5. Organization and Startup Costs

These costs include equipment modification and repair during startup, operator training, property taxes, insurance during construction, and materials consumed during startup. These costs are typically 5% of the plant facilities investment for nonelectric power plant facilities and 3% of investment for electric power plant facilities. Regulated industry startup costs are capitalized, while nonregulated industry startup costs are expensed.

3.6. Depreciable Investment

Regulated industry plant depreciable investment is the sum of plant facilities investment, interest during construction, paid-up royalties, and startup expenses, typical of a regulated utility. Nonregulated industry plant depreciable investment includes plant facilities investment and paid-up royalties, representative of an industrial venture.

A 20- to 30-yr life and straight line depreciation is normally used for regulated utilities. A 15- to 30-yr life and an accelerated depreciation method is frequently used for industrial (nonregulated) investments, depending upon several factors including the type and durability of construction materials.

3.7. Plant On-Stream Factor

After the startup period is completed, the biomass conversion plants are generally assumed to operate on-stream about 90% of the time, or 328.5 days/yr. Conversion plants producing electric power are assumed to operate about 80% of the time (292 days/yr), as would be typical of a new base-load power plant.

3.8. Conversion Plant Capacities

Commercial biomass conversion plant sizes are often referred to by feedstock input rate rather than by a product output rate (e.g., 25 million gallons of ethanol per year or 250 million scf per day of SNG) to reflect the viewpoint that biomass conversion technologies tend to be feedstock-limited. For commercial thermochemical conversion plants, a feedstock rate of 1,000 dry tons/day is often advocated with an upper limit of 3,000 dry tons necessitated by collection and preparation difficulties. For biochemical missions, the typical rate is also about 1,000 dry tons/day, but the possible commercial range of feedstock inputs may be greater. For example, some advocates of commercial kelp conversion plants have proposed anaerobic digestion facilities exceeding 6,000 tons/day and several small cattle manure conversion plants of 30–50 dry tons/day have been designed.

4. Feedstock Prices

Feedstock costs can generally be considered a function of availability in almost all regional markets. The greater the available supply, the lower the

TABLE 2. Cost Factors of Biomass Sources (1979)

Feedstock	Dollars per dry ton	
	base cost	sensitivity range
Woody plants	19	0–$36
Manure (from feedlots)	5	0–10
Cellulosic material (low moisture)	25	6–40
High sugar content plants (bagasse)	15	0–30
Marine crop (kelp)	60	30–120
High moisture crops	35	10–60

cost to the conversion facility owner. A series of supply curves has been prepared on an area basis covering nine U.S. regions.* In general, a doubling in feedstock price results in almost a doubling in the quantities of feedstock available, but feedstock prices can be highly variable depending upon location and market conditions. Table 2 provides information on cost factors used in recent national biomass fuel analysis.*

5. Operating and Maintenance Costs

Operating costs include maintenance material and supplies, operator labor, administration costs, taxes, insurance, and overhead expenses. Administrative and support labor is assumed to include the plant manager, process engineers, laboratory technicians, clerks, secretaries, telephone operators, janitors, guards, and firemen. Payroll burden is assumed to cover costs of health insurance, disability insurance, vacations, sick leave, and retirement payments.

The following factors are typical of those used for conversion plant operating labor:
- Plant operating labor ($/hr) $8–12
- Plant supervision (% of operating labor) 15–20%
- Maintenance labor (% of plant facilities investment) 2–3%
- Administrative labor (% of operating supervision and maintenance labor) 15–25%
- Payroll burden (% of all labor) 30–40%

The following factors are typical of those used for conversion plant operating materials:
- Maintenance materials and supplies (% of plant facilities investment) 2–3%
- Local property taxes and insurance (% of plant facilities investment) 2–3%

---

*See Reference 1, Volume 4, *Mission Analysis.*

- General administrative and overhead expenses (% of plant facilities investment) 1–3%
- Electric power (cents per kilowatt hour) 2.5–3.0
- Water (cents per 1000 gallons delivered to the plant) 50–100

General administrative and overhead expenses include head office expenses for accounting, purchasing, legal services, office supplies, communications, travel fees, and contracted services.

6. Calculation of Revenue Requirements

The calculation of revenue requirements involves the following:

1. Estimation of net expenditures annually (annual capital and operating costs less byproduct and other credits).
2. Determination of an average annual revenue using discounted cash flow (DCF) analysis described below.
3. Division of the average annual revenue amount by the annual product production unit estimate (gallons of ethanol, barrels of oil, etc.) to obtain an annual cost per productions unit.
4. Conversion of the annual cost per production unit into dollars per million Btu by dividing the cost per biomass fuel production unit by the millions of Btus in an average production unit. (Table 3 provides data

TABLE 3. Approximate Btu Content of Several Fossil Fuels and Biomass Sources

Fuel unit	Approximate Btu content
1 bbl of crude oil	5,800,000
1 ton of dry woody biomass	15,000,000–18,000,000
1 ton of dry agricultural residues	13,000,000–17,000,000
1 therm	100,000
1 ton of ammonia	18,000,000–19,000,000
1 gallon of gasoline	
regular	124,000
high octane	123,000
1 gallon of Fuel Oil #2	140,000
1 cubic foot of:	
Natural gas or SNG	900–1,100
Intermediate Btu gas	300–500
Low Btu gas	100–200
1 gallon of methanol	55,000–58,000
1 gallon of ethanol	78,000–81,000
1 kilowatt hour	3,400–3,420

on the approximate Btu content of several fossil fuels and biomass sources).

6.1. Revenue Required—Nonregulated Industry

Discounted cash flow (DCF) analysis is used to determine the revenue required to yield a desired rate of return on equity investment over the life of the project. A detailed discussion of DCF analysis[2] is not attempted in this text. Very briefly, the net cash flow for any year is the sum of positive and negative cash flows consisting of revenue received, operating costs, income tax, equity investment, and payments for debt principal and interest. The annual net cash flow is discounted each year by multiplying it by the appropriate discount factor defined by

$$d = \frac{1}{(1 + i)^n} \qquad (1)$$

where d is the discount factor (also referred to as P or present worth), i is the DCF rate of return on equity investment, and n is the year of project life to which cash flow applies.

The revenue required to yield the desired rate of return on equity investment is determined by iterative calculation such that the sum of all discounted positive and negative annual cash flows over the life of the project will be zero.

Equation (2) provides a method for determining the equivalent annual costs given P, the present worth of all annual expenditures:

$$R = P \frac{i(1 + i)^n}{(1 + i)^n - 1} \qquad (2)$$

where R is the equivalent annual cost, i is the minimum acceptable rate of return, n is the number of years of facility operation or life, and P is the present worth.

6.2. Revenue Required—Regulated Industry

The accounting methods permitted and required, as regulated U.S. utilities, vary somewhat with the regulatory agency. The method described here has been used by the U.S. Federal Power Commission.[3] With this method, organization and startup costs, interest during construction, and paid-up royalties are capitalized as part of the depreciable investment. Straightline depreciation is calculated for each year of the project. This depreciation is subtracted from the undepreciated investment each year to obtain the depreciated plant investment. The rate base in any year is the sum of the depreciated plant investment, the cost of land, and the working capital. The return on rate base in a weighted average of the return and the simple interest rate on debt is given by

$$RRB = d(i) + (1 - d)r \qquad (3)$$

where RRB is the return on rate base, d is the debt fraction (60–70%), i is the interest on debt (9–12%), $1 - d$ is the equity fraction, and r is the return on equity (15–18%) for 1979.

Byproduct credits, if any, are calculated separately, and the net product revenue is obtained by subtracting it from the total revenue required.

Typical revenue requirements are shown for several biomass conversion processes in Reference 1, Volume I: *Mission Analysis Study—Summary and Conclusions* (Tables 1 and 2). A few additional references[4-6] have been included which discuss economic considerations in further detail.

References

1. F. Schooley and R. L. Dickenson, Mission Analysis for the Federal Fuels from Biomass Program: Final Report, Vol. No. I–VII, prepared for The U.S. Department of Energy, Division of Distributed Solar Technology, Fuels from Biomass Systems Branch, SRI, Menlo Park, California (December, 1978 and January, 1979).
2. Eugene L. Grant and W. Grant Ireson, *Principles of Engineering Economy*, The Ronald Press Company, New York (1964).
3. Federal Power Commission, Final Report: The Supply Technical Advisory Task Force—Synthetic Gas—Coal, Appendix I (April, 1973).
4. W. D. Baasel, *Preliminary Chemical Engineering Plant Design,* Elsevier, New York (1976).
5. P. H. Jeynes, *Profitability and Economic Choice,* Iowa State University Press, Ames, Iowa (1968).
6. G. L. Wells, *Process Engineering with Economic Objective,* John Wiley & Sons, New York (1973).

Index